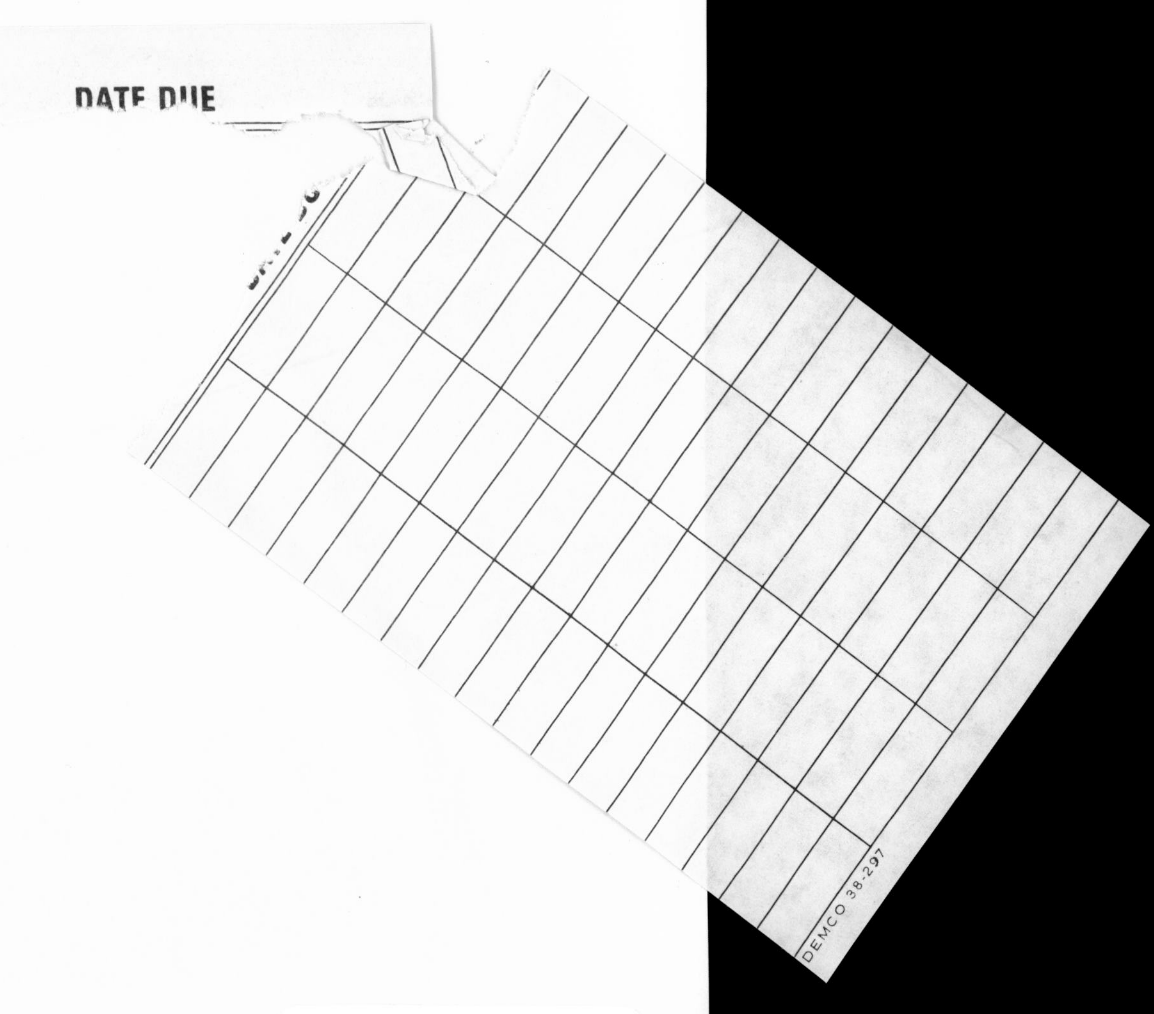
DEMCO 38-297

D1251437

PHYSICAL ACOUSTICS

Principles and Methods

VOLUME XIII

CONTRIBUTORS TO VOLUME XIII

Arthur Ballato
Ralph R. Gajewski
A. S. Nowick
Yih-Hsing Pao
Thomas L. Szabo
L. R. Testardi
R. S. Wagers

PHYSICAL ACOUSTICS

Principles and Methods

Edited by WARREN P. MASON

SCHOOL OF ENGINEERING AND APPLIED SCIENCE
COLUMBIA UNIVERSITY
NEW YORK, NEW YORK

and

R. N. THURSTON

BELL TELEPHONE LABORATORIES
HOLMDEL, NEW JERSEY

VOLUME XIII

1977

ACADEMIC PRESS New York San Francisco London

A Subsidiary of Harcourt Brace Jovanovich, Publishers

ACADEMIC PRESS, INC.
111 Fifth Avenue, New York, New York 10003

United Kingdom Edition published by
ACADEMIC PRESS, INC. (LONDON) LTD.
24/28 Oval Road, London NW1

Library of Congress Cataloging in Publication Data

Mason, Warren Perry, Date ed.
Physical acoustics.

Includes bibliographies.
1. Sound. 2. Ultrasonics. I. Thurston, Robert N., Joint ed. II. Title.
QC225.M42 534 63-22327
ISBN 0-12-477913-1 (v. 13)

PRINTED IN THE UNITED STATES OF AMERICA

CONTENTS

3

Plate Modes in Surface Acoustic Wave Devices

R. S. Wagers

4

Anisotropic Surface Acoustic Wave Diffraction

Thomas L. Szabo

5

Doubly Rotated Thickness Mode Plate Vibrators

Arthur Ballato

6

The Generalized Ray Theory and Transient Responses of Layered Elastic Solids

YIH-HSING PAO AND RALPH R. GAJEWSKI

CONTRIBUTORS

ARTHUR BALLATO
U.S. Army Electronics Technology & Devices Laboratory
Fort Monmouth, New Jersey

RALPH R. GAJEWSKI
Department of Civil Engineering
Engineering Mechanics and Materials
United States Air Force Academy
USAFA, Colorado

A. S. NOWICK
Henry Krumb School of Mines
Columbia University
New York, New York

YIH-HSING PAO
Department of Theoretical and Applied Mechanics
Cornell University
Ithaca, New York

THOMAS L. SZABO
Deputy for Electronic Technology
Hanscom AFB, Massachusetts

L. R. TESTARDI
Bell Laboratories
Murray Hill, New Jersey

R. S. WAGERS[1]
Central Research Laboratories
Texas Instruments Incorporated
Dallas, Texas

[1] Present address: Electrical Engineering and Computer Science Faculty, Princeton University, Princeton, New Jersey 08540.

PREFACE

This volume deals with a variety of topics in physical acoustics, including the use of ultrasonic waves in determining the effects of imperfections and structure, unwanted effects of plate modes and diffraction in surface acoustic wave devices, thickness mode vibrators with both stress and temperature compensation, and the application of ray theory to seismic phenomena and acoustic emission.

The first chapter, by A. S. Nowick, is essentially a review of the theory and application of anelasticity in studying various types of relaxations. These include point defect, grain-boundary, thermoelastic, phonon and electron relaxations, and magnetic relaxations. This chapter brings up to date the material in the book "Anelastic Relaxation in Crystalline Solids" by Nowick and Berry (Academic Press, 1972).

Similarly, Chapter 2, by L. R. Testardi, is a review and updating of the material on A-15 superconductors which was first covered in Volume X of this serial publication. Testardi's chapter discusses all the different methods that have been used in studying these very important Type II superconductors, which have produced the highest superconducting temperatures T_c of any crystal structures so far obtained, i.e., 23°K.

The third chapter, by R. S. Wagers, covers plate modes in surface acoustic wave devices such as transversal filters. Plate modes degrade the performance by limiting the out-of-band rejection or producing other undesired responses. The chapter develops the theory needed to understand plate modes in piezoelectric media, and eliminate or reduce their effect on the response.

The fourth chapter is "Anisotropic Surface Acoustic Wave Diffraction" by Thomas L. Szabo. This subject is important in the application of surface acoustic waves to signal processing devices and the nondestructive evaluation of materials. The chapter covers ways of predicting diffraction loss and phase distortion, and discusses the alleviation of diffraction effects by acoustic beam shaping, material selection and orientation, and alterations in the transducer structure.

The fifth chapter, on doubly rotated thickness mode plate vibrators, by Arthur Ballato, treats plate vibrators whose thickness direction has an arbitrary crystallographic orientation (in contrast to "singly rotated cuts" in which the thickness direction is normal to one of the crystal axes). By double rotation, it is possible to reduce simultaneously the effects of changes of temperature and stress, thereby obtaining reduced sensitivity to initial stress, acceleration, thermal or mechanical shock, and reduced non-linear response, intermodulation, and mode coupling effects. The chapter includes not only the tools for the analysis of the properties of doubly rotated cuts, but a wealth of specific information on such cuts in quartz, berlinite, lithium tantalate, and lithium niobate.

The final chapter, by Yih-Hsing Pao and Ralph R. Gajewski, discusses generalized ray theory and transient responses of layered elastic solids. It is pointed out that, while the theory was first developed for geophysical applications, it can be used for analyzing signals in acoustic emission. By following the ray paths of reflection and refraction, the theory of generalized rays allows one to obtain the shape at the source of the signal produced by various sources of acoustic emission.

The editors again wish to thank the many contributors who have made these volumes possible and the publishers for their unfailing help and advice.

WARREN P. MASON
ROBERT N. THURSTON

PHYSICAL ACOUSTICS

Principles and Methods

VOLUME XIII

—1—

Anelasticity: An Introduction

A. S. NOWICK

Henry Krumb School of Mines
Columbia University
New York, New York 10027

I. Introduction

The term anelasticity, although sometimes used quite loosely to refer to nonelastic behavior, was given a very specific definition by Zener (1948). Since that time Zener's definition has been widely accepted, and interest in anelastic phenomena and the physical origins of these phenomena has increased rapidly. The subject has been dealt with in detail in recent treatises (De Batist, 1972; Nowick and Berry, 1972). The present article is intended as a brief introductory overview.

Anelasticity may be treated on three levels. First, it may be considered in a formal way as a generalization of Hooke's law relating stress and strain, to allow for time-dependent effects. Section II of this article is concerned with this formal theory, and introduces the specific restrictions that serve to define anelasticity. Second, anelasticity may be considered in thermodynamic terms as a process by which one or more internal variables adjust (or undergo relaxation) to an equilibrium state determined by the applied stress. This viewpoint is given in Section III. Finally, the subject may be considered in terms of the physical origins or atomistic mechanisms that give rise to anelastic effects, as presented in Section IV. Although emphasis is placed on applications to crystalline materials, amorphous materials (or glasses) are also discussed. Omitted from consideration, however, are the physical mechanisms of anelasticity in polymeric materials. For this, the reader is referred to the book by McCrum *et al.* (1967). The formal and thermodynamic theory of Sections II and III is, of course, applicable to all materials.

II. Formal Theory of Anelasticity

A. The Concept of Anelasticity and the Standard Anelastic Solid

The formal theory of anelasticity begins as a departure from perfectly elastic behavior, as expressed by Hooke's law:

$$\varepsilon = J\sigma, \tag{1}$$

where σ is the stress, ε the conjugate strain, and J the corresponding compliance (equal to the reciprocal of the appropriate elastic modulus M). In anelasticity, Hooke's law is extended to include time as a variable in the relation between stress and strain. The "instantaneous" response and the "single-valuedness" inherent in Eq. (1) are thereby discarded. Two restrictions are retained however; these are: (a) *linearity*, in the sense that doubling the stress doubles the strain at each instant of time; and (b) a *unique equilibrium relationship*, which means that to every value of stress there corresponds a unique equilibrium value of strain which the material will attain if given sufficient time. Because of (b), it follows that regardless of the previous history of the sample, complete removal of the stress must be followed by a gradual return to a state of zero strain. Thus *total recoverability* is a feature of anelastic behavior which follows as a corollary to (b). This condition is placed on the anelastic process to rule out viscous creep behavior, which would not be consistent with a relaxation process. Figure 1 schematically illustrates anelastic behavior under conditions of *creep* (constant stress) followed by *elastic aftereffect* (or "creep recovery") when the stress is removed.

Adoption of the two restrictions (a) and (b) above makes it possible to generalize Eq. (1) to introduce time derivatives. In view of the requirement of

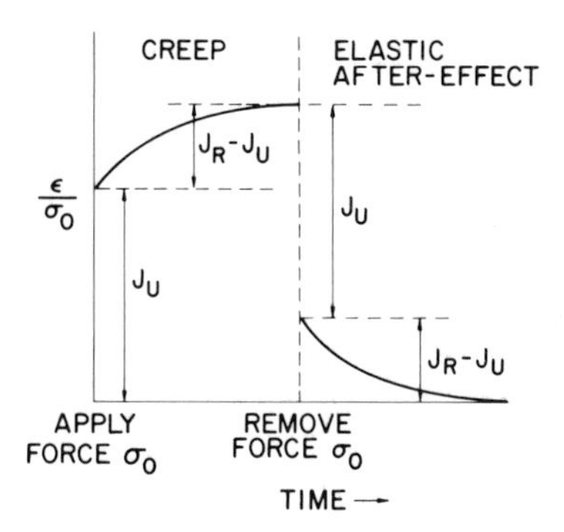

FIG. 1. Behavior of an anelastic solid upon application of a static stress σ_0 ("creep") and upon subsequent removal of the stress ("elastic aftereffect").

linearity each new term in the equation may have either σ or ε, or some time derivative of σ or ε, but products of such terms may not occur. The simplest generalization of Eq. (1) meeting this requirement and at the same time satisfying condition (b) is of the form

$$a\sigma + b\dot{\sigma} = c\varepsilon + d\dot{\varepsilon}. \tag{2}$$

This equation characterizes what we shall call the *standard anelastic solid*. Actually, only three of the constants appearing in Eq. (2) are independent since we may divide through by one of them. In anticipation of the significance of these constants, we rewrite Eq. (2) as

$$J_R\sigma + \tau_\sigma J_U\dot{\sigma} = \varepsilon + \tau_\sigma\dot{\varepsilon} \tag{3}$$

in terms of three new constants τ_σ, J_R, and J_U. Equation (3) may be solved under suitable conditions to obtain the response of a standard anelastic solid to different experimental situations. For example, under creep conditions we have

$$\sigma = \sigma_0, \qquad \dot{\sigma} = 0 \qquad (t > 0)$$
$$\varepsilon = J_U\sigma \qquad \text{at} \quad t = 0, \tag{4}$$

where J_U is the "unrelaxed compliance" which gives the instantaneous response of the solid to an applied stress (in this case applied at $t = 0$). If Eq. (3) is solved subject to conditions (4), the creep function $J(t) \equiv \varepsilon(t)/\sigma_0$ is obtained. The result is clearly

$$J(t) = J_U + (J_R - J_U)[1 - \exp(-t/\tau_\sigma)]. \tag{5}$$

The meaning of the constants J_R and τ_σ now becomes clear. J_R is the value of the compliance for $t \to \infty$, i.e., when the system has come to equilibrium. It is commonly called the "relaxed compliance." A quantity δJ defined by

$$\delta J \equiv J_{\mathrm{R}} - J_{\mathrm{U}} \tag{6}$$

and called the *relaxation of the compliance* measures the magnitude of the anelastic effect. Its significance is shown in Fig. 1. The quantity τ_σ is the time for the change in strain to reach $1/e$ of completion; it is commonly called the "relaxation time at constant stress." The whole process of the time-dependent adjustment of the strain to a new equilibrium value, after application of stress, is called *relaxation.*

Equation (3) can also be solved under conditions of constant strain, corresponding to a "stress relaxation" experiment. The conditions are

$$\varepsilon = \varepsilon_0, \qquad \dot{\varepsilon} = 0 \qquad (t > 0)$$

$$\varepsilon = M_{\mathrm{U}}\sigma \qquad \text{at} \quad t = 0, \tag{7}$$

where $M_{\mathrm{U}} \equiv 1/J_{\mathrm{U}}$ is the "unrelaxed modulus." It is easily shown that the solution for the stress relaxation function $M(t) \equiv \sigma(t)/\varepsilon_0$ is

$$M(t) = M_{\mathrm{R}} + (M_{\mathrm{U}} - M_{\mathrm{R}}) \exp(-t/\tau_\varepsilon), \tag{8}$$

where τ_ε, called the "relaxation time at constant strain," is given by

$$\tau_\varepsilon = \tau_\sigma(J_{\mathrm{U}}/J_{\mathrm{R}}) \tag{9}$$

and $M_{\mathrm{R}} \equiv 1/J_{\mathrm{R}}$ is the "relaxed modulus."

The behavior of a standard anelastic solid under dynamical conditions has been of great interest. Here a sinusoidal stress is applied; in complex notation

$$\sigma = \sigma_0 \exp(i\omega t), \tag{10}$$

where σ_0 is the stress amplitude and ω the circular frequency. For an anelastic material, the strain is in general not in phase with the stress, but lags behind by a phase angle ϕ, so that

$$\varepsilon = \varepsilon_0 \exp i(\omega t - \phi) \equiv (\varepsilon_1 - i\varepsilon_2) \exp(i\omega t), \tag{11}$$

where ε_0 is the strain amplitude while ε_1 and ε_2 are respectively the components of strain in phase with and 90° out of phase with the stress. The relationship between stress and strain in such an experiment may be expressed as $\varepsilon = J^*\sigma$, where J^* is called the *complex compliance* and is given by

$$J^*(\omega) = J_1(\omega) - iJ_2(\omega). \tag{12}$$

Note that the real part of J^*, $J_1 = \varepsilon_1/\sigma_0$, and the imaginary part, $J_2 = \varepsilon_2/\sigma_0$, are in general both functions of ω. The relationship of the phase angle ϕ to these quantities is readily seen from Eq. (11) to be

$$\tan\phi = \varepsilon_2/\varepsilon_1 = J_2/J_1. \tag{13}$$

Equations (10)–(13) are completely general, in the sense that they apply to any anelastic solid, not only to the standard solid. If now we substitute these equations into the differential equation (3) of the standard anelastic solid, we obtain

$$J_1(\omega) = J_U + \frac{\delta J}{1 + \omega^2 \tau_\sigma^2}, \tag{14}$$

$$J_2 = \delta J \frac{\omega \tau_\sigma}{1 + \omega^2 \tau_\sigma^2}. \tag{15}$$

These equations are the well-known "Debye equations," originally used to describe dielectric relaxation (where J_1 and J_2 are replaced by the components of the dielectric constant). The function $J_2(\omega)$, when plotted vs. log $\omega\tau_\sigma$ gives a symmetric peak centered about log $\omega\tau_\sigma = 0$ (i.e., $\omega\tau_\sigma = 1$), called the "Debye peak." On a similar plot the function $J_1(\omega)$ is antisymmetric about the point $J_1 = J_U + \frac{1}{2}\delta J$, log $\omega\tau_\sigma = 0$. Figure 2 shows these plots. An important characteristic of the Debye peak, i.e., of the function $\omega\tau/(1 + \omega^2\tau^2)$, is that its width at half-maximum $\Delta \log_{10} \omega\tau$ is just about one decade; specifically,

$$\Delta \log_{10} \omega\tau = 1.144. \tag{16}$$

The phase angle ϕ, or more often tan ϕ, is called the *internal friction* since it is a measure of the internal mechanisms that give rise to energy

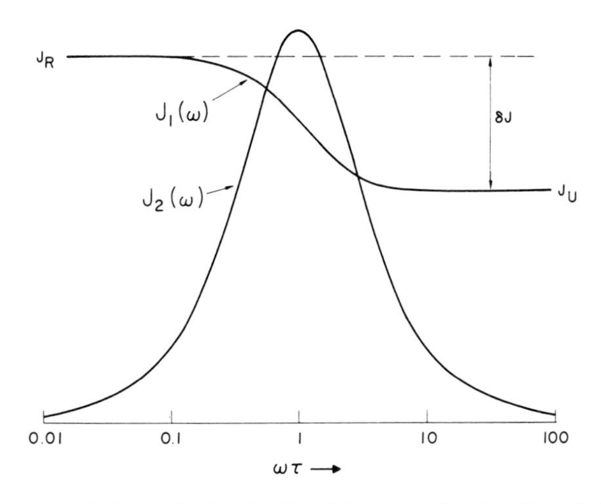

FIG. 2. Behavior of the standard anelastic solid upon application of a periodic stress with angular frequency ω. The real and imaginary parts J_1 and J_2 of the complex compliance are plotted against $\log_{10} \omega\tau$.

dissipation. There are several ways of measuring this energy dissipation (Beshers, 1976; Nowick and Berry, 1972). For example, if a bar of anelastic material is driven in forced vibration through one of its resonant modes of vibration, the width of the resonance curve (amplitude vs. frequency) at $1/\sqrt{2}$ of maximum expressed as a fraction of the resonant frequency ω_r is called Q^{-1}, i.e.,

$$(\omega_2 - \omega_1)/\omega_r \equiv Q^{-1} \tag{17}$$

by analogy to a resonant electric circuit. It is easily shown that $Q^{-1} = \tan\phi$ when $Q^{-1} \ll 1$. Similarly, in free vibration at the resonant frequency, one obtains the logarithmic decrement of free decay δ of the system, which is given by $\delta/\pi = \tan\phi$ when $\delta \ll 1$. Finally, one may propagate a high frequency periodic pulse of finite duration down the bar with velocity v, and measure the attenuation of the wave amplitude $A = A_0 \exp(-\alpha x)$ with distance x. The *attenuation* α is also related to $\tan\phi$, according to the relation

$$\tan\phi = 2v\alpha/\omega = Q^{-1} = \delta/\pi \tag{18}$$

which also includes the relations, involving Q^{-1} and δ, already mentioned above.

To obtain an expression for $\tan\phi$ for the standard anelastic solid, we must substitute Eqs. (14) and (15) into (13). By suitable manipulation, the resulting expression can be written in the form of a Debye peak, viz.,

$$\tan\phi = \frac{\Delta}{(1+\Delta)^{1/2}} \frac{\omega\bar{\tau}}{1+\omega^2\bar{\tau}^2}, \tag{19}$$

where

$$\Delta \equiv \delta J/J_{\mathrm{U}} \tag{20}$$

is a dimensionless parameter called the *relaxation strength*, while

$$\bar{\tau} \equiv \tau_\sigma(1+\Delta)^{-1/2} = (\tau_\sigma\tau_\varepsilon)^{1/2} \tag{21}$$

and τ_ε is given by Eq. (9). In other words, $\tan\phi$ also obeys the equation of a Debye peak, but this one is centered about $\omega\bar{\tau} = 1$, rather than at $\omega\tau_\sigma = 1$ as in the case of the J_2 peak. Since Δ is positive, $\bar{\tau} \leq \tau_\sigma$. Often the relaxation strength is small ($\Delta \ll 1$), in which case $\bar{\tau}$ and τ_σ are essentially indistinguishable.

The standard anelastic solid can also be described in terms of a three-component mechanical model consisting of two springs and a dashpot. Such models have been widely used in the description of anelasticity and viscoelasticity (Alfrey, 1948; Nowick and Berry, 1972, Chapter 3). For reasons of lack of space, however, we shall not employ this approach, but the interested reader should consult the references.

B. Relaxation Spectra

Several phenomena in solids are describable quite closely in terms of the equations of the standard anelastic solid. In many cases, however, this elementary model is insufficient to describe the behavior of the material. For example, the internal friction may show a superposition of two or more Debye peaks, instead of a single one. Or, there may be obtained a single peak that is broader than the theoretical width given by Eq. (16). To continue to treat such cases in terms of a linear theory, it is natural to introduce a multiplicity of standard anelastic solids, called a relaxation spectrum. Thus, instead of having a single relaxation term, involving a magnitude δJ and time constant τ_σ as in Eqs. (5), (14), and (15), we introduce a set of relaxation magnitudes $\delta J^{(i)}$ and corresponding relaxation times $\tau_\sigma^{(i)}$ which may be written $\{\delta J^{(i)};\ \tau_\sigma^{(i)}\}$. This set can be plotted as a series of spectral lines located on a scale of ln τ as shown in Fig. 3a. The corresponding functions $J(t)$, $J_1(\omega)$, and $J_2(\omega)$ are then summations over the expressions given in Eqs. (5), (14), and (15) for a single relaxation process. For example, Eq. (15) for $J_2(\omega)$ becomes

$$J_2(\omega) = \sum_i \delta J^{(i)} \frac{\omega\tau_\sigma^{(i)}}{1 + \omega^2\tau_\sigma^{(i)2}}. \tag{22}$$

Correspondingly, the stress relaxation function $M(t)$ becomes a summation over a set of relaxation times $\tau_\varepsilon^{(i)}$ derivable from $\tau_\sigma^{(i)}$ and $\delta J^{(i)}$. On the other hand, tan ϕ will, in general, no longer be expressible as a sum of Debye

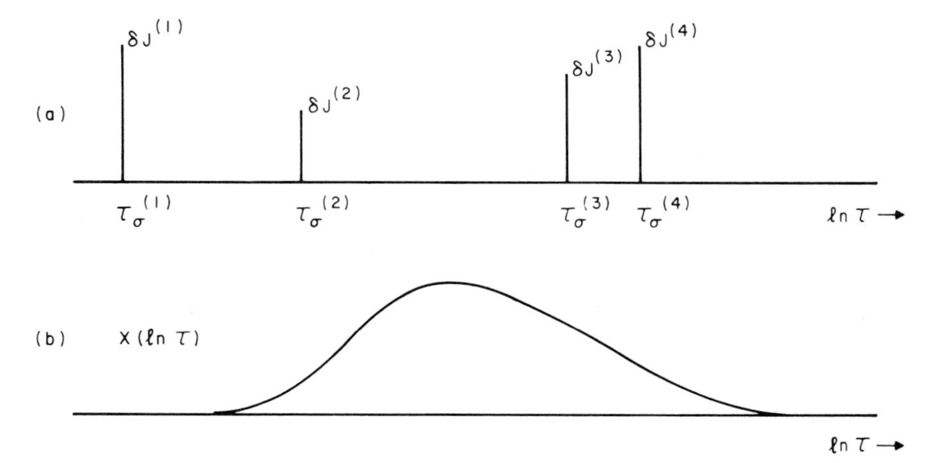

Fig. 3. Illustration of (a) a discrete relaxation spectrum and (b) a continuous relaxation spectrum.

peaks unless either the total relaxation strength is small, or the component relaxation times are widely separated. We refer to the case represented by Fig. 3a as a *discrete relaxation spectrum.* Another generalization of the standard anelastic solid takes the form of a *continuous spectrum* as shown in Fig. 3b. Here a distribution function $X(\ln \tau)$ is plotted, where $X(\ln \tau_\sigma)d \ln \tau_\sigma$ represents the relaxation magnitude in the range of relaxation time $d \ln \tau_\sigma$. The corresponding functions $J(t)$, $J_1(\omega)$, and $J_2(\omega)$ then become integrals, e.g.,

$$J_2(\omega) = \int_{-\infty}^{\infty} X(\ln \tau) \frac{\omega\tau}{1 + \omega^2\tau^2} d \ln \tau. \tag{23}$$

One of the objects of anelastic measurements is to obtain the relaxation spectrum from an experimentally measured function, e.g., $X(\ln \tau)$ from $J_2(\omega)$. In general this is a difficult problem, but it can be solved to various degrees of approximation (Alfrey and Doty, 1945; Gross, 1953; Staverman and Schwartzl, 1956). The problem is greatly simplified when the form of the distribution function is known. A useful form for $X(\ln \tau)$ is a Gaussian (or normal) distribution in $\ln \tau$, often called the *lognormal distribution*:

$$X(\ln \tau) = (\delta J/\beta\sqrt{\pi}) \exp[(-z/\beta)^2], \tag{24}$$

where $z = \ln(\tau/\tau_m)$ (τ_m being the most probable value of τ), β is a distribution parameter, and δJ is the total relaxation magnitude $\delta J = \int_{-\infty}^{\infty} X(\ln \tau)d \ln \tau$. The availability of tables of the functions $[J(t) - J_U]/\delta J$, $[J_1(\omega) - J_U]/\delta J$, and $J_2(\omega)/\delta J$ for different values of β (Nowick and Berry, 1961) have made this distribution function a very useful one to attempt to fit to experimental data. If the function fits well for some value of β, it can be concluded that Eq. (24) with that particular β value adequately represents the relaxation spectrum of the given material.

C. Dynamic Experiments as a Function of Temperature

It was shown earlier that for the standard anelastic solid under oscillatory stress, such experimentally observable functions as $J_1(\omega)$, $J_2(\omega)$, or $\tan \phi(\omega)$ take on particularly simple forms when plotted vs. $\log \omega\tau$.[1] In that discussion it was implied that ω is varied while τ remains constant. Although a continuous variation in vibration frequency is possible in some experimental methods, it is usually difficult to cover the two decades in frequency needed to trace out a Debye peak with a single specimen. There is, however, another way to trace out such a peak which is of great importance in practice, viz., by varying τ while keeping ω constant. The basis of this

[1] Henceforth, we shall drop the distinction between τ_σ and $\bar{\tau}$ on the grounds that the reader will be able to discern which is intended.

method is the knowledge that in many cases (in particular, those involving atomic migration) τ^{-1} is expressible by an Arrhenius equation of the form

$$\tau^{-1} = \tau_0^{-1} \exp(-H/kT), \tag{25}$$

where τ_0^{-1} is a frequency factor, H is the "activation energy," k is Boltzmann's constant, and T the absolute temperature. The fact that τ obeys such a relation makes it possible to vary τ over a wide range simply by changing the temperature. From Eq. (25) we may write

$$\ln \omega\tau = \ln \omega\tau_0 + \frac{H}{k}\left(\frac{1}{T}\right), \tag{26}$$

which shows that there is a linear relation between the quantity $\ln \omega\tau$ (plotted as abscissa in Fig. 2) and the reciprocal absolute temperature. Thus, the form of the plots of J_1, J_2, and $\tan \phi$ vs. T^{-1} are the same as the corresponding plots vs. $\ln \omega\tau$, except for a scale factor of H/k. The condition that $\omega\tau = 1$ at a Debye peak becomes, when the temperature is varied,

$$\ln \omega\tau_0 + \frac{H}{k}\left(\frac{1}{T_p}\right) = 0, \tag{27}$$

where T_p is the temperature at which the peak occurs. Equation (27) provides the basis for determining H from a series of peaks obtained at different frequencies. Specifically, if a series of frequencies are chosen and $\log_{10} \omega$ is plotted vs. $1/T_p$, a straight line is obtained whose negative slope is $H/2.303k$ and whose intercept gives τ_0. We shall later (in Fig. 6) see some examples of such plots.

From Eqs. (26) and (16) we may compute the width of a Debye peak at half-maximum when plotted vs. T^{-1} as

$$\Delta(T^{-1}) = (1.144)(2.303k/H) = 2.635k/H. \tag{28}$$

The peak width therefore varies inversely as the activation energy H. Equation (28) can be used to calculate H from the width if one knows that the material under study is a standard anelastic solid; more often, it is used to test whether the material is a standard solid, when H has been determined independently from experiments at different frequencies, using Eq. (27).

In the case of a spectrum of relaxation times dynamical experiments can again be performed as a function of temperature. Here the analysis is more complex because the distribution in $\ln \tau$ may be due to a distribution in the activation energy H, in the frequency factor τ_0^{-1}, or in both. In the particular case where a lognormal distribution is applicable one may distinguish among these possibilities by obtaining the distribution parameter β as a function of temperature. Under reasonable approximations (Nowick and

Berry, 1961, 1972) β is shown to vary as

$$\beta = |\beta_0 \pm \beta_H/kT|, \tag{29}$$

where β_0 and β_H are the distribution parameters for the quantities ln τ_0 and H, respectively.

D. Anisotropic Anelasticity

The analysis thus far has been based on generalization of Hooke's law given in the form of Eq. (1), where a single stress σ, its conjugate strain ε, and the corresponding compliance $J = \varepsilon/\sigma$ are considered. If σ and ε represent, respectively, tensile stress and strain, J is a reciprocal Young's modulus. On the other hand, for shear stress and strain, J would be a reciprocal shear modulus. In a crystal both Young's modulus and the shear modulus are functions of crystal orientation, i.e., they are anisotropic (Nye, 1957). It must be expected therefore that the relaxation δJ of such quantities is also anisotropic. In order to present a complete picture of the anelastic behavior of a crystal it is therefore necessary that the relaxation magnitude δJ be known as a function of orientation.

To illustrate the approach used we shall confine ourselves to the equations for cubic crystals, which are the ones that have been most widely studied. Elasticity of cubic crystals is describable in terms of the three well-known compliances S_{11}, S_{12}, and S_{44}. For present purposes, there is a considerable advantage to taking the following linear combinations of these constants:

$$S \equiv S_{44}, \qquad S' \equiv 2(S_{11} - S_{12}), \qquad S'' \equiv S_{11} + 2S_{12}. \tag{30}$$

The constant S is the shear compliance for shearing stress applied across the (100) plane in the [010] direction, while S' is the shear compliance for $(110)[1\bar{1}0]$ shear. On the other hand, S'' is one-third the hydrostatic compressibility.

In longitudinal or flexural vibration of a rod, the appropriate elastic constant is Young's modulus along the axis of the rod. The reciprocal of Young's modulus is the compliance J_E given in terms of the above constants by (Nye, 1957)

$$J_E = \tfrac{1}{3}(S' + S'') - (S' - S)\Gamma, \tag{31}$$

where Γ is an orientation factor, expressed in terms of the direction cosines γ_1, γ_2, and γ_3 of the sample axis relative to the three cube axes as

$$\Gamma = \gamma_1^2\gamma_2^2 + \gamma_2^2\gamma_3^2 + \gamma_3^2\gamma_1^2. \tag{32}$$

This orientation factor varies from a value $\Gamma = 0$ for a $\langle 100 \rangle$ oriented crystal to a value $\Gamma = \frac{1}{3}$ for a $\langle 111 \rangle$ orientation. The relaxation δJ_E is then expressible in terms of $\delta S'$ and $\delta S''$:

$$\delta J_E = \tfrac{1}{3}(\delta S' + \delta S'') - (\delta S'' - \delta S)\Gamma. \tag{33}$$

In some cases it is known from theoretical considerations that $\delta S'' = 0$. Under these conditions Eqs. (32) and (33) give

$$\delta J_{E\langle 100\rangle} = \delta S'/3, \qquad \delta J_{E\langle 111\rangle} = \delta S/3, \tag{34}$$

so that the shear relaxation $\delta S'$ may be obtained from an experiment carried out on a single crystal rod oriented in the $\langle 100\rangle$ direction while δS may be obtained from a study of a $\langle 111\rangle$ oriented crystal. All intermediate orientations give relaxation magnitudes that depend on both δS and $\delta S'$. In still other cases $\delta S = \delta S' = 0$ and only the hydrostatic constant S'' undergoes relaxation.

For torsional oscillations of a circular rod of a cubic crystal, the relaxation of the reciprocal shear modulus, i.e., δJ_G, is given by

$$\delta J_G = \delta S + 2(\delta S' - \delta S)\Gamma. \tag{35}$$

Thus, by combinations of experiments on longitudinal and torsional oscillations of selected crystals, the three basic relaxation magnitudes δS, $\delta S'$, and $\delta S''$, and their associated relaxation times can be studied. Figure 4 provides an example of such measurements on a Ag–Zn alloy.

For crystals of lower than cubic symmetry, similar relations exist, but they involve a larger number of constants to be determined (Nowick and Berry, 1972, Chapter 6).

The question of which (if any) of the basic relaxation magnitudes is zero is termed the "selection rules" for anelasticity. As already indicated above, the occurrence of such zero magnitudes is of profound importance in revealing the type of mechanism operating in a given anelastic phenomenon.

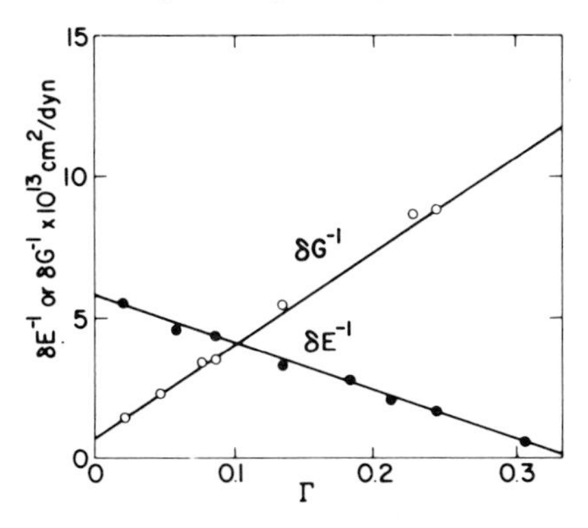

FIG. 4. Orientation dependence of the relaxations of the reciprocal Young's modulus (δE^{-1} or δJ_E) and torsional modulus (δG^{-1} or δJ_G) for the alloy Ag–26 at. % Zn at 350°C (from Seraphim and Nowick, 1961).

III. Internal Variables and the Thermodynamic Basis for Relaxation Spectra

In Section II we dealt with the purely formal approach to anelasticity, as a generalization of Hooke's law to include time-dependent effects but with certain restrictions. In this section we shall look into the origin of anelastic behavior, but still in general terms. It will be shown that the existence of a relaxation spectrum is a direct consequence of the existence of a set of internal variables which obey rather simple equations.

A. Case of a Single Internal Variable

Let us consider the simple case of a solid whose state is completely defined by the temperature, the state of stress (or of strain), and the value of one internal variable ξ. Such a variable is a parameter that may describe the internal state of the material, e.g., the distribution of a collection of defects or the electron distribution in the solid. Specific examples will be given in Section IV. Without loss of generality, we may define $\xi = 0$ as the equilibrium value of ξ at zero stress.

Since we seek a linear theory, all functional relationships between variables are given as Taylor expansions taken only to the linear approximation. Thus, treating σ and ξ as independent variables, and holding the temperature T as constant, we write the strain ε as

$$\varepsilon(\sigma, \xi) = J_{\mathrm{U}}\sigma + \kappa\xi. \tag{36}$$

In this way ε is defined as zero when $\sigma = 0$ and $\xi = 0$. The first assumption made here is that the strain depends on ξ, i.e., that there is a *coupling* between the internal variable and the elastic properties. The term $\kappa\xi$ is the anelastic strain, and the coefficient κ depends on the nature of the strain ε. As a second assumption, we require that there be a definite equilibrium value of ξ (denoted by $\bar{\xi}$) for each value of σ. Since $\bar{\xi} \equiv 0$ for $\sigma = 0$, the required linear relation is just

$$\bar{\xi} = \mu\sigma, \tag{37}$$

where μ depends on the type of stress applied and on T. For the final assumption, we require that with a change in stress, ξ approaches its equilibrium value according to the first-order kinetic equation

$$\frac{d\xi}{dt} = -\frac{1}{\tau}(\xi - \bar{\xi}), \tag{38}$$

where τ is a constant with dimensions of time. This approach to equilibrium is called "relaxation" of the internal variable.

The above three assumptions are sufficient to define a standard anelastic solid. By eliminating ξ, $\dot{\xi}$, and $\bar{\xi}$ from Eqs. (36)–(38), we readily obtain Eq. (3) with $J_R = J_U + \kappa\mu$ and τ_σ given by the quantity τ that appears in (38). Thus, a material describable by a single internal variable is completely equivalent to the standard anelastic solid.

B. Case of a Set of Coupled Internal Variables

We now consider the more general problem in which the internal state of the material is describable in terms of a finite number n of internal variables ξ_p ($p = 1, 2, \ldots, n$) and attempt to generalize Eqs. (36)–(38). Again, we define $\bar{\xi}_p = 0$ at $\sigma = 0$ for all values of p. Thus

$$\varepsilon = J_U\sigma + \sum_{p=1}^{n} \kappa_p \xi_p, \tag{39}$$

as the generalization of (36). Similarly, each ξ_p has an equilibrium value given by

$$\bar{\xi}_p = \mu_p \sigma. \tag{40}$$

The generalization of (38), however, is more complex. When the ξ_p are not in equilibrium, each ξ_p will, in general, approach equilibrium at a rate that depends on the deviation of *all* the variables from their equilibrium values. This behavior constitutes coupling of the internal variables. Thus, to maintain the linear approximation in its full generality, we must write

$$\dot{\xi}_p = -\sum_{q=1}^{n} \omega_{pq}(\xi_q - \bar{\xi}_q) \qquad (p = 1, 2, \ldots, n) \tag{41}$$

in terms of a set of coefficients ω_{pq}. Equations (41) constitute a coupled set of first-order differential equations in n unknowns. It can be shown (Meixner, 1949, 1954), however, that it is always possible to find a linear transformation to a new set of variables $\xi'_r = \sum_p B_{rp}\xi_p$, such that Eqs. (41) become

$$\dot{\xi}'_r = -\tau_r^{-1}(\xi'_r - \bar{\xi}'_r) \qquad (r = 1, 2, \ldots, n); \tag{42}$$

i.e., the new variables ξ'_r are decoupled. Further, all of the quantities τ_r, which may be obtained from the coefficients ω_{pq} are real and positive. The basis for the existence of such a transformation originates in classical thermodynamics and in the thermodynamics of irreversible processes. For further details, see Nowick and Berry (1972, §5.3).

The new variables ξ'_r are actually just as good a set of internal variables for describing the system as the original ξ_p. These new variables still obey Eqs. (39) and (40) but with different coefficients κ'_r and μ'_r, respectively. The important difference lies in Eq. (42), i.e., in the decoupling of these new internal variables. Because of the analogy to the problem of normal coordinates in the vibration of a system of many degrees of freedom, the set of new variables ξ'_r are called "normal internal variables."

The solution to Eq. (42) under constant stress is clearly

$$\xi'_r = \bar{\xi}'_r[1 - \exp(-t/\tau_r)], \tag{43}$$

which is the same form as for a single internal variable. Accordingly, we see that each normal internal variable ξ'_r undergoes relaxation independently of the others, so that each solution of the type (43) is referred to as a *relaxational normal mode*. From Eqs. (43) and the modified forms of (39) and (40) we readily obtain for the creep function

$$J(t) = J_{\mathrm{U}} + \sum_{r=1}^{n} \kappa'_r \mu'_r[1 - \exp(-t/\tau_r)]. \tag{44}$$

This behavior corresponds to a discrete relaxation spectrum of n lines $\{\delta J^{(r)}; \tau_\sigma^{(r)}\}$, with $\delta J^{(r)} = \kappa'_r \mu'_r$ and $\tau_\sigma^{(r)} = \tau_r$, while the total relaxation magnitude is

$$\delta J = \sum_r \delta J^{(r)} = \sum_r \kappa'_r \mu'_r. \tag{45}$$

Thus, each line of a discrete relaxation spectrum is related to one of the normal internal variables.

If n becomes very large and the τ values close together, the discrete spectrum will go over into a continuous spectrum, as already described in Section II. A continuous spectrum is then the result of a quasi-continuously varying set of normal internal variables.

IV. Physical Origins of Anelasticity

Having shown in Section III that anelasticity is a consequence of the relaxation of internal variables that are coupled to the elastic properties of the material, we are now ready to illustrate the most important examples of anelastic phenomena that have been studied. These examples fall into several general classes which will be considered in turn in the present section.

A. Point Defect Relaxations

The elementary point defects are the vacancy (missing atom), the interstitial (extra atom), and the impurity atom (either interstitial or substitutional). In addition, point defect complexes can be created as pairs or higher clusters of the elementary defects. The type of point defect that may give rise to anelasticity in a crystal is one that has lower symmetry than the crystal as a whole, so that there exists a number n_{d}, of crystallographically equivalent defect orientations. In the absence of stress, if the total concentration (i.e., mole fraction) of defects is C_0, the concentration in each equivalent orientation must be C_0/n_d. In the presence of stress the concentrations of defects that have different orientations need no longer be the same. If we write $\xi_p \equiv C_p - (C_0/n_d)$, C_p being the concentration in orientation p, it is clear that the ξ_p so defined constitute internal variables in the sense of the previous section. The quantities κ_p of Eq. (39) are now components of a tensor (called the $\boldsymbol{\lambda}$ tensor) which measures the distortion produced by the defect in question, specifically, the strain per unit defect concentration in orientation p (Nowick and Heller, 1963). This $\boldsymbol{\lambda}$ tensor, like any second-order symmetric tensor, can be described in terms of a tensor ellipsoid with three principal values measured along the three principal axes of the ellipsoid. The resulting expression for the relaxation magnitude of the appropriate compliance constant then takes the form

$$\delta J = (C_0 v_0/n_d kT)(\delta\lambda)^2, \tag{46}$$

where v_0 is the molecular volume and $\delta\lambda$ involves a particular combination of principal values of the $\boldsymbol{\lambda}$ tensor.

Studies of anelasticity due to point defects can give the following information about the defects involved:

(a) From the observation of which compliances do and which do not show anelastic behavior one can obtain considerable information concerning the local symmetry of the point defects that undergo relaxation. This involves the "selection rules" of anelasticity (Nowick and Heller, 1965) and requires that experiments be carried out on crystals cut in particularly simple orientations and vibrating in simple modes.

(b) From the measured relaxation magnitudes and a knowledge of the total concentration of defects, certain combinations of the principal values of the $\boldsymbol{\lambda}$ tensor can be obtained, specifically, the quantity $|\delta\lambda|$ in Eq. (46). This gives information on the directionality of the distortion produced by the defect under consideration. Conversely, if the principal values of the $\boldsymbol{\lambda}$ tensor are known, the relaxation magnitude can be used to obtain the total defect concentration C_0.

(c) From the measured relaxation times and a model of the defect in the crystal, one may determine the jump rates of the atoms or ions that change positions in the process of reorientation of the defect (Nowick, 1967). If relaxation times are measured as a function of temperature, appropriate activation energies and frequency factors may be obtained (see Section II,C). This often amounts to obtaining direct information on diffusion (mass transport) in a crystal at temperatures far below those at which diffusion can normally be measured.

Turning now to specific examples of point defect relaxations, we first

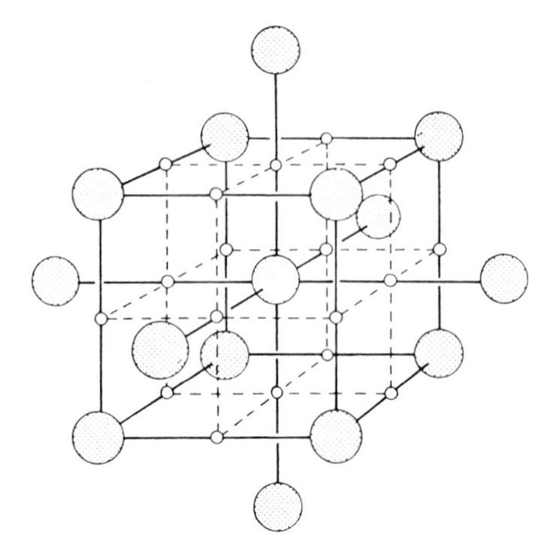

FIG. 5. The location of the octahedral interstitial sites (small circles) in the body-centered cubic lattice. The large circles represent the metal atoms.

mention the oldest and best understood example, the *Snoek relaxation.* In this case the point defects are interstitial solute atoms (e.g., C, N, or O) in body-centered cubic metals (e.g., Fe, Nb, Ta). The solute is located in the "octahedral" interstitial sites that are at the centers of the cell edges and in the face centers of the bcc lattice, as shown in Fig. 5. As one can see from this figure, there are three equivalent orientations of these interstitial sites depending on whether the two nearest neighbor solvent atoms lie along the x, y, or z directions. In this case the defect and therefore the lattice distortion that it produces have tetragonal symmetry. Because $\delta\lambda$ in Eq. (46) is now known for several choices of solute and solvent, measurement of the relaxation magnitude constitutes an excellent method of analysis for determining the interstitial content in solution down to very small concentrations. Since elements such as C, N, or O are difficult to determine chemically, this is a very important application. On the other hand, measurement of relaxation times has made it possible to obtain activation energies for migration of these solutes to a high precision. Figure 6 shows some of the Arrhenius plots for various interstitials in Ta, Nb, and V. For a more detailed discussion of the Snoek effect, see Nowick and Berry (1972, Chapter 9).

The *Zener relaxation* is a second example of a point defect relaxation. The responsible defects are substitutional solute atoms (s) in adjacent lattice sites, to form the so-called s-s pairs. In concentrated solid solutions, however, the concept of simple pairs breaks down and the relaxation is better treated in terms of the concept of directional short-range order (Le Claire and Lomer, 1954). Figure 7 shows the internal friction peaks in a

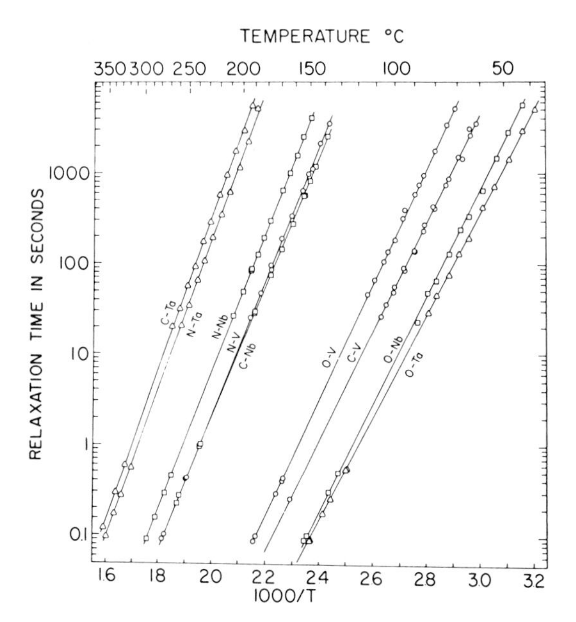

FIG. 6. Arrhenius plots of the Snoek relaxation time in several alloys (from Powers and Doyle, 1959).

series of Ag–Zn alloys. The peak heights in dilute solution vary in proportion to the square of the solute concentration, in accordance with the pair mechanism. The Zener relaxation effect has been observed in a wide variety of alloys involving fcc, bcc, and hcp structure as well as in ionic crystals of the NaCl structure (Nowick and Berry, 1972, Chapter 10). It generally is describable in terms of a narrow distribution of relaxation times, rather than as a standard anelastic solid. The kinetics (i.e., relaxation times) of the Zener effect have been widely used to study atomic mobility in solid solutions. Comparison of activation energies determined from anelasticity with high temperature diffusion measurements shows very good agreement, as summarized in Fig. 8.

A wide variety of other types of other point defect relaxations have been observed, including i-s, v-s, and i-i pairs (where v is a vacancy, while s is a substitutional, and i an interstitial impurity) as well as simple interstitial defects (Nowick and Berry, 1972, Chapter 11). Some well-studied examples of i-s, s-v, and i-i pairs are, respectively, the Li–B pair in silicon (Berry, 1970), the Ca^{++}–oxygen vacancy pair in ThO_2 (Wachtman, 1963), and the oxygen–oxygen pair in tantalum (Powers, 1955). Intrinsic defects in metals produced by irradiation have also been studied extensively (Nowick, 1977).

In noncrystalline substances many of the structural point defects that

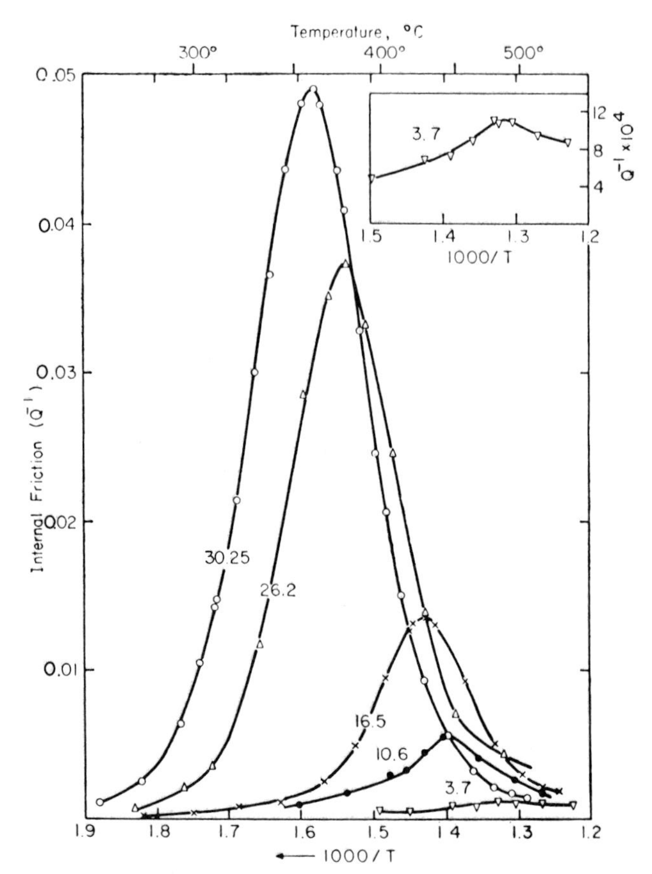

FIG. 7. Internal friction peaks (at ~500 Hz) of a series of similarly oriented single crystals of Ag–Zn alloys. The number on each curve is the at. % Zn (from Seraphim and Nowick, 1961).

are defined for crystals become meaningless. Nevertheless, relaxations attributable to impurity atoms or ions have been observed. In particular in oxide glasses two characteristic peaks are observed. The lower temperature peak is related to alkali ions, while the upper temperature peak is not well understood but may be due to the motion of hydrogen ions (Doremus, 1973).

In the case of amorphous metals or "metallic glasses," a detailed study of anelastic behavior of amorphous $Pd_{80}Si_{20}$ was recently reported by Berry (1977). In contrast to the relatively sharp Zener relaxation observed for crystalline alloys, behavior corresponding to an extremely wide distribution

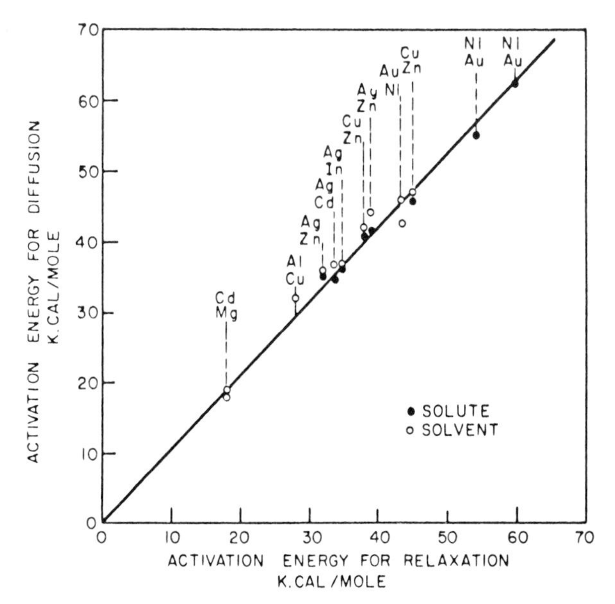

FIG. 8. Comparison of activation energy for self-diffusion of the two constituents of a binary solid solution with the activation energy for the Zener relaxation (from Seraphim, 1960).

of relaxation times was found for Pd–Si, with an internal friction that continues to rise exponentially as the temperature increases.

B. Dislocation Relaxations

Dislocations are the line defects in crystals that are responsible for plastic deformation. One of the characteristics of a dislocation is that shear stress (resolved on the slip plane) will set it into motion, and conversely, dislocation motion gives rise to shear strain. Accordingly, the coupling required by Eq. (36) is present when the internal variable is taken as the dislocation displacement. The type of displacement usually anticipated is the bowing out of the dislocation line between two anchor points, as shown in Fig. 9. In many cases, however, dislocation motion is so facile as to be

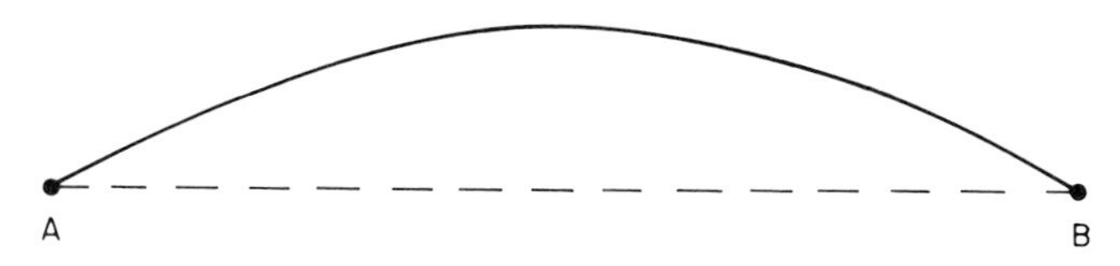

FIG. 9. Schematic illustration of the bowing out of a dislocation line between two anchor points under the influence of a shear stress. The dashed line is the position of the dislocation in the absence of stress.

always in phase with the applied stress, so that the relaxation behavior required by Eq. (38) is not present. Only when the dislocation motion is hindered can this relaxation condition obtain.

One type of dislocation that can be expected to show relaxation is that which lies in the "Peierls valley." This valley is a direction of minimum energy for the dislocation line in the crystal (usually, a close-packed direction); therefore, moving such a dislocation means going over an activation barrier to the next minimum energy position. Such motion is most easily accomplished by the generation and propagation of a pair of kinks, as shown in Fig. 10. A relaxation process in which the internal variable is the

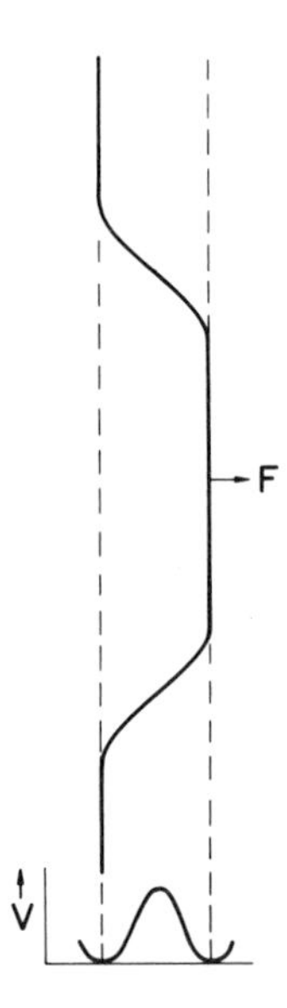

FIG. 10. Movement of a dislocation line that originally lies in a Peierls valley through the generation of a pair of kinks. The dashed lines show the location of the Peierls valleys and the bottom curve represents potential energy V vs. displacement.

number of such kink pairs that have been generated is therefore suggested. Such a relaxation process is believed to be responsible for the "Bordoni peaks" of internal friction which are observed in the range ~50–100°K for typical fcc metals at kilohertz frequencies after plastic deformation (Bordoni, 1960; Niblett, 1966; De Batist, 1972). Figure 11 shows that the peak is produced by plastic deformation and eliminated by annealing. It is noteworthy that this peak is considerably broader than that of a standard anelastic solid and, in addition, possesses a smaller satellite peak on the low tempera-

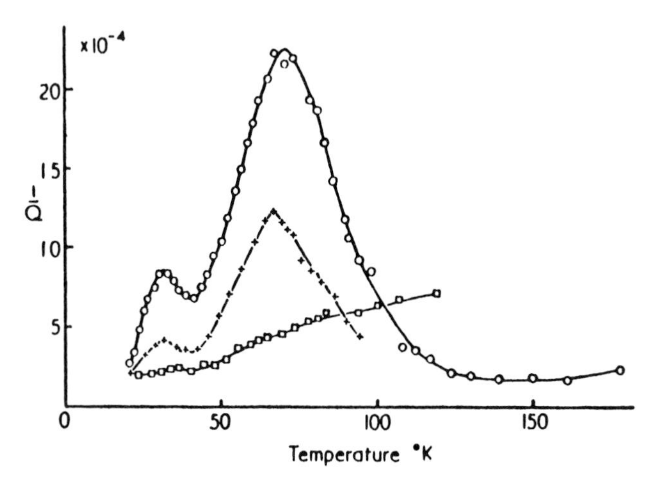

FIG. 11. The Bordoni peak in copper and the effect of annealing. Circles, after straining 8.4%; crosses, after 1-hr anneal at 180°C; boxes, after 1 hr at 350°C. Frequency is 1100 Hz (from Niblett and Wilks, 1957).

ture side. The interpretations of the various complexities of the Bordoni relaxation peaks has received much attention (Seeger and Schiller, 1966; De Batist, 1972; Beshers and Gottschall, 1975; Benoit *et al.*, 1976). Related relaxation peaks have also been observed in the bcc metals (Chambers, 1966), and their investigation has recently been reviewed (Seeger and Wüthrich, 1976).

Still another type of dislocation line that can undergo relaxation is one that lies in an arbitrary direction but is held up in its motion by point defects which serve as "pinning points." The internal variables can then be considered to be the extent of dislocation bowing, which is controlled by breaking away of the dislocation line from the pinning points. Such breaking away will constitute a thermally activated process at low stress levels, although it can become an athermal process at higher stresses (leading to "hysteresis" rather than relaxation behavior). As an alternative to breaking away, there is the possibility that the point defect may be dragged along with the moving dislocation. There are two known relaxation processes that may be due to mechanisms of the above types. The first involves internal friction peaks that occur below room temperature (but above the Bordoni peak) in plastically deformed metals. These peaks, discovered by Hasiguti and co-workers (1962), are often called "Hasiguti peaks." The second is found in deformed bcc metals containing interstitial solutes and occurs at somewhat higher temperatures. It has been called the Snoek–Köster or "cold work" peak

(Nowick and Berry, 1972; De Batist, 1972) and appears to be due to the bowing out of dislocation loops that drag along the solute atoms. It has been proposed (Schoeck and Mondino, 1963) that the activation energy is given by

$$H = H_0 + E_B,$$

where H_0 is the diffusion activation energy of the interstitial solute, and E_B its binding energy to the dislocation.

There are other relaxations due to dislocation motions, but the ones mentioned above are those that have been most extensively studied.

C. Grain-Boundary Relaxation

A fine grained polycrystalline metal shows a large and very broad relaxation peak at relatively high temperatures, which is not present for single crystals, as illustrated for aluminum in Fig. 12. This relaxation has been

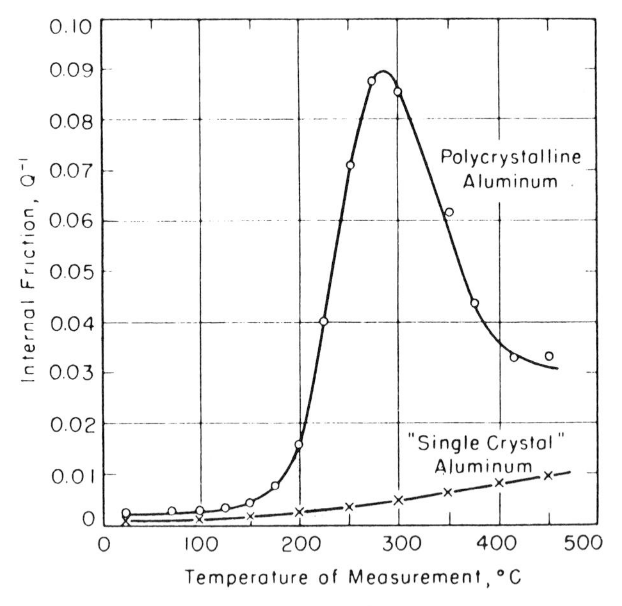

Fig. 12. Comparison of the internal friction of polycrystalline and "single crystal" (actually very coarse grained) aluminum at a frequency of 0.8 Hz (from Kê, 1947).

attributed to viscous sliding of one grain over another at the interface (Zener, 1948). The internal variables are then the shear displacements at the various boundaries. Due to the variety of structures possessed by boundaries that meet at different crystallographic orientations, a distribution in activation energies can be anticipated as the main cause of the breadth of the relaxation peak.

The above description is actually oversimplified since there are cases in which multiple peaks and other complexities are observed (Nowick and Berry, 1972, Chapter 15). It has also been suggested that boundary migration as well as sliding can contribute to the anelastic effects (Roberts and Leak, 1975).

D. Thermoelastic Relaxation

The ability of an applied stress to change the temperature of a solid is the basis for thermoelastic relaxation, in which heat flows between a specimen and its surroundings or between one part of a specimen and another. In this case we select the entropy s per unit volume as the internal variable, so that the quantity κ in Eq. (36) measures

$$\kappa = (\partial \varepsilon / \partial s)_\sigma = \alpha T / c_\sigma, \tag{47}$$

where α is the linear thermal expansion coefficient and c_σ the specific heat per unit volume at constant stress. The last equality follows from the thermodynamic reciprocity relations and shows that both κ and α are manifestations of thermoelastic coupling. Thus, if a sample, initially in thermal equilibrium with its surroundings, is deformed in tension, heat will flow into it, giving rise to thermoelastic relaxation. The driving force for the relaxation is the change in temperature T, which is conjugate to the internal variable s.

The most important manifestations of thermoelastic relaxation occur under inhomogeneous deformation, as for example, in the flexure of a thin reed. In this case relaxation occurs by heat flow from the hotter (compressed) layers to the cooler (extended) layers of the sample. The behavior is very nearly that of a standard anelastic solid, where the relaxation time is inversely proportional to the thermal diffusivity of the material (Zener, 1948).

E. Phonon Relaxation

The thermoelastic effect is a thermal relaxation which is generally observed at relatively low frequencies. On the other hand, for high frequency (megahertz or higher) phenomena involving ultrasonic waves, the lattice thermal vibrations are best regarded as quasi-particles called phonons which, in equilibrium, obey the well-known Planck distribution law (Ziman, 1960). It was first suggested by Akhieser (1939) that an ultrasonic wave traversing the crystal would change the frequencies of the thermal phonons and thereby disturb the Planck distribution. This leads to a phonon relaxation, or Akhieser effect, which may be regarded as a flow of heat, not in real space as in the thermoelastic effect, but among phonons having different polarization and propagation directions (Woodruff and Ehrenreich, 1961; Mason and Bateman, 1964). Standard solid anelastic behavior is obtained, but only in a range for which $\omega\tau \ll 1$, where τ is the time for establishment of phonon equilibrium. When $\omega\tau \gtrsim 1$, the process becomes one of collision between the ultrasonic waves and the thermal phonons and is no longer a relaxation effect (Beyer and Letcher, 1969, Chapter 8; Truell *et al.*, 1969, Chapter 3, Part VI; Klemens, 1965).

F. Magnetic Relaxations

In a ferromagnetic material, the coupling between elasticity and magnetism occurs via magnetostriction. The appropriate internal variable is now the change in magnetic induction ΔB. A sudden application of stress produces an instantaneous change in the intensity of magnetization I. This in turn gives rise to eddy currents, which flow in such a way as to preserve instantaneously the flux density B at its original value. As time proceeds, the eddy current field collapses and B relaxes toward an equilibrium value appropriate to the applied stress σ.

There are two types of such magnetic relaxations, one due to macroeddy currents and the other to microeddy currents. The macroeffect is illustrated by a long rod axially magnetized to an intermediate induction B. Application of a uniaxial stress along the axis gives rise to circumferential eddy currents. The behavior corresponds to a discrete (but infinite) relaxation spectrum (Zener, 1948; Nowick and Berry, 1972, §18.2), and the relaxation strength is zero for either a demagnetized or fully (saturation) magnetized sample. Figure 13 shows some recent data that demonstrates this effect. The microeddy current effect, on the other hand, can be observed

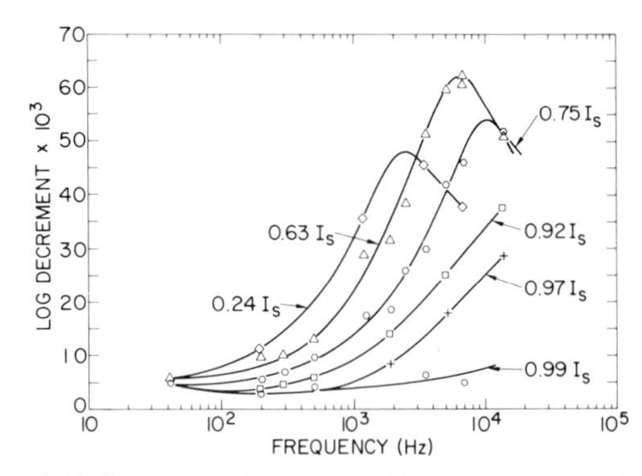

Fig. 13. Internal friction produced by macroeddy currents in a reed of pure Ni, 0.060 in. thick, magnetized to the levels shown (unpublished data of Berry and Pritchet).

in demagnetized samples. It reflects the existence of a domain structure in a ferromagnetic material. The movements of domain walls induced by stress give rise to local eddy currents on the scale of the domain size (typically $\sim 10^{-2}$ cm), the losses due to which are additive. This effect occurs at higher frequencies than the macroeffects and is very sensitive to internal stress (which retards the motion of domain walls). The linear dependence on ω at low frequencies and the elimination of the effect through cold working, which introduces internal stress, are illustrated in Fig. 14.

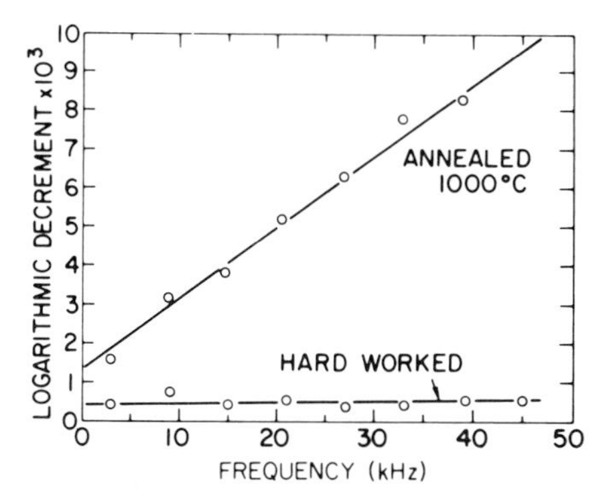

FIG. 14. Frequency dependence of the internal friction is demagnetized Permalloy in both the annealed and cold worked conditions (from Bozorth, 1951).

Magnetoelastic relaxation effects of a different nature occur in ferromagnetic alloys in which the state of short-range order is a function of the state of magnetization. Here the motion of domain boundaries under stress is not instantaneous, but is controlled by the rate at which the local atomic order can undergo a change in orientation. This drag on the domain boundary motion gives rise to relaxation peaks in alloys such as Fe–Al and Fe–Co (Nowick and Berry, 1972, §18.5).

G. ELECTRONIC RELAXATIONS

Stress may induce a change in the electronic configuration of a crystal provided that a redistribution of electrons can give rise to a corresponding strain. Such relaxation effects occur in metals, semiconductors, and insulators, but due to the differences in electronic structure of these three types of materials, the interpretation of each requires different concepts.

The electrons in a metal form a highly degenerate gas that obeys Fermi–Dirac statistics and fills a volume in momentum space (or wave-vector k space) up to the "Fermi surface." Application of stress in general results in an instantaneous promotion of some electrons located very close to the Fermi surface to higher energy states and demotion of others to lower

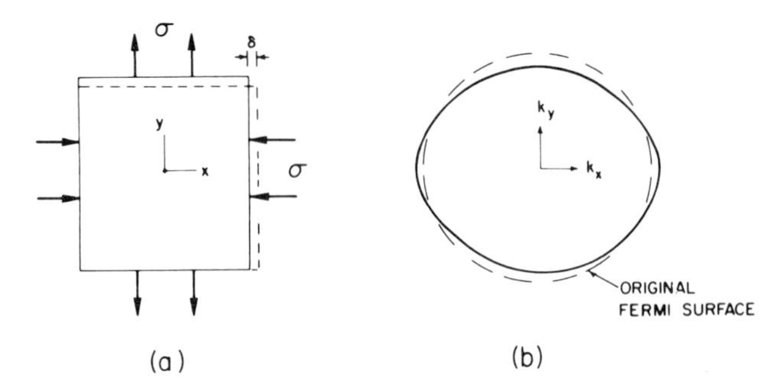

FIG. 15. (a) Deformation of a sample by the application of shear stress. (b) The corresponding instantaneous deformation of the Fermi surface for an ideal free-electron metal. (Relaxation then occurs such as to return to the spherical Fermi surface.)

energy states (Beyer and Letcher, 1969, Chapter 10). This is illustrated in Fig. 15 for a free-electron metal, where the Fermi surface is a sphere. The subsequent relaxation of the electron distribution toward equilibrium with the stress-displaced energy states causes the metal to behave as a standard anelastic solid. The relaxation time is $\tau = l/v_F$, where l is the mean free path and v_F the magnitude of the velocity of electrons at the Fermi surface. Just as in the case of phonon relaxation, standard-solid behavior is obtained only when $\omega\tau$ is extremely small. At higher values of $\omega\tau$ electron–phonon collisions must be considered, and the resulting behavior ceases to be an anelastic relaxation process (Beyer and Letcher, 1969; Truell *et al.*, 1969, Chapter 3, Part V).

In semiconductors there are two ways by which the electronic configuration can couple to the elastic properties of the crystal. The first mechanism applies to so-called "multivalley" semiconductors (Mason, 1966). This case is very analogous to the distortion by stress of the Fermi surface in metals, except that in a semiconductor there are relatively few electrons, which are located in pockets of wave-vector space called "valleys." In germanium these valleys lie along $\langle 111 \rangle$ directions of the crystal. Accordingly, only the elastic shear constant S shows this relaxation effect and not the constant S' [see Eq. (30)]. The second coupling mechanism between stress and electrons occurs in semiconductors that are piezoelectric, so that a local strain gives rise to an internal electric field that is proportional to the strain. When an ultrasonic wave is passed through the material, mobile charge carriers move toward regions of potential minimum in the piezoelectrically induced electric field. In general, this is not an anelastic process, but in the simple case in which diffusion of the charge carriers can be neglected, standard-solid behavior is obtained (Truell *et al.*, 1969).

Finally, in insulating crystals relaxation effects have been observed, which can be attributed to the redistribution of bound electrons among two or more equivalent sites in much the same way as that which occurs for point defects (Section IV,A). One set of examples involves crystals containing transition metal ions in two valence states, as in the case of magnetite (Fe_3O_4) and various related compounds. A second set of examples occurs for insulating crystals that possess trapped holes or electrons. A well-studied example is the case of lithium-doped NiO, in which an electronic hole accompanies each Li^+ ion and is bound to that ion, moving about it by a hopping mechanism (van Houten and Bosman, 1964). A review of these various relaxation effects is given in Nowick and Berry (1972, §19.3).

Acknowledgment

This article was prepared as part of a research program supported by the National Science Foundation under grant NSF DMR75-09603.

References

Akhieser, A. (1939). *J. Phys. (Moscow)* **1**, 277.
Alfrey, T. (1948). "Mechanical Behavior of High Polymers." Wiley (Interscience), New York.
Alfrey, T., and Doty, P. (1945). *J. Appl. Phys.* **16**, 700.
Benoit, W., Fantozzi, G., and Esnouf, C. (1976). *Nuovo Cimento B* **33**, 1.
Berry, B. S. (1970). *J. Phys. Chem. Solids* **31**, 1827.
Berry, B. S. (1977). *In* "Metallic Glasses." Am. Soc. Metals, Metals Park, Ohio (in press).
Beshers, D. N. (1976). *In* "Techniques of Metals Research" (R. F. Bunshah, ed.), Vol. 7, Part 2. Wiley, New York.
Beshers, D. N., and Gottschall, R. J. (1975). *In* "Internal Friction and Ultrasonic Attenuation in Crystalline Solids" (D. Lenz and K. Lucke, eds.), Vol. 2, p. 134. Springer-Verlag, Berlin and New York.
Beyer, R. T., and Letcher, S. V. (1969). "Physical Ultrasonics." Academic Press, New York.
Bordoni, P. G. (1960). *Nuovo Cimento* **17**, Suppl. 1, 43.
Bozorth, R. M. (1951). "Ferromagnetism." Van Nostrand-Reinhold, Princeton, New Jersey.
Chambers, R. H. (1966). *Phys. Acoust.* **3A**, 123.
De Batist, R. (1972). "Internal Friction of Structural Defects in Crystalline Solids." North-Holland Publ., Amsterdam.
Doremus, R. H. (1973). "Glass Science," Chapter 11. Wiley, New York.
Gross, B. (1953). "Mathematical Structure of the Theories of Viscoelasticity." Hermann, Paris.
Hasiguti, R. R., Igata, N., and Kamoshita, G. (1962). *Acta Metall.* **10**, 442.
Kê, T. S. (1947). *Phys. Rev.* **71**, 533.
Klemens, P. G. (1965). *Phys. Acoust.* **3B**, 201.
Le Claire, A. D., and Lomer, W. M. (1954). *Acta Metall.* **2**, 731.
McCrum, N. G., Read, B. E., and Williams, G. (1967). "Anelastic and Dielectric Effects in Polymeric Solids." Wiley, New York.
Mason, W. P. (1966). *Phys. Acoust.* **4A**, 299.
Mason, W. P., and Bateman, T. B. (1964). *J. Acoust. Soc. Am.* **36**, 644.
Meixner, J. (1949). *Z. Naturforsch., Teil A* **4**, 594.
Meixner, J. (1954). *Z. Naturforsch., Teil A* **9**, 654.
Niblett, D. H. (1966). *Phys. Acoust.* **3A**, 77.

Niblett, D. H., and Wilks, J. (1957). *Philos. Mag.* [7] **2**, 1427.

Nowick, A. S. (1967). *Adv. Phys.* **16**, 1.

Nowick, A. S. (1977). *J. Nucl. Mater.* (in press).

Nowick, A. S., and Berry, B. S. (1961). *IBM J. Res. Dev.* **5**, 297 and 312.

Nowick, A. S., and Berry, B. S. (1972). "Anelastic Relaxation in Crystalline Solids." Academic Press, New York.

Nowick, A. S., and Heller, W. R. (1963). *Adv. Phys.* **12**, 251.

Nowick, A. S., and Heller, W. R. (1965). *Adv. Phys.* **14**, 101.

Nye, J. F. (1957). "Physical Properties of Crystals." Oxford Univ. Press, London and New York.

Powers, R. W. (1955). *Acta Metall.* **3**, 135.

Powers, R. W., and Doyle, M. V. (1959). *J. Appl. Phys.* **30**, 514.

Roberts, G., and Leak, G. M. (1975). *In* "Internal Friction and Ultrasonic Attenuation in Crystalline Solids" (D. Lenz and K. Lucke, eds.), Vol. 1, p. 370. Springer-Verlag, Berlin and New York.

Schoek, G., and Mondino, M. (1963). *J. Phys. Soc. Jpn.* **18**, Suppl. I, 149.

Seeger, A., and Schiller, P. (1966). *Phys. Acoust.* **3A**, 361.

Seeger, A., and Wüthrich, G. (1976). *Nuovo Cimento B* **33**, 38.

Seraphim, D. P. (1960). *Trans. Metall. Soc. AIME* **218**, 485.

Seraphim, D. P., and Nowick, A. S. (1961). *Acta Metall.* **9**, 85.

Staverman, A. J., and Schwartzl, F. (1956). *In* "Die Physik der Hochpolymeren" (H. A. Stuart, ed.), Chapter 1. Springer-Verlag, Berlin and New York.

Truell, R., Elbaum, C., and Chick, B. B. (1969). "Ultrasonic Methods in Solid State Physics." Academic Press, New York.

van Houten, S., and Bosman, A. J. (1964). *Proc. Buhl Int. Conf. Transition Metal Compounds, 1st*, p. 123.

Wachtman, J. B., Jr. (1963). *Phys. Rev.* **131**, 517.

Woodruff, T. O., and Ehrenreich, H. (1961). *Phys. Rev.* **123**, 1533.

Zener, C. (1948). "Elasticity and Anelasticity of Metals." Univ. of Chicago Press, Chicago, Illinois.

Ziman, J. M. (1960). "Electrons and Phonons." Oxford Univ. Press, London and New York.

—2—

Structural Instability of A-15 Superconductors[1]

L. R. TESTARDI

Bell Laboratories
Murray Hill, New Jersey 07974

I. Introduction

Since the publication of the original two review articles on the A-15 superconductors (Testardi, 1973; Weger and Goldberg, 1973) a number of other reviews have appeared (Hein, 1973; Izyumov and Kurmaev, 1974; Dew-Hughes, 1975; Testardi, 1975a,b). The field has remained sufficiently active, however, that many new experimental and theoretical developments have occurred too recently to be included in the prior review articles. In what follows we shall summarize a number of these findings with particular emphasis on experimental work.

II. Metallurgical and Chemical Studies

The translation of the book on superconducting materials by Savitskii *et al.* (1973) has provided a wealth of new information to English speaking mater-

[1] This article has been solicited by L. Gor'kov for the Mir Publishing Co. of Moscow to accompany the Russian translation of a previous review chapter ("Physical Acoustics," Vol. X).

ials scientists. The book contains considerable data on the preparation, analysis, and properties of superconductors including the A-15 compounds.

Johnson and Douglass (1974b) have compiled the reported A-15 phases and their lattice parameters. An enlarged set of Geller radii, predicted lattice parameters, and a discussion of the systematic behavior are given.

Wang (1973, 1974) has discussed some chemical aspects of A-15 compounds based on the atomic orbitals of the constituent atoms and has reported a correlation between T_c and the solid solution range of the A-15 structure. Hartsough (1974) has reviewed much of the previous work on the stability of A-15 compounds including data for 60 known phases and has developed stability criteria based on Engel–Brewer correlations.

Studies of T_c versus composition have been made in $(Nb_{1-y}M_y)_3Al_{1-x}Ge_x$ with M = Ti, Zr, and Hf by Cadieu (1970), and in a large number of previously unreported A-15 Nb alloys by Johnson and Douglass (1974a) using sputtered films. Studies of Cr and Mo based A-15s by Flukiger *et al.* (1974) show that the maximum T_c does not always occur at the 3/1 composition, but that T_c does correlate with the degree of order. The Knight shift, Seebeck coefficient, electrical resistivity, and T_c of the ternary alloys $V_3(Ge_{1-y}Ga_y)$ and $V_3(Ge_{1-y}Al_y)$ have been reported by Surikov *et al.* (1972). The homogeneity range and single crystal growth of V_3Si was discussed by Seeber and Nickl (1973). The effect on T_c of minor additions of Cr, Mo, Zr, Hf, and Ti in Nb_3Sn was studied by Ronami and Berezina (1974).

Vieland *et al.* (1975) studied the structure and properties of Nb_3SnH_x. For $x > 0.1$, the low temperature structural transformation disappears; and for $x \sim 1$, T_c falls below 4.2K. The H atoms were found to lie at the alternate Nb positions, crosslinking the orthogonal Nb chains.

A number of interesting papers with new results on the preparation and properties of A-15 superconductors can be found in the Proceedings of the 1974 Applied Superconductivity Conference, *IEEE Transactions on Magnetics*, Vol. MAG-11, No. 2, March, 1975.

An anisotropy in Hc_2 for V_3Si of $\sim$ 1–5% has been reported by Reed *et al.* (1967), Kramer and Knapp (1975), and Pulver (1972a). The anisotropy of the critical current J_c and the uniaxial stress dependence of T_c and J_c were also investigated by Pulver (1972a,b). Foner and McNiff (1976) have measured the anisotropy of H_{c_2} in Nb_3Sn and V_3Si and find no evidence of a highly anisotropic Fermi surface.

III. Magnetic Susceptibility, NMR, and Electrical Resistivity

Kodess (1973) has found an anomalous increase in susceptibility on cooling for $Nb_3(Al_{0.7}Ge_{0.3})$ which decreases in samples with lower T_c's.

Fradin and Williamson (1974) have found from an NMR study that for $V_3Ga_{1-x}Y_x$ with Y = Si or Sn the ratio T_c/θ_d (θ_d = Debye temperature) is a universal function of the bare electronic density of states at the Fermi level.

Marchenko (1973) measured the electrical resistivity of V_3Si from 17 to 1100K. From 17 to 28K he finds a quadratic T dependence of ρ which he ascribes to electron–electron interactions. The well known $\rho(T)$ anomaly at T between 30 and 300K is explained by electronic effects arising from fine structure in the density of states (cf. Cohen *et al.*, 1967).

A detailed study of $\rho(T)$ from V_3Si has been made by Milewits *et al.* (1976). However, they interpret the anomalous curvature as due to selective scattering between two pieces of the Fermi surface by [100] TA phonons in both V_3Si and Nb_3Sn. On the other hand, Allen *et al.* (1976) explain the behavior as due to anharmonicity and have obtained semiquantitative agreement in a calculation of $\rho(T)$ for V_3Si. In a new theory of A-15 compounds Gor'kov (1974) suggests that because of a singularity in the density of states, ρ should be proportional to ln T_c—a result that can be supported by experiment. Offering still another explanation, Fisk and Webb (1976) have argued that the negative curvature occurs because the mean free path of the electrons at high temperature is approaching the (limiting) value of an interatomic separation. The anomalous $\rho(T)$ of A-15 superconductors has now been found in a wide variety of relatively high T_c materials (Fisk and Lawson, 1973).

It is clear from the above that no single and accepted explanation of this behavior is available. An old experimental finding of Sarachik *et al.* (1963) is sometimes ignored with the implication that this anomaly is characteristic of the high T_c materials. These workers, who were the first to observe the anomalous $\rho(T)$, found that the behavior in V_3Si ($T_c \sim 17K$) was nearly identical to that in V_3Ge ($T_c \sim 6K$). (Both have the A-15 structure.) Thus the anomaly is not always associated with high T_c's and a meaningful explanation must at least allow and, one hopes, "predict" this. To accentuate the breadth of this problem, we shall discuss in Section VIII the very marked effect of defects on the anomalous resistivity of A-15 superconductors.

IV. The Batterman–Barrett Transformation

A low temperature structural transformation (~80–130K) determined from X-ray studies has been reported for a wide range of A-15 Nb–Ge–Al alloys by Kodess (1975). Viswanathan *et al.* (1974) have observed a structural transformation in specific heat studies of V_3Ga and Nb_3Al.

Fung *et al.* (1975) studied the depression of the structural transformation temperature T_m in high magnetic fields for different crystallographic directions in V_3Si. The results show that the degeneracy of the $m_l = \pm 2$

bands [in the Weger–Labbe–Friedel (WLF) model] is removed and the orbital moment is quenched. The WLF model has been extended to account for finite lifetimes of the d electrons by Williamson *et al.* (1974). This formalism would account for a variation in the behavior of crystals differing, for example, in their resistance ratios.

V. X-Ray Studies

In a study of the temperature dependence of the X-ray diffraction Kodess (1974) has observed that the intensity of the (440) reflection in transforming and nontransforming V_3Si decreases between 300 and 80K (i.e., a "negative" Debye–Waller factor) while the (320) and (321) lines grew stronger as usually expected. This result is both surprising and important since 80K is four times the structural transformation temperature. Thus the crystal is showing extraordinary precursor effects (e.g., mode softening and anharmonicity), presumably even for the high frequency phonons, and at temperatures where present theory and understanding presume quasinormal behavior. Kodess also observed: (i) that the anomalous softening occurred in a (440) diffraction involving both V and Si atoms but not in the (320) diffraction which is determined by the V atoms only; and (ii) that the unusual behavior was far less pronounced in V_3Ge ($T_c \sim 6$K).

From detailed and accurate X-ray diffraction data Staudenmann *et al.* (1976) have obtained the charge distribution in V_3Si. From these excellent data, part of which are given in Fig. 1, the authors conclude that the main interaction in this compound is the strong convalent bonding between adjacent V atoms. A charge transfer $\sim$ 1.8–2.4 electrons from the Si to V atoms is also found. In a more recent work (P. Coppens, private communication) the degree of covalency between the Cr atoms in the isostructural Cr_3Si is found to be considerably less.

VI. Elastic Moduli

Dieterich and Stollhoff (1974) continued earlier theoretical work, showing how in Labbé–Friedel type models, the shape of the electronic density of states near E_f can be determined from the magnetic field dependence of the sound velocity. Cheeke *et al.* (1972) and Fukase *et al.* (1974) have studied the experimental behavior for V_3Si following the original work of Maita and Bucher.

The pressure dependence of the shear modulus in V_3Si has been calculated using the Labbé–Friedel model by Barsch and Rogowski (1973) and Ting and Ganguly (1974). Barsch (1974) has also calculated the nonlinear

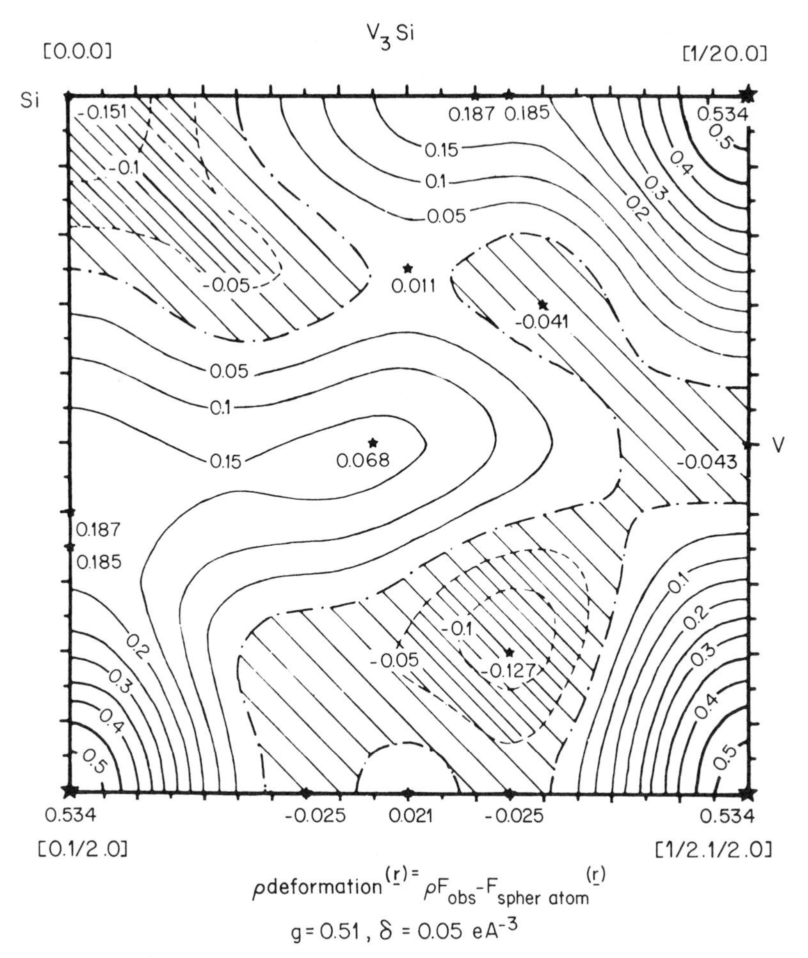

FIG. 1. Charge deformation density (observed minus spherical atom) in (000) plane of V_3Si. Contours at 0.05 e A^{-3}. Negative contours broken. (After Staudenmann *et al.*, 1976.)

stress–strain relation in V_3Si.

Noolandi and Varma (1975) have shown theoretically how the difference in shear moduli between transforming and nontransforming V_3Si could arise from a larger defect concentration in the latter.

Ultrasonic attenuation studies in V_3Si have continued. Using these methods Ishibashi and Akao (1976) find that for the tetragonal phase H_{c2} is $\sim 16\%$ less than for the cubic phase and T_c is ~ 0.1K smaller. Large attenuation and velocity peaks have been observed in the mixed state by Fukase *et al.* (1976), who explain the effect as due to tetragonal domain reorientation driven by the sound wave stress and the flux anisotropy. Finlayson *et al.* (1975) have studied internal friction in some A-15 and C-15 superconductors which undergo structural transformations above T_c and find loss peaks for HfV_2 and ZrV_2 but not for Nb_3Sn or V_3Si.

VII. Stress Behavior

The pressure dependence of the structural (T_m) and superconducting (T_c) transition temperatures have been obtained in V_3Si (Chu and Testardi, 1974) and Nb_3Sn (Chu, 1974). The results, given in Figs. 2 and 3 show that in Nb_3Sn dT_c/dP and dT_m/dP each have opposite signs to that in V_3Si and that, for either material, dT_c/dP has a sign opposite to dT_m/dP. Testardi (1973) has previously argued that increased instability would lead to increased T_c. No matter what their microscopic origins, the pressure effects support this suggestion by showing that as T_m approaches (V_3Si) or departs (Nb_3Sn) from T_c the instability (at T_c) increases or decreases, respectively, and leads to positive or negative values of dT_c/dP.

Chu and Vieland (1974) find that T_c of tetragonal Nb_3Sn is only several tenths K below that of the cubic state—a result previously shown for V_3Si.

The pressure dependence of T_c in $V_3(Si_{1-x}Ge_x)$ has been measured by Rabinkin *et al.* (1973) who do not find the behavior predicted by Testardi for the volume changes. Pupp *et al.* (1974) have studied the uniaxial stress

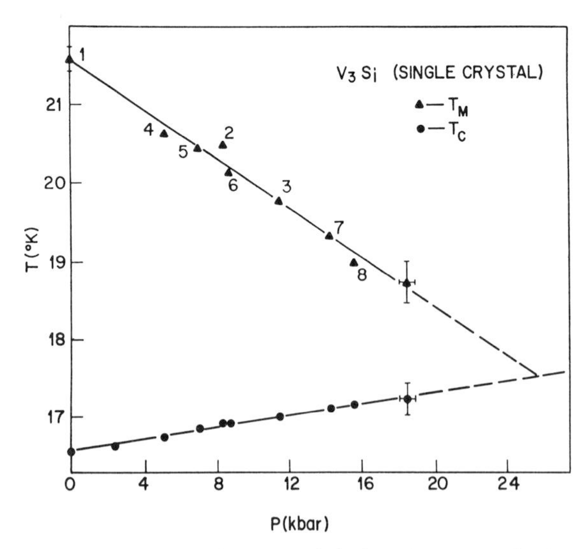

FIG. 2. Pressure dependence of structural (T_m) and superconducting (T_c) transition temperatures in V_3Si. (After Chu and Testardi, 1974.)

dependence of T_c for Nb_3Sn films and find linear as well as nonlinear behavior.

Blaugher *et al.* (1974) obtained one of the most unexpected pressure effects in V_3Si in showing that hydrostatic pressure applied at room temperature does not lead to the expected lattice parameter reduction. Their data (see Fig. 4) show that between 0 and 10 kbar the lattice parameter reduction with static pressure is considerably larger than that predicted from the ultrasonic sound velocity results. (At pressures of 20–110 kbar the expected behavior is observed.) Similar behavior has been observed in Nb_3Al by Galev and Rabin'kin (1974), and confirming the results of Blaugher *et al.*, in V_3Si by Testardi and Chu (1976). The data indicate a new mechanism for mechanical compliance in A-15 crystals with a relatively long ($> 10^{-7}$ sec) time scale. Varma *et al.* (1974) have suggested the discrepancy is due to vacancy motion. Further discussion on the nature of the defect is given in Section VIII. The Blaugher result, however, does indicate that the assumption in the interpretation of most previous pressure measurements (T_c, T_m, and elastic moduli) of only a normal lattice parameter change may be in error since the pressure was applied at room temperature. Shelton and Smith (1976) have found a break in dT_c/dP for V_3Si at ~ 16 kbar and suggest a connection with the Blaugher *et al.* finding.

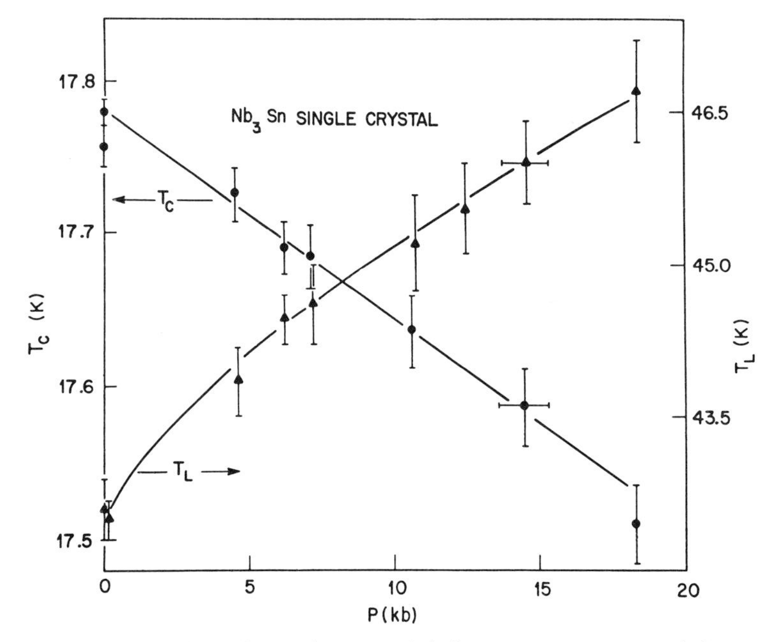

FIG. 3. Pressure dependence of structural (T_L) and superconducting (T_c) transition temperatures in Nb_3Sn. (After Chu, 1974.)

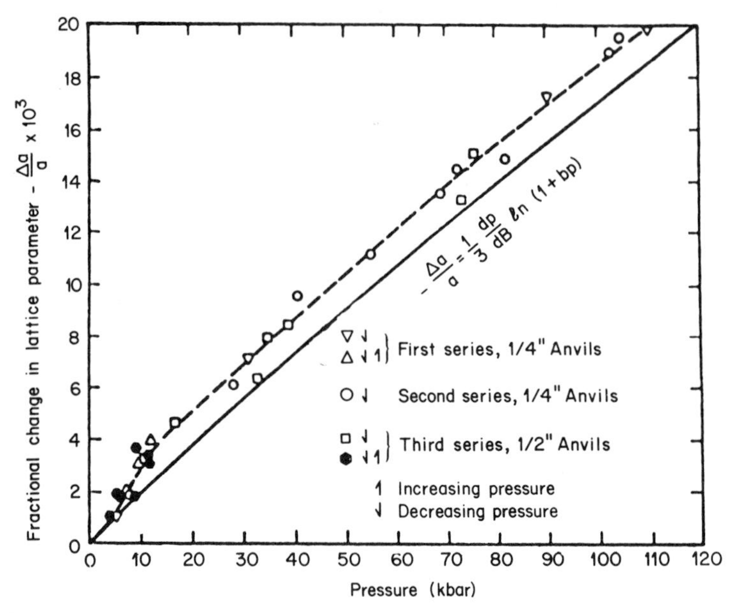

FIG. 4. Reduction of V_3Si lattice parameter with pressure at 300K. The expected behavior, based on ultrasonic measurements, is shown by the dashed lines. (After Blaugher *et al.*, 1974.)

VIII. Composition, Disorder, and Defects: Their Effects on T_c and Electrical Resistivity

Hot substrate sputtering has been lately used to obtain metastable phases with enhanced T_c's (Testardi *et al.*, 1971, 1974a). Recently Gavaler (1973) obtained the significant finding of T_c (onset) = 22.3K in Nb_3Ge sputtered onto hot substrates in high argon pressures. This was subsequently increased to T_c onsets at ~23K (Testardi *et al.*, 1974b; Gavaler *et al.*, 1974).

One of the benefits of this finding was the increased metallurgical flexibility of the sputtering process, which allows greater compositional variations and alloying range than that obtained by conventional pyrometallurgical techniques. In addition to this, Rutherford backscattering techniques have been used (Testardi *et al.*, 1975) to yield chemical composition analyses accurate to ~3% (absolute) with no standards required, and further yielding the compositional profile through the depth of the film.

A study of Nb–Ge films by Testardi *et al.* (1975) led to the following results relevant to the question of which metallurgical parameters were crucial to achieving the high T_c's in this A-15 superconductor. (1) The achievement of exact stoichiometry was not crucial in obtaining the high T_c's.

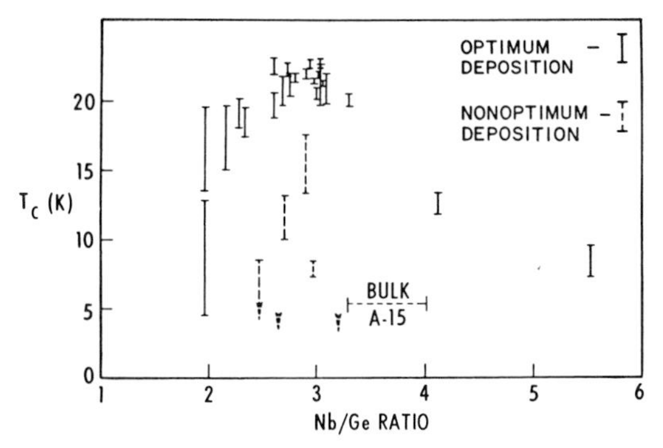

FIG. 5. Variation of T_c with Nb/Ge ratio for optimal and nonoptimal deposition temperatures. (After Testardi *et al.*, 1975.)

Figure 5, which gives the variation in T_c with Nb/Ge ratio, shows that high T_c's ($\sim$20K onsets) can be achieved over a compositional range exceeding 20%. X-ray data show that for Nb/Ge > 2.6 the samples are predominantly A-15 single phase and critical current measurements (Testardi, 1975c) clearly demonstrate that the bulk of the sample exhibits superconductivity. (2) Antisite defects (A atoms on B sites and vice versa) are expected to have their greatest effect when Nb/Ge < 3.0 since this should put Ge atoms interrupting the Nb chains. The data, however, show that neither T_c nor the electrical resistivity (or the resistance ratio) are much affected in this range of nonstoichiometry. (Impurity levels were $\lesssim 1\%$ in high T_c films). (3) A simple correlation of T_c with resistance ratio $\rho(300\text{K})/\rho(25\text{K})$ was found (see Fig. 6) which was independent of almost all deposition conditions and chemical composition in the A-15 phase range 2.6 < Nb/Ge < 5.5. It was later shown (Poate *et al.*, 1975) that by introducing defects only at constant chemical composition using ^{4}He particles (which do not stop in the film) in an initially high T_c film it was possible to retrace the T_c-resistance ratio correlation for the as-grown film by varying the defect concentration only (see Fig. 6).

It was concluded that the crucial factor for a high T_c in this material is not exact stoichiometry but the absence of a peculiarly detrimental defect to which the A-15 structure is susceptible. Subsequent work (Poate *et al.*, 1976) has shown that many of the above findings (and presumably the conclusion as well) apply to Nb_3Sn and V_3Si.

The importance of defects in A-15 materials was shown in two other experimental areas. The finding of an anomalous compressibility in V_3Si by Blaugher *et al.* (1974) discussed above lead Varma *et al.* (1974) to suggest that the normal A-15 lattice was stabilized by several precent vacancies. Sweedler and colleagues (1974) were the first to show that neutron included damage in A-15 materials leads to large universal T_c reductions in most of

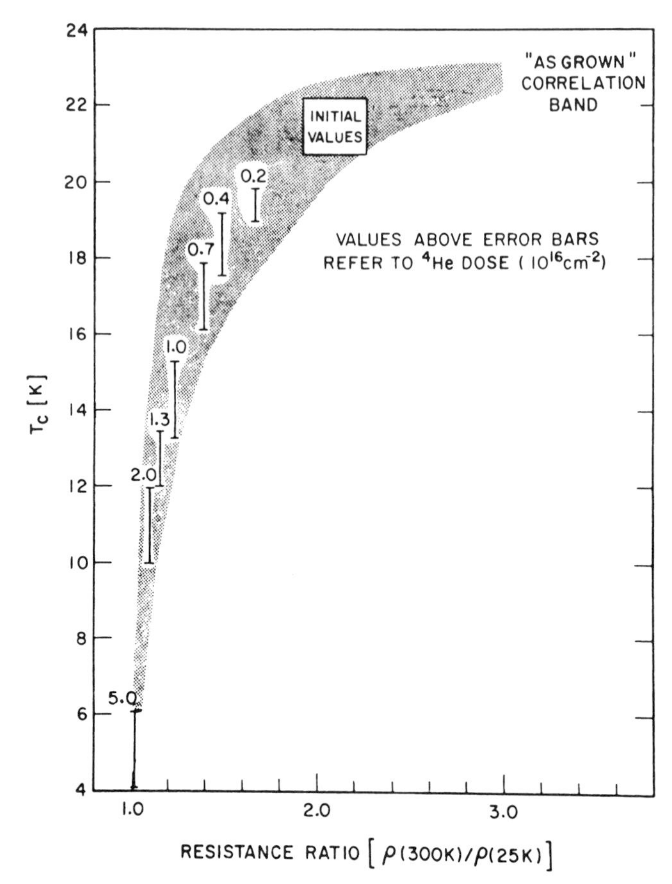

FIG. 6. Correlation of electrical resistance ratio with T_c for 130 films of Nb–Ge. Data are also given for an initially high T_c film following various doses of 2 MeV ^{4}He damage producing particles. (After Testardi *et al.*, 1975.)

the high T_c phases (see Fig. 7). This has been interpreted (Sweedler *et al.*, 1974; Sweedler and Cox, 1975) as due to the production of antisite defects which disrupt the chains of transition metal atoms.

The effect of defects on some normal state properties are now being investigated. Sweedler and Cox (1975) and Poate *et al.* (1975) both find that defects expand the room temperature lattice parameter of high T_c A-15 compounds. Noolandi and Testardi (1976) have shown a correlation in the reduction of T_c and the increase in lattice parameter for a wide variety, both chemical and processing, of Nb_3X superconductors which in many (if not most) cases is due to defect formation (see Fig. 8). The anomaly in the normal state resistivity of A-15 materials (V_3Si, V_3Ge, Nb_3Ge) is also found

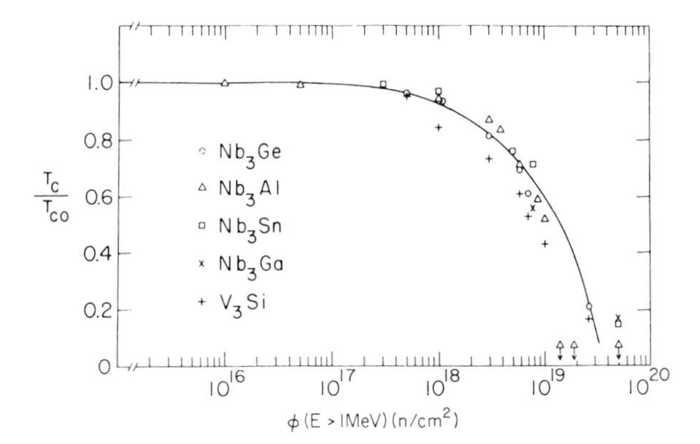

FIG. 7. Normalized reduction in T_c versus (>1 MeV) neutron dose for some A-15 compounds. (After Sweedler *et al.*, 1974.)

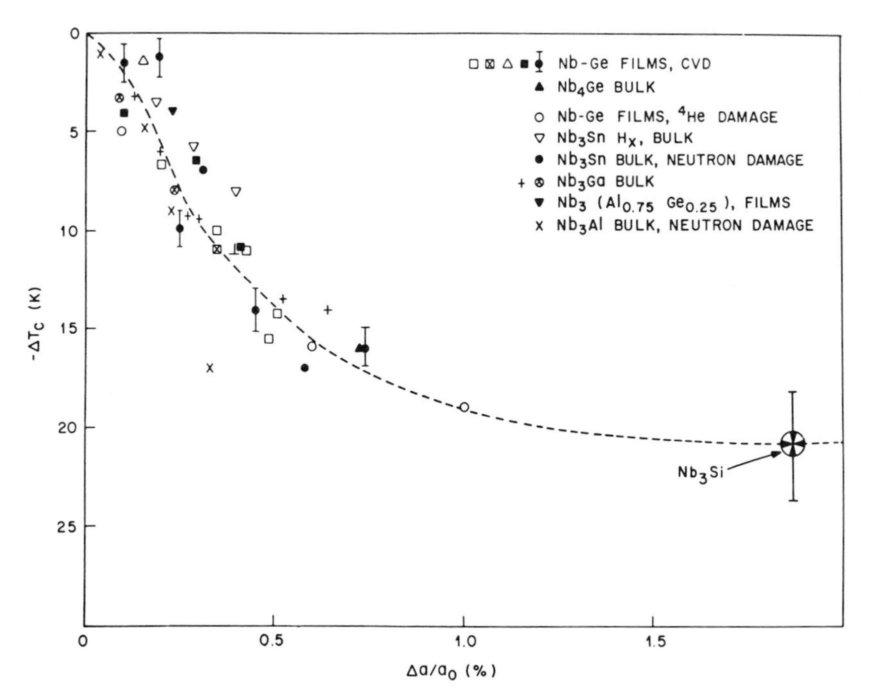

FIG. 8. Reduction in T_c versus increase in lattice parameter—both relative to the highest T_c state—for Nb–X A-15 superconductors. (After Noolandi and Testardi, 1976.)

to disappear (see Fig. 9) with increasing defect density (Testardi *et al.*, 1977).

It has been shown (Testardi *et al.*, 1977) that the A-15 materials are altered in their behavior by defect producing mechanisms at a rate which is, perhaps, one hundred times faster than that for Nb. The wealth of very recent data now indicates that the low temperature Batterman–Barrett transformation is not the only structural manifestation of the A-15 instability. It appears that even at room temperatures these materials are highly

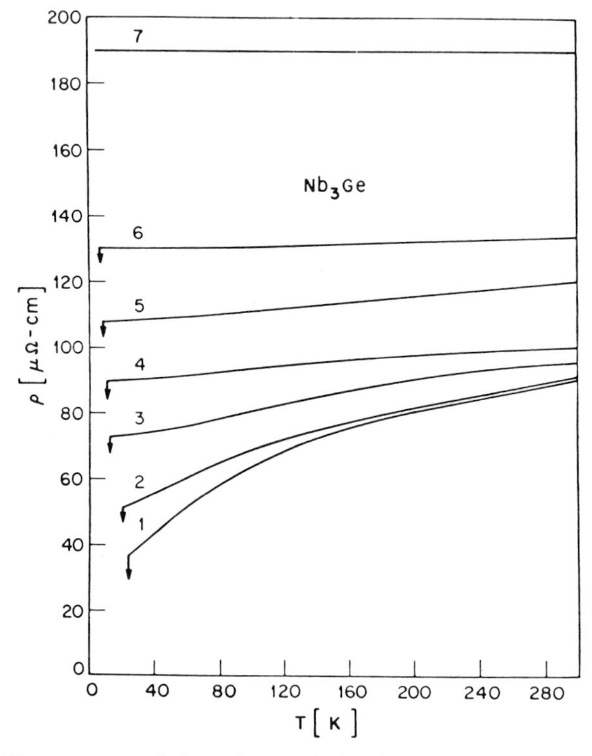

FIG. 9. Disappearance of thermal part of electrical resistivity with increasing defect concentration in Nb_3Ge. Samples 1–7 have T_c's decreasing from 23 to 3.6K due to increasing defects. (After Testardi *et al.*, 1977.)

susceptible to defects which, in turn, have a profound effect on the superconducting and normal state anomalies. Experiments in progress indicate that neither simple antisite disorder nor any "point" type (bond breaking) imperfection is the crucial defect. The defect appears to be the gradual and continuous collapse of the structure in which the translational symmetry of the lattice is lost in a manner tending toward an amorphous structure. The collapse of the structure (or whatever the defect) is as much an A-15 characteristic as the numerous anomalies so extensively studied in the past. Like the high T_c it is the consequence and the compromise of the structural instability.

IX. High Frequency Phonon Behavior

Extensive specific heat studies of a number of A-15 compounds have been reported by Gel'd *et al.* (1974), Junod (1974), and Knapp *et al.* (1975). The first two groups have analyzed their data using the density of states peak models. Junod's thesis in addition, contains considerable data on A-15 pseudobinary alloys and nonstoichiometric compounds. Knapp *et al.* give evidence of unusually large anharmonicity of the high T_c compounds and propose a model that relates the anharmonic heat capacity to the electron–phonon mass enhancement (and T_c).

Axe and Shirane (1973) have extended their previous inelastic neutron scattering data in Nb_3Sn. They describe the general mode softening and the appearance of the "central peak" on approaching the structural transformation temperature from above. Below T_c changes in phonon linewidths are observed for $\hbar\omega$ less than the superconducting gap.

Shen (1972) and Vedeneev *et al.* (1972) have observed some of the phonon peaks in Nb_3Sn by tunneling. Strong low frequency peaks in the range 6–9 meV have been found.

In a series of successful and highly informative experiments the Karlsruhe group (Nücker *et al.*, 1974-1975) have obtained the total phonon density of states for V_3Si, V_3Ge, V_3Ga, Nb_3Al, and Nb_3Sn at both room and low temperatures. Some, but not considerable mode softening is found in the total density of states (seven-eighths of which is from optic modes) with some evidence of optic as well as acoustic mode softening. Average values of $\langle\omega^2\rangle$ are reduced by $\sim 7\%$ to 13% on cooling. For Nb_3Sn, the data have been combined with tunneling results to yield the electron–phonon coupling parameter $\alpha^2(\omega)$ between 0 and 25 meV. The results for the phonon density of states, α^2F (from tunneling), and α^2 for Nb_3Sn are shown in Fig. 10.

Tanner and Sievers (1973) have obtained the superconducting energy gap of V_3Si from the far infrared reflectivity. They found a spread in energy gap values $2\Delta/kT_c = 1$ to 3.8. Perkowitz *et al.* (1976) measured the far infrared reflectivity of V_3Si over a wider energy range and report unusual structure between 6 and 25 meV in the superconducting state (see Fig. 11). Assuming a Holstein absorption process as the cause would indicate an unexpected peak in α^2F at 6 ± 2 meV. No change in reflectivity due to the structural transformation is observed between 6 and 25 meV.

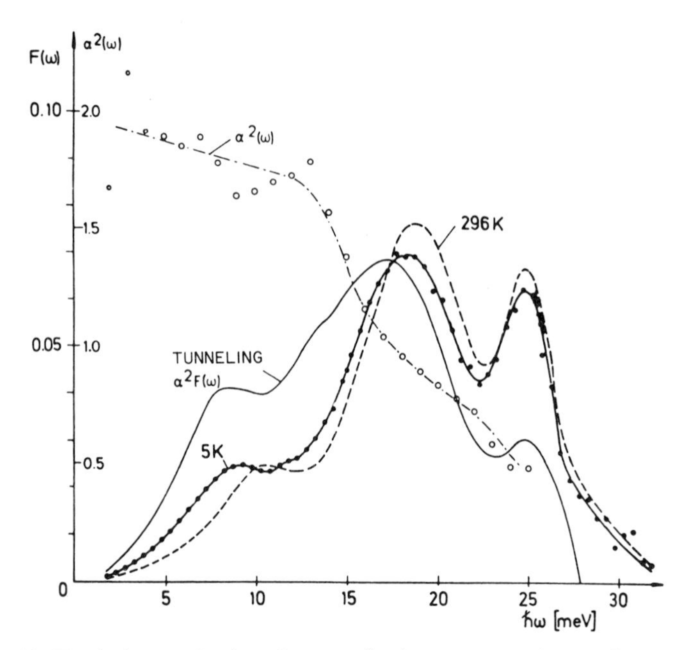

FIG. 10. Total phonon density of states $\alpha^2 F$ (from tunneling) and α^2 (extracted) for Nb_3Sn. (After Nücker *et al.*, 1975.)

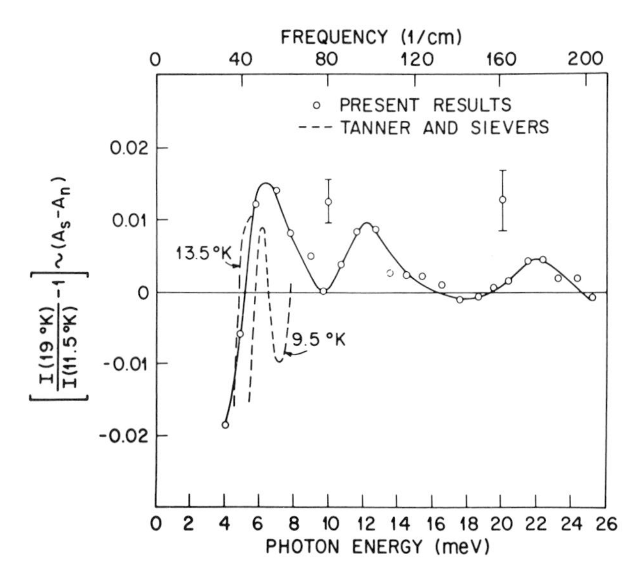

FIG. 11. Ratio of reflectivities at 19 and 11.5K (or superconducting minus normal state absorptivities) between 4 and 25 meV in V_3Si. (After Perkowitz *et al.*, 1976.)

X. Theory

Noolandi and Sham (1973) have treated the theory of the structural phase transition in Nb_3Sn using the Cohen, Cody, and Halloran density of states step model but accounting for the sublattice distortion as well. Good agreement is obtained with a number of measured properties using several adjustable parameters.

Using a Labbé–Friedel model Ting *et al.* (1973) predict that a discontinuous structural transformation in V_3Si will occur in an applied field of sufficient strength from a state with $c/a > 1$ to one with $c/a < 1$ below 17.4K.

Dieterich (1972) calculates large Gruneisen constants in V_3Si with a change in magnitude occurring for nontransforming samples in the superconducting state. Calculations of the soft mode phonon dispersion are reported by Saub and Barisic (1972).

Gor'kov (1974) and Gor'kov and Dorokhov (1976) have provided a new and extensive theory of the behavior of A-15 compounds based on the (symmetry allowed) electronic linear dispersion law at the X point in the Brillouin zone. The work provides a better justification for the assumptions of one dimensionality and the Fermi level positions used in previous theories. Logarithmic temperature dependences are found for all quantities above the structural transition temperature (consistent with much of the experimental data). Finally, the theory does provide a connection between structural instability and superconductivity by showing these to be two aspects of the basic instability of the electron spectrum against interactions involving attraction for one-dimensional systems.

Using the Gor'kov model Noolandi (1974) predicts a gap in the d-electron spectrum occurring between T_c and T_m. A Landau theory of the structural phase transition has been developed by Bhatt and McMillan (1976) based on the Gor'kov model and general agreement with much of the experimental behavior is obtained. A theoretical description of the effect of the structural transformation on superconductivity, again using the Gor'kov model, has been given by Bilbro and McMillan (1976). Ting and Birman (1975) have used the Gor'kov model to show that the onset of superconductivity can arrest the softening of $(C_{11} - C_{12})/2$ shear mode.

Bongi *et al.* (1974) present a phenomenological model for the A-15s based on the local character and charge transfer of the constituent atoms. Allen (1973, 1974) has discussed the effect of soft phonons and anharmonicity on the superconductivity.

Weger and Barak (1974) and Barak *et al.* (1975) have extended their earlier tight binding calculations of A-15 compounds.

Finally, Mattheiss (1975) has obtained the most reliable band calculations of the A-15 compounds to date using APW–LCAO methods. Although insufficient precision is obtained to yield the density of states on an meV scale he finds that (i) corrections to the Labbé–Friedel and Weger–Goldberg models are sufficient to wash out fine structure in the density of states on an meV scale and (ii) there is no evidence for describing the electronic structure of A-15 compounds as one-dimensional. The calculated density of states in V_3Si showing the various subband contributions is given in Fig. 12.

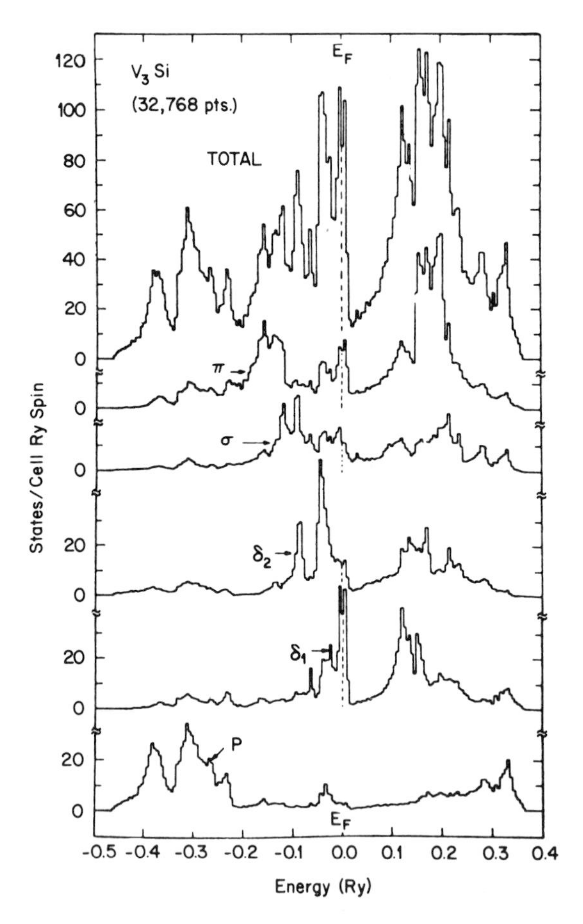

FIG. 12. Calculated density of states in V_3Si showing various subband contributions. (After Mattheiss, 1975.)

REFERENCES

Allen, P. B. (1973). *Solid State Commun.* **12**, 379.
Allen, P. B. (1974). *Solid State Commun.* **14**, 937.
Allen, P. B., Hui, J. C. K., Pickett, W. E., Varma, C. M., and Fisk, Z. (1976). *Solid State Commun.* **18**, 1157.
Axe, J. D., and Shirane, G. (1973). *Phys. Rev. B* **8**, 1965.
Barak, G., Goldberg, I. B., and Weger, M. (1975). *J. Phys. Chem. Solids* **36**, 847.
Barsch, G. R. (1974). *Solid State Commun.* **14**, 983.
Barsch, G. R., and Rogowski, D. A. (1973). *Mater. Res. Bull.* **8**, 1459.
Bhatt, R. N., and McMillan, W. L. (1976). *Phys. Rev.* **B14**, 1007.
Bilbro, G., and McMillan, W. L. (1976). *Phys. Rev.* **B14**, 1887.
Blaugher, R. D., Taylor, A., and Ashkin, M. (1974). *Phys. Rev. Lett.* **33**, 292.
Bongi, G., Fischer, O., and Jones, H. (1974). *J. Phys. F* **4**, L259.
Cadieu, F. J. (1970). *J. Low Temp. Phys.* **3**, 393.
Cheeke, J., Mallie, H., Roth, S., and Seeber, B. (1972). *C. R. Hebd. Séances Acad. Sci.* **275**, 661.
Chu, C. W. (1974). *Phys. Rev. Lett.* **33**, 1283.
Chu, C. W., and Testardi, L. R. (1974). *Phys. Rev. Lett.* **14**, 766.
Chu, C. W., and Vieland, L. J. (1974). *J. Low Temp. Phys.* **17**, 25.
Cohen, R. W., Cody, G. D., and Halloran, J. J. (1967). *Phys. Rev. Lett.* **19**, 840.
Dew-Hughes, D. (1975). *Cryogenics* **15**, 435.
Dieterich, W. (1972). *Z. Phys.* **254**, 464.
Dieterich, W., and Stollhoff, G. (1974). *Z. Phys.* **266**, 185.
Finlayson, T. R., Thomson, K. W., and Smith, T. F. (1975). *J. Phys. F* **5**, L225.
Fisk, Z., and Lawson, A. C. (1973). *Solid State Commun.* **13**, 277.
Fisk, Z., and Webb, G. W. (1976). *Phys. Rev. Lett.* **36**, 1084.
Flukiger, R., Paoli, A., and Muller, J. (1974). *Solid State Commun.* **14**, 443.
Foner, S., and McNiff, E. J., Jr. (1976). *Phys. Lett.* **58A**, 318.
Fradin, F. Y., and Williamson, J. D. (1974). *Phys. Rev. B* **10**, 2803.
Fukase, T., Vema, K., and Muto, Y. (1974). *Phys. Lett. A* **49**, 129.
Fukase, T., Tachiki, M., Toyota, N., and Muto, Y. (1976). *Solid State Commun.* **18**, 505.
Fung, H. K., Williamson, S. J., Ting, C. S., and Sarachik, M. P. (1975). *Phys. Rev.* **B11**, 2053.
Galev, V. N., and Rabin'kin, A. G. (1974). *Sov. Phys.—JETP (Engl. Transl.)* **38**, 525; *Zh. Eksp. Teor. Fiz.* **65**, 1061 (1973).
Gavaler, J. R. (1973). *Appl. Phys. Lett.* **23**, 480.
Gavaler, J. R., Janocko, M. A., and Jones, C. K. (1974). *J. Appl. Phys.* **45**, 3009.
Gel'd, P. V., Kalishevich, G. I., Surikov, V. I., Shtol'ts, A. K., and Zagryazhskii, V. L. (1974). *Sov. Phys.—Dokl. (Engl. Transl.)* **19**, 230; *Dokl. Akad. Nauk SSSR* **215**, 833 (1974).
Gor'kov, L. P. (1974). *Sov. Phys.—JETP (Engl. Transl.)* **38**, 830; *Zh. Eksp. Teor. Fiz.* **65**, 1658 (1973).
Gor'kov, L. P., and Dorokhov, O. N. (1976). *J. Low Temp. Phys.* **22**, 1.
Hartsough, L. D. (1974). *J. Phys. Chem. Solids* **35**, 1691.
Hein, R. A. (1973). *Sci. Technol. Supercond. Proc. Summer Course, 1971.* Vol. 1, p. 333.
Ishibashi, K., and Akao, F. (1976). *J. Phys. Soc. Jpn.* **40**, 396.
Izyumov, Yu. A., and Kurmaev, Z. Z. (1974). *Sov. Phys.—Usp. (Engl. Transl.)* **17**, 356; *Usp. Fiz. Nauk* **113**, 193 (1974).
Johnson, G. R., and Douglass, D. H. (1974a). *J. Low Temp. Phys.* **14**, 516.
Johnson, G. R., and Douglass, D. H. (1974b). *J. Low Temp. Phys.* **14**, 565.

Junod, A. (1974). Ph.D. Thesis, Faculty of Science, University of Geneva.
Knapp, G. S., Bader, S. D., Culbert, H. V., Fradin, F. Y., and Klippert, T. E. (1975). *Phys. Rev.* **11**, 4331.
Kodess, B. N. (1973). *Sov. Phys.—Solid State* (*Engl. Transl.*) **15**, 844; *Fiz. Tverd. Tela* **15**, 1252 (1973).
Kodess, B. N. (1974). *Sov. Phys.—Solid State* (*Engl. Transl.*) **16**, 782; *Fiz. Tverd. Tela* **16**, 1216 (1974).
Kodess, B. N. (1975). *Solid State Commun.* **16**, 269.
Kramer, E. J., and Knapp, G. S. (1975). *J. Appl. Phys.* **46**, 4595.
Marchenko, V. A. (1973). *Sov. Phys.—Solid State* (*Engl. Transl.*) **15**, 1261; *Fiz. Tverd. Tela* **15**, 1893 (1973).
Mattheiss, L. F. (1975). *Phys. Rev.* **12**, 2161.
Milewits, M., Williamson, S. J., and Taub, H. (1976). *Phys. Rev.* (to be published).
Noolandi, J. (1974). *Phys. Rev. Lett.* **33**, 1151.
Noolandi, J., and Sham, L. J. (1973). *Phys. Rev. B* **8**, 2468.
Noolandi, J., and Testardi, L. R. (1976). To be published.
Noolandi, J., and Varma, C. (1975). *Phys. Rev. B* **11**, 4743.
Nücker, N., Reichardt, W., Rietschel, H., Schneider, E., Schweiss, P., and Trioadus, V. (1974–1975). Karlsruhe Group. Progress Report KFK 2054, KFK 2183. Gesellschaft für Kernforschung, M. B. H., Karlsruhe, Germany.
Perkowitz, S., Merlin, M., and Testardi, L. R. (1976). *Solid State Commun.* **18**, 1059.
Poate, J. M., Testardi, L. R., Storm, A. R., and Augustyniak, W. M. (1975). *Phys. Rev. Lett.* **35**, 1290.
Poate, J. M., Dynes, R. C., Testardi, L. R., and Hammond, R. H. (1976). "Superconductivity in d- and f-band Metals (1976)" (D. H. Douglass, ed.). Plenum, New York.
Pulver, M. (1972a). *Z. Phys.* **257**, 22.
Pulver, M. (1972b). *Z. Phys.* **257**, 261.
Pupp, W. A., Sattler, W. W., and Saur, E. J. (1974). *J. Low Temp. Phys* **14**, 1.
Rabinkin, A. G., Galev, V. N., and Laukhin, V. N. (1973). *Sov. Phys.—JETP* (*Engl. Transl.*) **37**, 870; *Zh. Eksp. Teor. Fiz.* **64**, 1724 (1973).
Reed, W. A., Fawcett, E., Meincke, P. P. M., Hohenberg, P. C., and Werthamer, N. R. (1967). *Proc. Int.* Conf. *Low Temp. Phys., 10th, 1966*, Vol. IIA, p. 368.
Ronami, G. N., and Berezina, V. P. (1974). *Fiz. Met. Metalloved.* **37**, 872.
Sarachik, M. P., Smith, G. E., and Wernick, J. H. (1963). *Can. J. Phys.* **41**, 1542.
Saub, K., and Barisic, S. (1972). *Phys. Lett. A* **40**, 415.
Savitskii, E. M., Baron, V. V., Efimov, Yu. V., Bychokova, M. I., and Myzenkova, L. F. (1973). "Superconducting Materials." Plenum, New York.
Seeber, B., and Nickl, J. (1973). *Phys. Status Solidi* **15**, 73.
Shelton, R. N., and Smith, T. F. (1976). To be published.
Shen, L. Y. (1972). *Phys. Rev. Lett.* **29**, 1082.
Staudenmann, J.-L., Coppens, P., and Muller, J. (1976). *Solid State Commun.* **19**, 29.
Surikov, V. I., Pryadein, V. I., Shtol'ts, A. K., Stepanov, A. P., Prekul, A. F., and Gel'd, P. V. (1972). *Fiz. Met. Metalloved.* **34**, 724.
Sweedler, A. R., and Cox, D. E. (1975). *Phys. Rev. B* **12**, 147.
Sweedler, A. R., Schweitzer, D. G., and Webb, G. W. (1974). *Phys. Rev. Lett.* **33**, 168.
Tanner, D. B., and Sievers, A. J. (1973). *Phys. Rev. B* **8**, 1978.
Testardi, L. R. (1973). In "Physical Acoustics: Principles and Methods" (W. P. Mason and R. N. Thurston, eds.), Vol. 10, p. 193. Academic Press, New York.
Testardi, L. R. (1975a). *Comments Solid State Phys.* **6**, 131.
Testardi, L. R. (1975b). *Rev. Mod. Phys.* **47**, 637.

Testardi, L. R. (1975c). *Solid State Commun.* **17**, 871.
Testardi, L. R., and Chu, C. W. (1977). *Phys. Rev.* **B15**, 146.
Testardi, L. R., Hauser, J. J., and Read, M. H. (1971). *Solid State Commun.* **9**, 1829.
Testardi, L. R., Wernick, J. H., Royer, W. A., Bacon, D. D., and Storm, A. R. (1974a). *J. Appl. Phys.* **45**, 446.
Testardi, L. R., Wernick, J. H., and Royer, W. A. (1974b). *Solid State Commun.* **15**, 1.
Testardi, L. R., Meek, R. L., Poate, J. M., Royer, W. A., Storm, A. R., and Wernick, J. H. (1975). *Phys. Rev. B* **11**, 4304.
Testardi, L. R., Poate, J. M., and Levinstein, H. J. (1977). *Phys. Rev.* **B15**, 2570.
Ting, C. S., and Birman, J. L. (1975). *Phys. Rev. B* **12**, 1093.
Ting, C. S., and Ganguly, A. K. (1974). *Phys. Rev. B* **9**, 2781.
Ting, C. S., Ganguly, A. K., Zeyer, R., and Birman, J. L. (1973). *Phys. Rev. B* **8**, 3665.
Varma, C. M., Phillips, J. C., and Chui, S.-T. (1974). *Phys. Rev. Lett.* **33**, 1223.
Vedeneev, S. I., Golovashkin, A. I., and Motulevich, G. P. (1972). *JETP Lett.* (*Engl. Transl.*) **16**, 152; *Zh. Eksp. Teor. Fiz.* **63**, 1010 (1972).
Vieland, L. J., Wicklund, A. W., and White, J. G. (1975). *Phys. Rev.* **11**, 3311.
Viswanathan, R., Wu, C. T., Luo, H. L., and Webb, G. W. (1974). *Solid State Commun.* **14**, 1051.
Wang, F. E. (1973). *J. Solid State Chem.* **6**, 365.
Wang, F. E. (1974). *J. Phys. Chem. Solids* **35**, 273.
Weger, M., and Barak, G. (1974). *Phys. Lett. A* **48**, 319.
Weger, M., and Goldberg, I. B. (1973). *Solid State Phys.* **28**, 1.
Williamson, S. J., Ting, C. S., and Fung, H. K. (1974). *Phys. Rev. Lett.* **32**, 9.

—3—

Plate Modes in Surface Acoustic Wave Devices

R. S. WAGERS[1]

Central Research Laboratories
Texas Instruments Incorporated
Dallas, Texas

I. Introduction

A principal reason for current examination of acoustic plate modes is that their unintentional excitation in Rayleigh wave devices produces a degradation of the devices' terminal properties. The most serious detrimental effects of these modes appear in Rayleigh wave transversal filters. Their presence on the high-frequency side of the pass band in these devices tends to limit the out-of-band rejection achievable; consequently, one finds that surface acoustic wave filter designers have put forth substantial efforts toward

[1] *Present address:* Electrical Engineering and Computer Science Faculty, Princeton University, Princeton, New Jersey 08540.

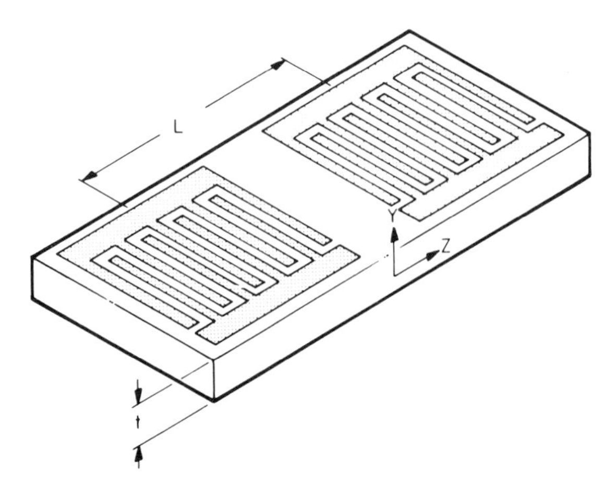

FIG. 1. Orientation of coordinates with respect to a piezoelectric plate and the two interdigital transducers disposed on its top surface.

understanding plate mode characteristics in order to selectively discriminate against the modes.

In Rayleigh wave filters, mode excitation is accomplished by depositing interdigital electrodes on the free upper surface of the piezoelectric crystal as illustrated in Fig. 1. The aperture of the transducer is made sufficiently wide that approximately plane wave excitation takes place. Since the transducer interacts with a mode through its electric potential, it is effective in exciting those modes of the piezoelectric crystal having an electric potential associated with them. For a given power flow, the plate modes usually have a weaker electric field than the Rayleigh mode so their excitation is weaker, but they are excited to some level nevertheless.

Many investigators have considered the problem of spurious mode excitation by interdigital transducers. Some have viewed the problem as bulk wave excitation where the bulk wave reflects off the top and bottom surfaces of the crystal ultimately reaching the output transducer. Others have sought to show that the spurious modes are pseudosurface waves—waves that are weakly bound to the top surface of the crystal and radiate energy into the bulk as they propagate. In a third category of analysis, the spurious mode problem has been treated by considering the modes of a piezoelectric plate. If the physical system one wishes to describe is in fact a plate, then the former two approaches are contained in the plate mode treatment because: (1) correct satisfaction of the boundary conditions for bulk wave excitation produces plate wave solutions; and (2) there are no pseudosurface wave or leaky wave solutions in a plate. Those solutions exist only in a half-space. Good guides to the literature of these approaches are the bibliographies of Mitchell and Reed (1975) and Wagers (1976a).

In this chapter the properties of plate modes of piezoelectric media will be developed. Because many good works (Meeker and Meitzler, 1964; Auld, 1973) have already been published on the plate modes of nonpiezoelectric media, their review will not be attempted. The analysis begins with a derivation of the stiffened Christoffel equation and its reduction to an explicit algebraic form. This reduction is extremely important in the development of effective numerical procedures for the solution of the boundary value problem. Without this simplification many redundant calculations occur in the iterative search for a normal mode of the plate.

One dispersion diagram for a thin plate of lithium niobate ($LiNbO_3$) is presented and interpreted. A dispersion diagram, or ω-β diagram, is the familiar graphical presentation of the propagation characteristics of modes of all types of wave guides. However, this type of presentation becomes extremely complicated for practical surface acoustic wave substrates because they typically are ten or more wavelengths thick. Also, one rarely cares about the complete dispersion relation; only the range within the bandwidth of the transducer has a bearing on device performance. Consequently, in the remainder of the chapter, the format for presentation of plate mode characteristics is mode coupling charts. They are line graphs of the midband coupling of the modes versus the velocity of the modes. Plate mode excitation by interdigital transducers is more directly interpreted from these charts.

Quantitative predictions of the terminal characteristics of surface acoustic wave filters are carried out in Section III by obtaining the interdigital transducer impedance characteristics for each plate mode of the substrate. In the example presented 96 plate modes are evaluated and the filter's insertion loss is accurately predicted over an octave range of frequency. Section III,C is devoted to qualitative illustrations of the positions and strengths of the plate modes of cuts of $LiNbO_3$. The theoretical calculations are for substrates on the order of 10λ thick. Because for thick plates the characteristics of the modes do not change rapidly as a function of plate thickness, the illustrations serve as useful guidelines for substrates of other thicknesses. The final section demonstrates the effectiveness of gross disruption of the bottom surface of the plate by sandblasting as a means of suppressing the plate modes' contributions to the terminal transfer characteristics of a Rayleigh wave filter.

II. Analysis of Plate Mode Propagation

A. Derivation of Stiffened Christoffel Equation

The analysis procedure employed here in the description of freely propagating plate modes is to seek partial wave solutions to the equations of motion and then to require that sums of these solutions give the prescribed behavior at the plate boundaries. The partial wave solutions are plane waves that would propagate in an infinite medium; and for a specified phase velocity along the free surface, a piezoelectric plate has eight solutions. Four of the waves propagate upward through the plate, and the other four propagate downward. This section will be concerned only with finding these waves. Their combination for the satisfaction of boundary conditions will be addressed in Section II,C.

The geometry under consideration is illustrated in Fig. 1. A piezoelectric plate has its length along the z axis and thickness t along the y axis. Propagation is assumed to be along the z axis, and the modal variations are taken to be independent of x; thus, $\partial/\partial x = 0$. At the top and bottom surfaces of the plate $y = \pm t/2$; the plate is assumed to be stress free. As a practical matter, the plate could be taped down to a supporting test fixture without inducing large perturbations in the modes. (A common test procedure is to wax the crystal down. In such a case, at frequencies of ~ 50 MHz and for $\sim 200\lambda$ of propagation, mode velocities are not appreciably perturbed, but attenuations of 10 dB or more are observed.)

Newton's second law requires that a divergence of the stress give rise to an acceleration of the body

$$\nabla \cdot \mathbf{T} = \rho \, \partial^2 \mathbf{u}/\partial t^2, \tag{1}$$

where $\mathbf{u}$ is the particle displacement field in the crystal, $\mathbf{T}$ is the tensor stress field, and ρ is the mass density. Considering only linear effects, strain is given by the symmetric portion of the gradient of the displacement field

$$\mathbf{S} = \nabla_{\mathrm{s}} \mathbf{u}. \tag{2}$$

In abbreviated subscript notation (Auld, 1973), which will be used here, $\mathbf{S}$ and $\mathbf{T}$ have matrix representations as six-element column vectors; $\mathbf{u}$ has a representation as a three-element column vector; and in rectangular coordinates, the operators $\nabla \cdot$ and ∇_{s} have the matrix representation

$$[\nabla \cdot] = [\nabla_{\mathrm{s}}]^{\mathrm{T}} = \begin{bmatrix} \partial/\partial x & 0 & 0 & 0 & \partial/\partial z & \partial/\partial y \\ 0 & \partial/\partial y & 0 & \partial/\partial z & 0 & \partial/\partial x \\ 0 & 0 & \partial/\partial z & \partial/\partial y & \partial/\partial x & 0 \end{bmatrix}. \tag{3}$$

Throughout the chapter, bold-faced symbols will be used to designate vectors, tensors, and their matrix representations. The position and use of the symbols in equations will specify which form is being considered.

In addition to the equations of motion [Eqs. (1)–(2)], equations of state are required to completely specify the interrelation of the field quantities in the plate. For piezoelectric media, these equations are

$$\mathbf{D} = \boldsymbol{\varepsilon}^{\mathrm{s}} \cdot \mathbf{E} + \mathbf{e} : \mathbf{S} \tag{4}$$

and

$$\mathbf{T} = \mathbf{c}^{\mathrm{E}} : \mathbf{S} - \mathbf{e} \cdot \mathbf{E}, \tag{5}$$

where $\mathbf{D}$ and $\mathbf{E}$ are the electric displacement and electric field, respectively; $\mathbf{e}$ is the piezoelectric stress tensor; and $\boldsymbol{\varepsilon}^{\mathrm{s}}$ and $\mathbf{c}^{\mathrm{E}}$ are the dielectric and stiffness tensors at constant strain and constant electric field, respectively. Finally, magnetic fields are neglected and the electric intensity is obtained from the quasistatic approximation

$$\mathbf{E} = -\nabla\phi, \tag{6}$$

where ϕ is the electric potential of the wave.

Harmonic variations in time, along the propagation direction, and along the thickness direction are assumed in seeking plane wave solutions to Eqs. (1)–(2) and Eqs. (4)–(6). Thus, the overall variation of the field quantities is given by

$$\exp[j(\omega t - \beta z - k_{\mathrm{t}} y)] = \exp[j\omega t - j\beta(z + py)],$$

where the transverse propagation constant k_{t} is normalized to be a fraction p of that along the substrate length. Only real values for the propagation constant β are considered, but p may be complex. Real values of p result when the phase velocity along the plate surface is greater than the bulk wave velocity in that direction. The corresponding partial wave is a purely propagating bulk wave with a propagation vector at some angle from the plate surface. Complex or imaginary values of p occur when the z velocity is less than the bulk wave velocity. These solutions have mode amplitudes that fall off exponentially away from the free surface. Rayleigh wave solutions are obtained from a superposition of four partial waves with complex p values.

With these assumptions for the spatial variations, matrix representations for the vector and tensor operators in Eqs. (1)–(3), (6) can be written. Thus, for the vector operation ∇ we have

$$[\nabla] = -j\beta \begin{bmatrix} 0 \\ p \\ 1 \end{bmatrix} = -j\beta\mathbf{L}. \tag{7}$$

The tensor divergence operator of Eq. (3) takes on a similarly simple form:

$$[\nabla \cdot] = -j\beta \begin{bmatrix} 0 & 0 & 0 & 0 & 1 & p \\ 0 & p & 0 & 1 & 0 & 0 \\ 0 & 0 & 1 & p & 0 & 0 \end{bmatrix} = -j\beta\mathbf{P}. \tag{8}$$

Since the plate is assumed to be charge free, the vector divergence of Eq. (4) vanishes:

$$\nabla \cdot \mathbf{D} = -j\beta\mathbf{L} \cdot \boldsymbol{\varepsilon}^{\mathrm{s}} \cdot \mathbf{E} - j\beta\mathbf{L} \cdot \mathbf{e} : \mathbf{S} = 0. \tag{9}$$

If the electric field from Eq. (6) and the strain from Eq. (2) are substituted in the above result,

$$\nabla \cdot \mathbf{D} = -(-j\beta)^2\mathbf{L}^{\mathrm{T}}\boldsymbol{\varepsilon}^{\mathrm{s}}\mathbf{L}\phi + (-j\beta)^2\mathbf{L}^{\mathrm{T}}\mathbf{e}\mathbf{P}^{\mathrm{T}}\mathbf{u} = 0$$

is obtained, where with the introduction of the superscript T for transpose it is clear that matrix representations are being considered. Final rearranging produces an expression for the electric potential in terms of the particle displacements and the (perhaps complex) ratio of the components of the propagation constant

$$\phi = \mathbf{L}^{\mathrm{T}}\mathbf{e}\mathbf{P}^{\mathrm{T}}\mathbf{u}/\mathbf{L}^{\mathrm{T}}\boldsymbol{\varepsilon}^{\mathrm{s}}\mathbf{L}. \tag{10}$$

Equation (10) shows that for a plane wave propagating in a piezoelectric medium, once the scalar constant p is specified and the particle displacement field found, then the electric potential associated with that particle displacement is uniquely known. Poles of Eq. (10) correspond to solutions to Laplace's equation. These arise when $\mathbf{L}^{\mathrm{T}}\boldsymbol{\varepsilon}^{\mathrm{s}}\mathbf{L}$ vanishes. In such cases the electric potential is uncoupled from the particle displacement $\mathbf{u}$.

After substituting for the electric field and strain from Eqs. (6) and (2), the tensor divergence of Eq. (5) can be written

$$\nabla \cdot \mathbf{T} = \nabla \cdot \mathbf{c} : \nabla_{\mathrm{s}}\mathbf{u} - \nabla \cdot \mathbf{e} \cdot \nabla\phi.$$

Substituting from Eq. (1) for the left-hand side above and using the matrix representations from Eqs. (7)–(8), we obtain

$$-\rho\omega^2\mathbf{u} = (-j\beta)^2\mathbf{P}\mathbf{c}^{\mathrm{E}}\mathbf{P}^{\mathrm{T}}\mathbf{u} + (-j\beta)^2\mathbf{P}\mathbf{e}^{\mathrm{T}}\mathbf{L}\phi.$$

Now, if Eq. (10) is used to replace the electric potential above, the resulting equations have the particle displacements as the only unknown field quantities. Algebraic manipulation and rearrangement of the terms then yields the stiffened Christoffel equation

$$\left\{\mathbf{P}\left[\mathbf{c}^{\mathrm{E}} + \frac{(\mathbf{e}^{\mathrm{T}}\mathbf{L})(\mathbf{L}^{\mathrm{T}}\mathbf{e})}{\mathbf{L}^{\mathrm{T}}\boldsymbol{\varepsilon}^{\mathrm{s}}\mathbf{L}}\right]\mathbf{P}^{\mathrm{T}} - \rho V^2\right\}\mathbf{u} = 0, \tag{11}$$

where V has been written for ω/β, the phase velocity along the z direction.

Equations (11) are a system of three homogeneous equations in the unknown particle displacements u_x, u_y, and u_z. Additionally, there are two other unknown parameters in Eq. (11), the phase velocity $V = \omega/\beta$ and the ratio of propagation constant components $p = k_t/\beta$. The system can be solved by specifying either V or p and obtaining the other parameter as an eigenvalue of the determinant. If phase velocity is taken as the specified variable, then

$$p = p(V).$$

The fractional propagation constant p is a continuous function of the phase velocity V. There are eight solutions p_n to the system [Eq. (11)] for every value of V. A unique set of values for the p_n and V is not obtained until the boundary conditions are specified.

B. Reduction of Stiffened Christoffel Equation from Matrix Form to Explicit Algebraic Form

In arbitrarily anisotropic and piezoelectric media the full algebraic complexity of Eq. (11) must be dealt with. Examination of the coefficients of the u_i in Eq. (11) shows that each one is a polynomial in V and p. Because of the complexity of the coefficients, analytic solutions to the equations usually cannot be written down. Since numerical procedures must be introduced to find the partial waves, it is important that the determinant of Eq. (11) be reduced as much as possible. Considerable savings in numerical computation time can be obtained by casting the secular equation in a form that avoids redundant calculations of intermediate steps. When writing the determinant, which will be a polynomial in p and V just as each of the matrix elements is, it is important to avoid a formulation where

$$\mathrm{Det} = \sum_n a_n(c_{IJ}, \varepsilon_{ij}, e_{iJ}, \rho, V)p^n = 0$$

in which the constants a_n are calculated anew for each value of velocity considered. Instead a form such as

$$\mathrm{Det} = \sum_m \sum_n b_{mn}(c_{IJ}, \varepsilon_{ij}, e_{iJ}, \rho)p^n V^m = 0 \tag{12}$$

should be obtained. In this form the coefficients of the determinant of Eq. (11) b_{mn} are functions only of the material constants of the plate and thus need be calculated only once for all the modes of a plate of any thickness. [The plate thickness does not appear in Eq. (11).]

We begin by defining the matrix column vector

$$\boldsymbol{\eta} = \mathbf{P}\mathbf{e}^{\mathrm{T}}\mathbf{L} \tag{13}$$

for the piezoelectric coupling term, and by defining the constant

$$\Lambda = \mathbf{L}^{\mathrm{T}}\boldsymbol{\varepsilon}^{\mathrm{s}}\mathbf{L}. \tag{14}$$

Substituting these definitions into Eq. (11), the Christoffel equation becomes

$$\{\Lambda[\mathbf{P}\mathbf{c}\mathbf{P}^{\mathrm{T}} - \rho V^2] + \boldsymbol{\eta}\boldsymbol{\eta}^{\mathrm{T}}\}\mathbf{u} = 0. \tag{15}$$

If the nonpiezoelectric portion of Eq. (15) is written as

$$N_{ij} = P_{iJ} c_{JK} P_{Kj} - \delta_{ij} \rho V^2, \tag{16}$$

where the subscripts take on the values $i, j = 1, 2, 3$ and $J, K = 1, 2, 3, 4, 5, 6$ and where δ_{ij} is the Kronecker delta function, then the elements of Eq. (15) can be expressed as $M_{ij} = \Lambda N_{ij} + \eta_i \eta_j$, and the determinant of the system becomes

$$|M_{ij}| = |\Lambda N_{ij} + \eta_i \eta_j| = 0. \tag{17}$$

Equation (17) can be expanded to separate the nonpiezoelectric portion of the Christoffel determinant from those parts coupled through piezoelectricity:

$$\begin{aligned}
&\Lambda^3 |N_{ij}| + \eta_1 \eta_1 \begin{vmatrix} M_{22} & M_{23} \\ M_{32} & M_{33} \end{vmatrix} - \eta_2 \eta_1 \begin{vmatrix} M_{12} & M_{31} \\ M_{32} & M_{33} \end{vmatrix} + \eta_3 \eta_1 \begin{vmatrix} M_{12} & M_{31} \\ M_{22} & M_{23} \end{vmatrix} \\
&- \Lambda \eta_1 \eta_2 \begin{vmatrix} N_{21} & M_{23} \\ N_{31} & M_{33} \end{vmatrix} + \Lambda \eta_2 \eta_2 \begin{vmatrix} N_{11} & M_{13} \\ N_{31} & M_{33} \end{vmatrix} - \Lambda \eta_2 \eta_3 \begin{vmatrix} N_{11} & M_{13} \\ N_{21} & M_{23} \end{vmatrix} \\
&+ \Lambda^2 \eta_1 \eta_3 \begin{vmatrix} N_{21} & N_{22} \\ N_{31} & N_{32} \end{vmatrix} - \Lambda^2 \eta_2 \eta_3 \begin{vmatrix} N_{11} & N_{12} \\ N_{31} & N_{32} \end{vmatrix} \\
&+ \Lambda^2 \eta_3 \eta_3 \begin{vmatrix} N_{11} & N_{12} \\ N_{21} & N_{22} \end{vmatrix} = 0,
\end{aligned} \tag{18}$$

where the M_{ij} in Eq. (18) are the elements of the stiffened Christoffel matrix [Eq. (15)] and the N_{ij} are the elements of the unstiffened Christoffel matrix. Writing out the 2×2 determinants in Eq. (18), much algebraic simplification takes place through cancellation of terms. Performing these operations, and making use of the fact that $N_{ij} = N_{ji}$ for a lossless medium, one arrives at the final result for the reduced determinant of the Christoffel equations

$$\begin{aligned}
&\Lambda |N_{ij}| + \eta_1^2 \begin{vmatrix} N_{22} & N_{23} \\ N_{23} & N_{33} \end{vmatrix} + \eta_2^2 \begin{vmatrix} N_{11} & N_{13} \\ N_{13} & N_{33} \end{vmatrix} + \eta_3^2 \begin{vmatrix} N_{11} & N_{12} \\ N_{12} & N_{22} \end{vmatrix} \\
&- 2\eta_2 \eta_3 \begin{vmatrix} N_{11} & N_{13} \\ N_{12} & N_{23} \end{vmatrix} + 2\eta_1 \eta_3 \begin{vmatrix} N_{12} & N_{13} \\ N_{22} & N_{23} \end{vmatrix} - 2\eta_1 \eta_2 \begin{vmatrix} N_{12} & N_{13} \\ N_{23} & N_{33} \end{vmatrix} = 0.
\end{aligned} \tag{19}$$

If explicit representations for the N_{ij}, η_i, and Λ are written out, typical multiplications for the 2×2 determinants and the 3×3 determinant can be carried out to determine the order of the equation. One finds that the equation is eighth order in the complex scalar constant p, and that sixth-order terms in phase velocity appear. All the coefficients of the powers of p in the equation are real. Consequently, the roots of p are either real or appear in complex conjugate pairs. Once a value for the phase velocity V has been specified, Eq. (19) is simply an eighth-order expression in p. Because the velocity only appears in the N_{ij} as V^2, the velocity in Eq. (19) occurs only in even powers of V, whereas all powers of p from 0 to 8 are present. Indexing the coefficients of Eq. (12) b_{mn} with a single subscript, the equation can be written

$$\begin{aligned}\text{Det} = {} & b_1 p^8 + b_2 p^7 + [b_3 + b_4 V^2]p^6 + [b_5 + b_6 V^2]p^5 \\ & + [b_7 + b_8 V^2 + b_9 V^4]p^4 \\ & + [b_{10} + b_{11} V^2 + b_{12} V^4]p^3 \\ & + [b_{13} + b_{14} V^2 + b_{15} V^4 + b_{16} V^6]p^2 \\ & + [b_{17} + b_{18} V^2 + b_{19} V^4 + b_{20} V^6]p \\ & + b_{21} + b_{22} V^2 + b_{23} V^4 + b_{24} V^6 = 0. \qquad (20)\end{aligned}$$

While it is known that Eq. (19) can be reduced to the form of Eq. (20), finding explicit expressions for the 24 b_n by pen-and-paper methods is virtually impossible due to the algebraic complexity of the b_n. Instead computer procedures are used to perform the required algebraic manipulations and produce the b_n. The first effort to obtain these coefficients appears to have been made by Solie (1971) using standard IBM subroutines. The version of the coefficients employed by the author was developed by L. Wigton at Stanford University from specially written subroutines. The latter program was written because very little cancellation takes place when the terms of Eq. (19) are multiplied out; thus to obtain compact expressions for the b_n, it is advantageous to have computer procedures that attempt to group calculations rather than to seek cancellations. The expressions generated by Wigton's programs are still lengthy however, requiring over 250 cards of Fortran source coding to define the 24 b_n.

C. Solution of Boundary Value Problem

When the medium is nonpiezoelectric, all the η_i in Eq. (19) are zero, and the equation reduces to a much simpler form

$$\Lambda\,|N_{ij}| = 0. \qquad (21)$$

Equation (21) is satisfied by either $\Lambda = 0$ or

$$|N_{ij}| = 0. \qquad (22)$$

The former case corresponds to solutions to Laplace's equation; and because the medium is nonpiezoelectric, the electric potential is uncoupled from the particle displacement.

Equation (22), which is the Christoffel equation for a nonpiezoelectric medium, is sixth order in the unknown p. There are consequently six solutions for each value of phase velocity specified. Half of these solutions are for waves propagating upward through the medium, i.e., $\exp[-j|\mathrm{Re}\, k_t|y]$, and the other half are for waves propagating downward, i.e., $\exp[j|\mathrm{Re}\, k_t|y]$. From each group of three modes propagating either upward or downward, one is a longitudinal mode and the other two are shear modes.

In a piezoelectric medium some of the η_i are nonzero; Eq. (19) shows that in such a case the two dispersion relations, $\Lambda = 0$ and Eq. (22), no longer vanish individually but are coupled together by the η_i and 2×2 subdeterminants of Eq. (22). Equation (19) is eighth order in p and thus has eight solutions—these eight being perturbed forms of the solutions to $\Lambda = 0$ and Eq. (22).

An example showing the six basically acoustic branches for a piezoelectric crystal is illustrated in Fig. 2. The plot is an inverse velocity plot where $1/V$ is plotted along the abscissa and p/V is plotted along the ordinate. Only those portions of the solutions where p is real are shown; they are given by the inner three curves. (The outer curve shows the Rayleigh wave velocity as a function of θ. The substrate illustrated at the origin of coordinates gives the normal and propagation direction assumed for the Rayleigh wave calculation.) The crystal plane illustrated in Fig. 2 corresponds to the sagittal plane of a commonly used surface wave device orientation, ST-cut quartz. The abscissa is along the crystal's x axis, while the ordinate lies between the y axis and z axis, 42.75° from the y axis. For $\theta = 0$, the substrate shown at the origin assumes an ST-cut orientation.

Considering the three inner curves, one can see that for a sufficiently large V, say at the first tick mark along the abscissa, there are six real solutions. All of the solutions have a k_x value of $(8500)^{-1}\omega$. For these solutions, the propagation constants along the ordinate can be read directly from the figure as $\pm 1.08\omega(8500)^{-1}$, $\pm 1.7\omega(8500)^{-1}$, and $\pm 2.0\omega(8500)^{-1}$. Moving to the right along the abscissa, eventually the inner curve is passed and only four real roots to Eq. (19) can be found. The point at which the inner curve drops out corresponds to the longitudinal velocity along the crystal's x axis. At 5400 m/s, there are still four distinct real solutions, but just to the right of this only two real (propagating) solutions can be found. Finally, at a velocity less than that of the slowest shear wave, where the eight p_n are complex, the Rayleigh wave solution is encountered.

The two roots of Eq. (19) associated with Laplace's equation are complex for all values of phase velocity along the plate's surface. For large phase

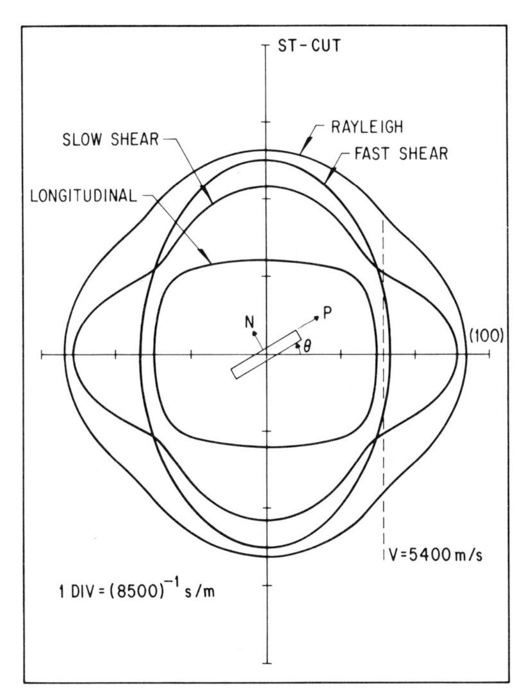

FIG. 2. Inverse velocity curves for α-quartz. The crystal's x axis and the normal to an ST-cut substrate lie in the plane shown. The three inner curves are for bulk wave propagation. The relative labeling of the two shear wave branches as fast or slow was made on the basis of their velocity along the x axis. The outer curve is for Rayleigh wave propagation as a function of the angle θ. The substrate shown at the origin of coordinates gives the normal (N) and propagation direction (P) assumed for the Rayleigh wave calculation (after Wagers, 1976a).

velocity, the values of p are negligibly different from those obtained by solving $\Lambda = 0$. However, as the velocity is decreased, the acoustic coupling can significantly perturb these two solutions. Thus for velocities near the Rayleigh velocity, the distinction between electrostatic and acoustic solutions is not clear. The mechanical displacements and electric potentials from the several branches are all comparable.

Another example of the bulk mode branches of a commonly used piezoelectric material is given in Fig. 3. Shown is the (yz) plane of $LiNbO_3$. Here, not only is there no symmetry about the major axes, but there is not any symmetry axis in the plane of the page. The plate modes of the crystal are accordingly more complex than those of ST-cut quartz. Because of the symmetry shown in Fig. 2, the partial waves of a plate mode can be associated in pairs, one propagating upward in the crystal at an angle θ from the z axis and the other propagating downward at $-\theta$. In contrast to this; Fig. 3

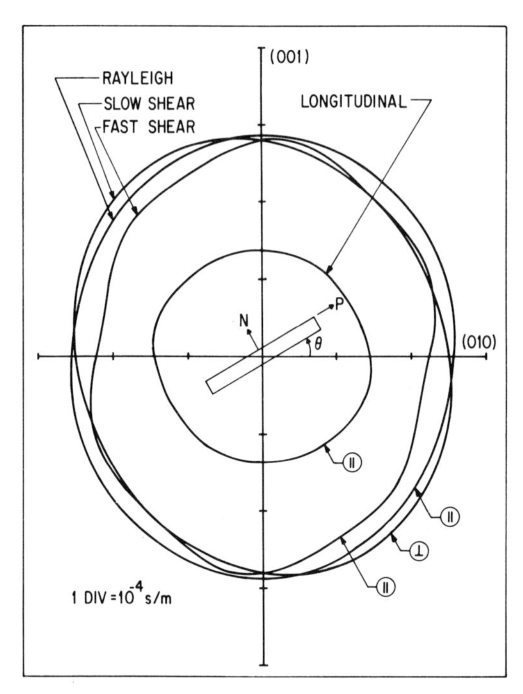

FIG. 3. Inverse velocity curves for the (zy) plane of $LiNbO_3$. The particle motion polarization symbols $\perp$ and $\|$ refer to motion with respect to the plane illustrated. Thus the slow shear mode has only x-directed particle motion (after Wagers, 1976a).

shows that, if the component of phase velocity along the abscissa is specified, then the angles θ_i to the branches with this component of z-axis velocity are all different. While this asymmetry complicates the mode patterns, a simplification comes from the fact that the bulk mode polarizations are either perpendicular or parallel to the (yz) plane. Because the branch labeled $\perp$ has particle displacements perpendicular to the sagittal plane, pure SH modes exist for plates of (yz) $LiNbO_3$. Additionally, the electric potential of these modes is zero.

Plate mode solutions are obtained by finding a sum of the eight modes specified by Eq. (19) satisfying the boundary conditions at the top and bottom of the plate. The conditions applied here are stress-free surfaces, $\phi = 0$ at the bottom surface, and continuity of ϕ and the normal component of electric displacement at the top surface. In the coordinate system used here, this means that the stress components T_1, T_5, and T_6 vanish at the surfaces. The electrical boundary conditions imposed here not only are the usual experimental ones, but in addition they are computationally fortuitous as well. Degenerate modes (with an electric potential) are split by the asymmetry of the electrical boundary conditions, making them more easily detected by numerical procedures.

Indexing by m the solutions $(\mathbf{u}_m, p_m)$ of the stiffened Christoffel equation, Eqs. (2) and (8) can be used to find $\mathbf{S}_m$, and Eq. (10) used to find ϕ. $\mathbf{S}_m$ and ϕ_m can then be substituted in Eqs. (4) and (5) to find $\mathbf{D}_m$ and $\mathbf{T}_m$. Finally, introducing complex coefficients C_m for the relative proportion of the mth partial wave in the plate mode, seven of the required boundary conditions can be written:

$$T_K(y = \pm t/2) = \sum_{m=1}^{8} C_m T_{Km}(y = \pm t/2) = 0, \tag{23}$$

and

$$\phi(y = -t/2) = \sum_{m=1}^{8} C_m \phi_m(y = -t/2) = 0, \tag{24}$$

where T_{Km} is the Kth component of the mth partial wave and $K = 1, 5$, and 6.

The eighth equation is obtained by matching the fields above the crystal to those of the crystal at the top surface. The electric field in the vacuum above the crystal is found from Laplace's equation. Thus, when the electric potential is required to vanish at $y = \infty$, the solution is

$$\phi = \phi_0 e^{-\beta y}.$$

The normal component of the electric displacement above the film is

$$D_y = \varepsilon_0 E_y = \varepsilon_0 \beta \phi_0 e^{-\beta y}.$$

Evaluating ϕ and D_y at the top of the plate and requiring that these fields be continuous with those of the crystal, one obtains

$$\phi_0 = e^{\beta t/2} \sum_{m=1}^{8} C_m \phi_m(y = t/2) \tag{25}$$

and

$$\phi_0 = \frac{e^{\beta t/2}}{\beta \varepsilon_0} \sum_{m=1}^{8} C_m D_{ym}(y = t/2). \tag{26}$$

Eliminating ϕ_0 between these two equations then yields the eighth homogeneous equation in the partial wave amplitudes C_m. An 8×8 boundary value determinant can then be formulated:

$$|B_{mn}| = 0. \tag{27}$$

Solutions to Eqs. (27) are found iteratively. First, a velocity is chosen; then Eq. (20) is solved for the eight p_m; then the elements of Eq. (27) are constructed and the boundary value determinant evaluated. If Eq. (27) is satisfied, then a plate mode has been found; if not, a new velocity is chosen and the process repeated.

D. Dispersion Characteristics of a Thin $LiNbO_3$ Plate

Applying the above procedure to a (yz) $LiNbO_3$ plate, the dispersion relations of Fig. 4 were obtained. Two different families of modes are indicated. Those marked $\perp$ have particle motions perpendicular to the sagittal plane and are thus true SH modes, while those designated $\parallel$ have particle motion in the sagittal plane and are piezoelectrically stiffened Lamb waves. The curves are very similar to those of isotropic media (Auld, 1973).

There is no electric potential associated with the SH modes of the (yz) $LiNbO_3$ plate. Because piezoelectricity is uncoupled from the equations of motion of these modes, the mathematics is simple enough that they can be solved for analytically without use of a computer. One finds that the dispersion relation for these modes is

$$\beta_n^2 = \frac{\omega^2}{V_s^2} + \frac{c_{66}}{c_{44}}\left(\frac{c_{56}}{c_{66}}\right)^2 - \frac{c_{66}}{c_{44}}\left(\frac{n\pi}{t}\right)^2 \tag{28}$$

where $V_s^2 = c_{44}/\rho$. The equivalent expression for an isotropic medium is

$$\beta_n^2 = \frac{\omega^2}{V_s^2} - \left(\frac{n\pi}{t}\right)^2, \tag{29}$$

which Eq. (28) reduces to when the stiffness constants assume the proper symmetry, i.e., $c_{56} = 0$, $c_{66} = c_{44}$. Equation (28) shows that the number of eigenvalues is infinite and for sufficiently large n, β_n becomes purely imaginary.

The Lamb modes are piezoelectrically active with some of them having extremely large wave impedances. At $\beta t = 1.0$, the wave impedances, $Z = \phi^2/2P$, where ϕ is the electric potential at the top surface of the crystal and P is the total power flow, are 67.8, 232, 3273, 4764, and 131 Ω m from the lowest frequency mode to the highest frequency mode. For comparison, note that the wave impedance of a Rayleigh wave of a (yz) $LiNbO_3$ half-space is only 175 Ω m. Thus, while the first two modes (the modified Rayleigh modes) and the fifth mode have impedances comparable to that of the half-space Rayleigh mode, the third and fourth modes have extraordinarily large impedances. The fraction of power flow in the electrical form is also large for these modes; it is 13 and 43% of the mechanical power flow for the third and fourth modes, respectively. A Rayleigh wave on a (yz) $LiNbO_3$ half-space has only 4% electrical power flow.

E. Plate Mode Completeness Considerations

In Section III the excitation of plate waves in a piezoelectric medium will be addressed. As is the usual case in that work, it will be assumed that the normal modes form a complete set for the expansion of the fields of an interdigital transducer. While completeness is commonly assumed, nevertheless proofs of completeness, examination of convergence rates, and definitions of the allowed properties of arbitrary driving functions are subjects on which there is little or no literature for acoustic plate wave guides. The subject is so little addressed that one finds no conspicuous references to proofs of completeness even for the SH modes of isotropic media although several different proofs for those modes are readily produced.

Lamb wave completeness is another matter; even for the isotropic case, the mathematical complexity obscures the prediction of a tractable approach. A major deficiency when dealing with these modes is the absence of a suitable variational expression. The commonly used Lagrangian, valid in the cross section for propagating modes (Lagasse, 1972), is not valid for evanescent modes. And completeness is not possible without the cut-off modes.

Several different proofs can be formulated to show that the SH modes of isotropic plates are a complete basis for expansion of an arbitrary transverse particle displacement function at a cross section. Perhaps the most direct method is to solve the differential equations for the eigenmodes and note that the variation in the cross section can be written as an infinite sum over the eigenmodes, each of which has either a sinusoidal or cosinusoidal variation. Grouping the sums and normalizing the arguments of the sinusoids, the series can be compared term by term to a Fourier trigonometric series which is known to be complete on the interval $[-\pi, \pi]$. In doing this, one finds that the sine function arguments have half-integer order indexing, which requires some additional arguments to show that $\sin[(n - \frac{1}{2})\theta]$ is complete for odd functions on $[-\pi, \pi]$.

An alternative procedure for the SH modes involves another comparison to well-established results. One begins by writing the differential equation for SH modes in an isotropic medium

$$c_{44} \frac{d^2 u_x}{dy^2} + \rho\omega^2 u_x - \beta_n^2 c_{44} u_x = 0$$

$$c_{44} \left. \frac{du_x}{dy} \right|_{y = \pm t/2} = 0, \tag{30}$$

where variation as $\exp[j(\omega t - \beta_n z)]$ has been assumed. Equation (30) is valid for all modes, cut off or not. Additionally, Eq. (29) shows that for the cut-off modes, $-\beta_n^2$ is a positive number tending to infinity with increasing n. If one makes the definitions $p = c_{44}$, $q = \rho\omega^2$, $r = c_{44}$, and $\lambda_n = -\beta_n^2$, Eq. (30) can be written in the standard form of a Sturm–Liouville problem

$$\frac{d}{dy}\left(p\frac{du_x}{dy}\right) + qu_x + \lambda_n r u_x = 0$$

$$\left.\frac{du_x}{dy}\right|_{y=\pm t/2} = 0 \tag{31}$$

with homogeneous Neumann boundary conditions. This is a problem that has been studied extensively, and completeness proofs can be found in standard texts such as Morse and Feshbach (1953).

III. Excitation of Plate Modes in Piezoelectric Substrates

A. General

Interdigital transducers are effective in exciting plate modes for the same reason that they work for Rayleigh waves: the electric field of the transducer, which is periodic in space, interacts synchronously with the plate mode driving it through its piezoelectric coupling. The transducer thus excites only those modes with wavelengths within its spatial Fourier spectrum.

In Fig. 4 the shaded region illustrates the effective range for a transducer with a 10% fractional bandwidth centered at $\beta t = 5.5$. One can see that as the frequency is increased from zero, excitation of the first mode begins at $\omega t/V_R \simeq 4.9$ and continues until $\omega t/V_R \simeq 5.4$. The next $\|$ polarized mode is not excited until the frequency reaches $\omega t/V_R \simeq 6.35$, and so forth. Because the modes are relatively widely spaced, there is never more than one mode excited at any frequency ω satisfying $0 < \omega t/V_R < 10$.

An experimental result showing the individual modes of a (yz) $LiNbO_3$ plate is given in Fig. 5. The plate with $\beta t \simeq 25$ was approximately five times thicker than the largest described in Fig. 4. One would expect that at $\beta t = 25$ on the abscissa of Fig. 4 the mode spacing would be getting sufficiently dense that more than one mode would be excited at any frequency. And, in fact, on the right-hand side of Fig. 5 the frequency response does not effectively separate the modes. However, for the slower modes, the transducer bandwidth was narrow enough to resolve the individual modes.

Returning to Fig. 4, it can be seen that if a narrow-band transducer is used, then the region of interest on the dispersion diagram is only the shaded portion. All characteristics to either the right or left of the shaded zone have no effect on the propagating modes. What is more, to the Rayleigh wave

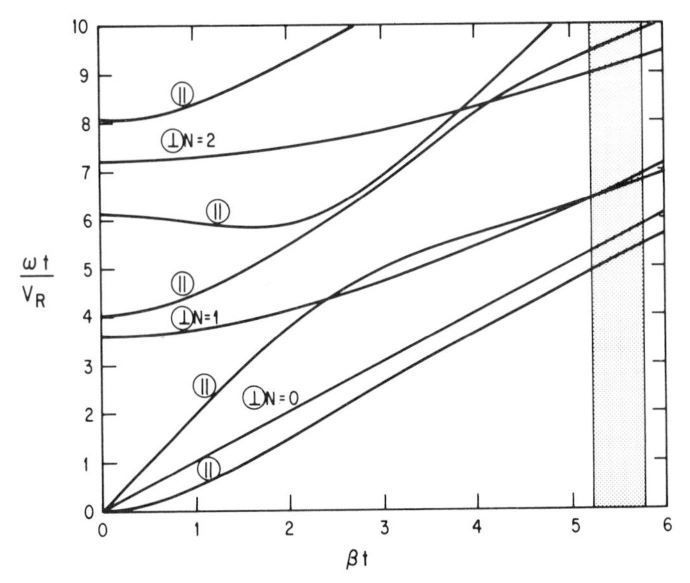

FIG. 4. Dispersion relations for a thin (yz) $LiNbO_3$ plate. Particle motion polarization symbols ⊥ and ‖ refer to motion relative to the sagittal plane. The ordinate has been normalized by the Rayleigh wave velocity of a (yz) $LiNbO_3$ half-space.

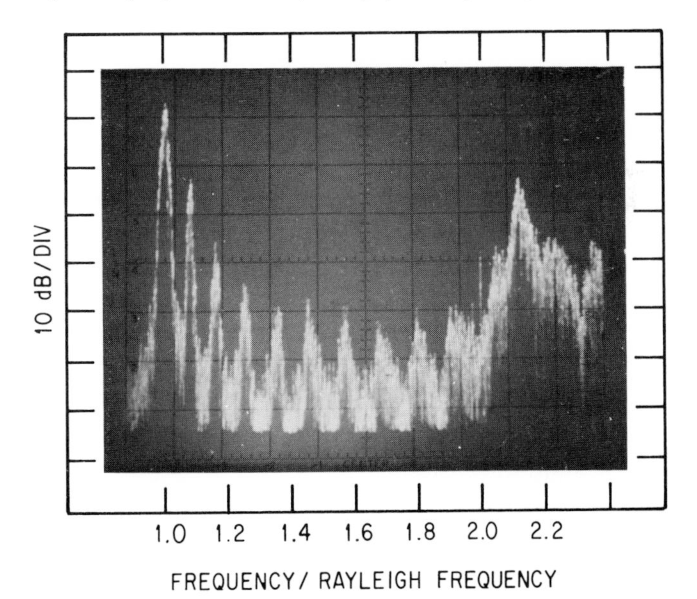

FIG. 5. Frequency response for a delay line configuration like Fig. 1 when the plate is (yz) $LiNbO_3$ with midband $\beta t \simeq 25$. The Rayleigh response is at 33.2 MHz. Between the Rayleigh mode frequency and two times that frequency, nine distinct modes can be seen.

device designer, the time delay of the modes (slope of the curves) is not usually of interest. All that matters to him is the midband frequencies of the plate modes and how strongly they are coupled to the transducer. Consequently, if the midband βt product is specified, the information of interest can be obtained by calculating the value of ω (or ω/β) and the coupling for each mode.

The standard measure of coupling for Rayleigh mode devices is $\Delta V/V$, the fractional change in velocity when the surface on which the transducer is to be applied is short-circuited. This is also an effective measure for the plate modes. It is the coupling parameter plotted in Fig. 6 that shows the allowed

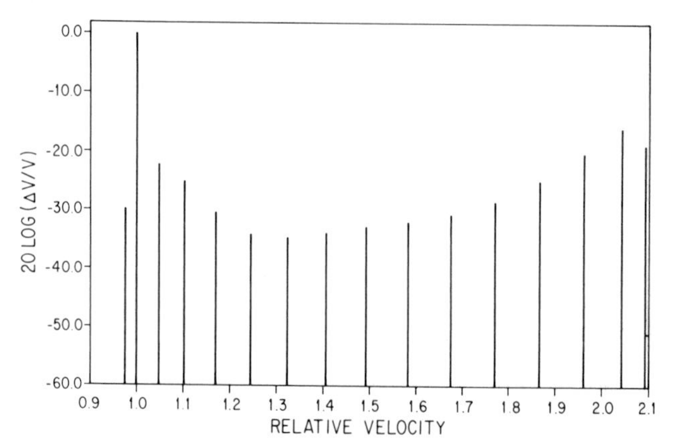

FIG. 6. Plate mode coupling plot for a (yz) $LiNbO_3$ plate with $\beta t = 30$. Mode velocities indicated along the abscissa are shown relative to the Rayleigh mode velocity. The relative couplings for the modes are also referenced to the Rayleigh mode on the top surface. The mode shown with a relative velocity less than unity is a Rayleigh mode on the bottom surface of the plate.

modes of a (yz) $LiNbO_3$ plate with midband $\beta t = 30$. The velocities relative to the Rayleigh velocity are plotted along the abscissa, while the ordinate is 20 times the common logarithm of the ratio of $\Delta V/V$ for a plate mode to that of the Rayleigh mode. The mode shown with a velocity less than that of the Rayleigh mode is also a Rayleigh mode; it is the Rayleigh mode on the short-circuited bottom surface of the crystal. Coupling to the Rayleigh mode on the lower surface occurs through the weak but nonzero field of the wave 4.77λ away from its guiding surface.

The Rayleigh mode of the lower surface does not appear in Fig. 5 because that surface was waxed down to the test fixture. The wax also damps the plate modes appearing at frequencies higher than the Rayleigh mode but not nearly so thoroughly. Between the Rayleigh frequency and two times that in Fig. 5, nine distinct modes appear, whereas Fig. 6 indicates 12 modes in the same velocity range. The difference is due to two different values of βt for the experimental and theoretical results. Fewer modes fit in the frequency range for the thinner plate. This effect is graphically illustrated by considering Fig. 4. At $\beta t = 1.25$, the number of sagitally polarized modes above the Rayleigh mode but within an octave of it is one. However, at $\beta t = 3.0$, an octave range almost encompasses the first three modes.

B. Plate Mode Excitation by Interdigital Transducers

Very narrow-band surface acoustic wave transducers on piezoelectric plates provide excellent illustrations of the modes of the plate. In the experiment described here, the ST-cut quartz substrate was $\sim 24\lambda$ thick ($\beta t = 152$) and over 400λ long. A thickness of $\beta t = 152$ brought many modes within the first octave above the Rayleigh response; however, because of the length of the substrate, very narrow-band transducers could be fabricated.

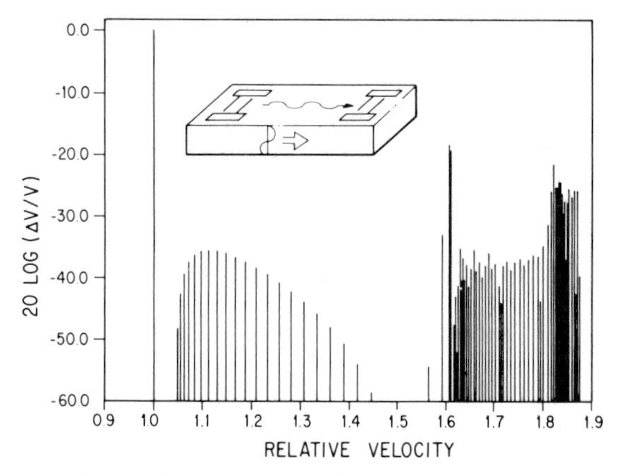

FIG. 7. Plate mode coupling plot for an ST-cut quartz plate with midband $\beta t = 152$. In the range shown 96 allowed modes exist. Bulk wave velocities along the plate surface are ~ 1.04, ~ 1.6, and ~ 1.8 times the Rayleigh wave velocity for the two quasishear and quasi-longitudinal waves, respectively (after Wagers, 1976a).

Equations (19) and (27) were solved for the first 96 modes of the substrate. Their relative velocities and couplings are shown in Fig. 7. Nearly all of the modes are electrically active and have relative couplings $\Delta V/V$ within 60 dB of the Rayleigh mode. Comparing the positions of the modes of

Fig. 7 to the curves in Fig. 2, one can see that those modes with $1.04 < V/V_R < 1.45$ are a family of shear modes associated with the slow shear wave branch of Fig. 2. For $V/V_R > 1.6$, the density of modes in Fig. 7 increases as another family of modes, associated with the fast shear wave branch of Fig. 2, is reached. Finally, for $V/V_R > 1.8$, the mode density of Fig. 7 can be seen to increase again with the introduction of a family of quasilongitudinal plate modes.

The transducers used in the experiment were each $\sim 125\lambda$ long; their combined response had very low near-in side lobes and was only $\sim 4\%$ wide at the -70-dB points of the main lobe. As shown in the experimental curve of Fig. 8, the individual slow shear modes are well resolved by the trans-

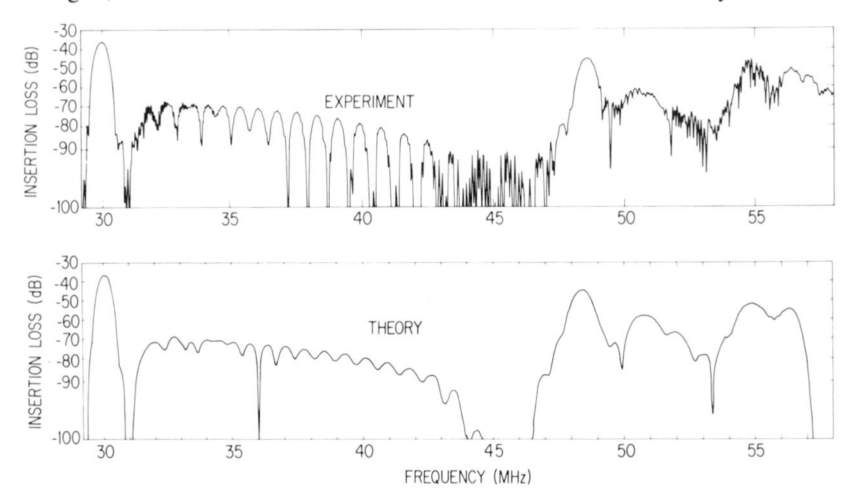

FIG. 8. Experimental and theoretical results for the terminal characteristics of a delay line like that in Fig. 1 where the plate is ST-cut quartz with midband $\beta t = 152$. Very narrow-band transducers were used in order to resolve the individual modes. Plate mode coupling calculations for the substrate tested are shown in Fig. 7 (after Wagers, 1973).

ducers. For frequencies above 45 MHz, the response is essentially continuous as would be expected from the Fig. 7 mode plot. Side-lobe responses of the transducers appear in Fig. 8 as a rapid modulation of the envelope of the response near strongly coupled modes. It can be seen on the high-frequency side of the Rayleigh mode as interference with the first four or five shear plate modes. Also above and below the strong plate mode at 48.5 MHz, side-lobe interference is present.

Quantitative theoretical prediction of the terminal response of the ST-cut quartz delay line requires a characterization of the impedance properties of each of the modes interacting with an interdigital transducer. One of the clearest presentations of the mode impedances is that of Auld

(1973) in the derivation of the radiation conductance of the Rayleigh mode. Fortunately, that approach can be applied to the multimoded quartz delay line with essentially no modifications; it is necessary only to establish that the range of validity of the method extends to the plate modes as well as the Rayleigh mode.

The starting point for the analysis is the complex reciprocity theorem. [The theorem is derived in Chapter 10 of Auld (1973).] If one of the solutions in the theorem is taken to be the nth free mode of the plate, and if the fields induced by the transducer are written as an infinite sum over all modes, then a general differential equation for the nth plate mode excited by the transducer can be obtained.

$$4P_{nn}\left(\frac{d}{dz} + j\beta_n\right) a_n(z) = -j\omega\phi_n^*(t/2)D_y(t/2, z), \tag{32}$$

where P_{nn} is the power flow of the nth mode, $\phi_n(t/2, z)$ is the electric potential of the nth mode at the upper surface of the crystal, and $D_y(t/2, z)$ is the normal component of electric displacement induced by the transducer just inside the crystal. Only two approximations were made in developing Eq. (32): (1) electric fields outside the crystal were neglected for the normal modes; and (2) it was assumed that the modes could be described by a weak coupling, or adiabatic, approximation in which each mode was written as $a_n(z)\psi_n(y)$, where $\psi_n(y)$ is the cross-sectional variation of a field quantity of the nth normal mode. Fortunately, these approximations are usually valid, and in particular they are valid for all the modes of the quartz delay line of Fig. 8.

Equation (32) can be integrated for the nth and $-n$th modes to obtain

$$|a_{\pm n}(\pm l)| = \frac{|\phi_n|}{4P_n}\left|\int e^{\pm j\beta_n\xi} j\omega D_y(t/2, \xi)\, d\xi\right|, \tag{33}$$

where $P_n = |P_{nn}|$, $a_{\pm n}(\pm l)$ is evaluated at the outputs of the transducer, and the ξ integration is over the entire D_y field of the transducer. Written in terms of the mode amplitudes at the transducer outputs, the total power input to the nth modes by the transducer takes the form

$$P_{\text{RAD}} = [|a_n(l)|^2 + |a_{-n}(-l)|^2]P_n. \tag{34}$$

We observe that for real D_y in Eq. (33), the integral is the same for either $\pm j\beta_n$. Then using the relation between input power and radiation conductance

$$P_{\text{RAD}} = \tfrac{1}{2}G|V|^2, \tag{35}$$

the radiation conductance for the nth plate mode becomes

$$G_n = S(\beta_n)\frac{\omega^2}{4}\frac{|\phi_n|^2}{P_n}, \tag{36}$$

where

$$S(\beta_n) = \left| \int e^{j\beta_n \xi} \frac{D_y(t/2, \xi)}{V} \, d\xi \right|^2. \tag{37}$$

Equation (36) shows that the radiation conductance for a mode is dependent only on its wave impedance $Z_n = |\phi_n|^2/2P_n$, the applied frequency, and a spectrum function $S(\beta_n)$ *which is the same for all modes.* If the propagation constant $\beta_n = \omega/V_n$ of the normal mode is within the bandwidth of the transducer, the mode receives the same array effect as every other mode; differences in energy radiation into the modes come only through the other factors of Eq. (36) and changes in the impedance match. Of course, ω in Eq. (36) is not independent of β_n; ω and β_n must jointly satisfy the dispersion relation for the plate.

It should be noted that only the real part of the admittance of the modes has been calculated. Because the coupling to the modes of the quartz delay line of Fig. 8 is so weak, all susceptances, except that of the static capacitance, can be neglected. However, the normal mode method can be applied to obtain the B_n. It is algebraically complicated, but for the standard N-finger-pair transducer, the above approach ultimately yields the correct expression. As a practical matter though, if the susceptances are required for an analysis, they can more easily be obtained by taking the Hilbert transform of the real part of the admittance (Guillemin, 1957).

Equation (36) is written in terms of the wave impedance $|\phi_n|^2/2P_n$, whereas the more common measure of coupling is $\Delta V/V$. To obtain a more conventional formulation, one makes use of a result of Kino and Reeder (1971), namely that, when the modes are not too densely spaced,

$$\frac{|\phi_n|^2}{2P_n} = \frac{2}{\omega[\bar{\varepsilon} + \varepsilon_0]} \left(\frac{\Delta V}{V} \right)_n, \tag{38}$$

where

$$\bar{\varepsilon} = [\varepsilon_{22}^S \varepsilon_{33}^S - \varepsilon_{23}^S]^{1/2}.$$

[To be theoretically rigorous, Eq. (36) needs to be adjusted for the fact that the normal mode expansion employed modes with $D_y(t/2) = 0$ instead of $D_y(t/2)$ continuous through the surface. However, the differences between the two modal expansions is usually negligibly small. See Chapter 10 of Auld (1973).]

Before numerical computations for the radiation conductances of Eq. (36) can be carried out, Eq. (37) must be evaluated for the particular transducers used in the experiment. For the complex transducer designs used in narrow-band surface wave filters, this is an exceedingly difficult task. Compounding the problem in the case of the Fig. 8 spectrum is the fact that the transducers at each end of the device were different from each other. This leads to an overall frequency response that is proportional to a product of functions like Eq. (37). Because of the difficulty of analytically evaluating Eq. (37) and because there is a simpler and more accurate way to obtain the actual effective product of spectrum functions, one generally does not attempt to find the transducer D_y field in order to perform the integration of Eq. (37).

Instead, since the spectrum function $S(\beta_n)$ is the same for all modes, one simply measures the effective product of the transducer responses at a mode that is distinct from the others. In Fig. 8 the Rayleigh mode serves this purpose well. For a constant drive voltage magnitude at the input transducer, the output voltage magnitude across a load resistor was measured for the Rayleigh mode. From this, the portion of the frequency response corresponding to Eq. (37) was extracted and stored as a function of β_n in a computer subroutine.

Obtaining the total frequency response meant summing the signals due to all the plate modes at the output transducer. With $S(\beta_n)$ known from the Rayleigh mode measurements, the rest of the factors appearing in Eq. (36) were added to the responses of the modes, and their phases at the output port were obtained from $\exp[-j\beta_n L]$, the free mode propagation delay for transducers L meters apart. All 96 modes of Fig. 7 were used in the summation which is shown as the theoretical curve of Fig. 8. [In fact, a series impedance model was used (Wagers, 1976a). However, for high Q modes, it is totally equivalent to the above parallel formulation.]

To facilitate visual comparison of experiment and theory, the theoretical results of Fig. 8 were plotted with a nonlinear ordinate that matched the characteristics of the spectrum analyzer. General features of the experimental and theoretical curves are in good agreement with each other. In particular, the general signal level across the octave bandwidth is correct, and the shear modes between 31 and 44 MHz are distinct with midband frequencies corresponding to the experimental curves. The strong response at 48.5 MHz comes from the two modes shown in Fig. 7 at that position acting in phase.

One obvious difference between experiment and theory is the absence of zeros between the shear mode predictions in the 31–44-MHz range. Also, the experimental response shows that, not only are there zeros between shear modes, but each shear mode response is narrower than the Rayleigh mode response. Both effects are a consequence of dispersion, which was neglected in the analysis. Looking back at Fig. 4, one can see examples of how ω/β_n varies throughout the shaded zone. However, the use of center-of-the-band velocities everywhere imposes a dispersion like the solid line shown in Fig. 9. In fact most often the dispersion would be more like the dashed curve—faster at low frequencies and slower at high frequencies. The projections over

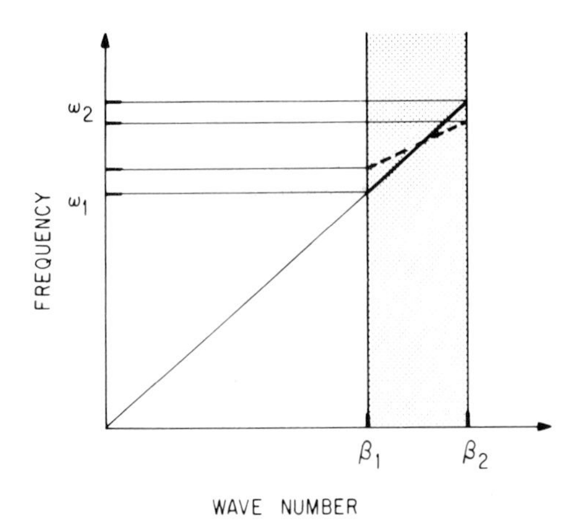

FIG. 9. Illustration of how the use of the midband velocity throughout the coupling range is equivalent to assuming a dispersion relation like the solid curve. A mode with an actual dispersion, such as the dashed curve, has a narrower frequency response than the dispersionless case as the projections over to the ω axis show.

to the ω axis of Fig. 9 show that an actual dispersion would narrow the frequency response of the mode.

At any frequency within the dense mode range above 46 MHz in Fig. 8, many modes are within the bandwidth of the transducers. With the output voltage being a phasor sum of the individual voltages of the contributing modes, the phase of each mode becomes critically important. It is in this range of the frequency response that approximations in the theory have their most serious effect. One can see however that in spite of neglecting dispersion and using Eq. (38) in a densely spaced mode range, the agreement between experiment and theory is quite good.

C. Examples of Plate Mode Coupling to Interdigital Transducers

Much research on the plate modes of various cuts of the commercially available piezoelectric crystals has been performed by those associated with Rayleigh wave filtering technology. Materials considered include lithium niobate, lithium tantalate, quartz, bismuth germanium oxide, and bismuth silicon oxide. Examples of the mode coupling diagrams (like Fig. 7) for common cuts of all of these materials except lithium tantalate can be found in Wagers (1976a). On some of these substrates, experimental results using narrow-band transducers have been obtained. The responses provide additional graphic demonstration of the mode spectra of piezoelectric plates.

In the following sequence of results the objective was to find a cut of $LiNbO_3$ in which the first electrically coupled plate mode occurred at a much higher velocity than that of the Rayleigh mode. Such a cut could advantageously be applied to narrow-band filtering applications because the distortion introduced by the spurious plate modes would be separated from the Rayleigh response, allowing it to be discriminated against by external circuitry.

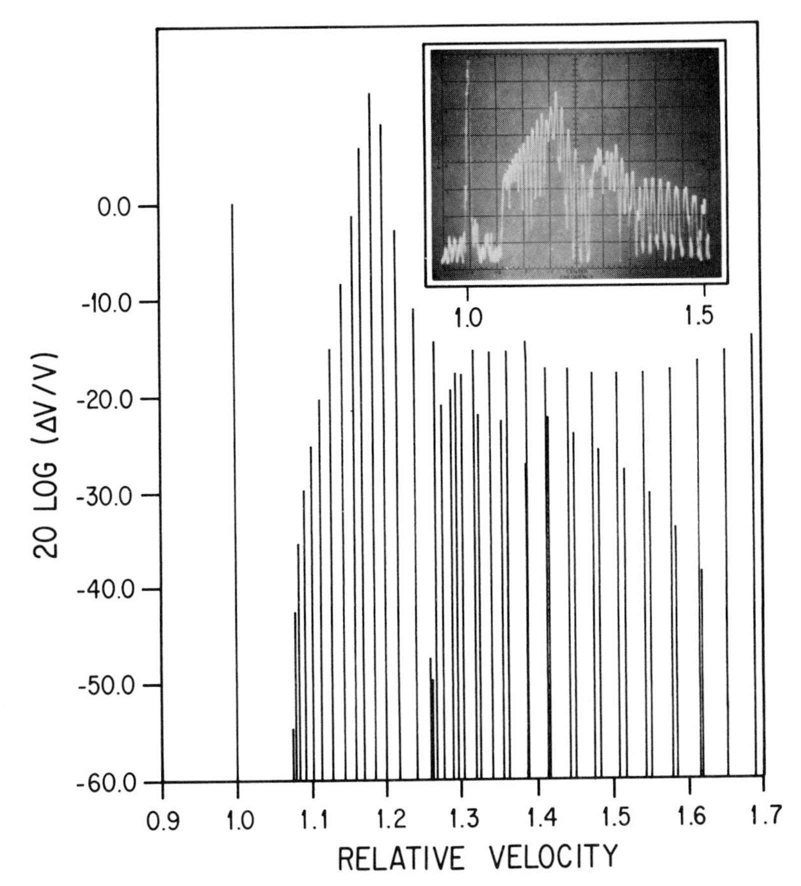

FIG. 10. Experimental and theoretical results for a (zx) $LiNbO_3$ delay line. The calculations were for $\beta t = 60$, while the actual plate tested was much thicker than that (after Wagers, 1974).

Figure 10 shows both theoretical mode coupling calculations and experimental frequency response characteristics for a (zx) $LiNbO_3$ plate. The Rayleigh response is at ~80 MHz, and the substrate was ~0.040-in. thick. While the calculations were for a plate ~9.5λ thick, the actual plate thickness corresponded to more like 20λ, so quantitative comparison of experiment and theory is not possible. However, the train of modes that starts at $V/V_R \simeq 1.08$, independent of plate thickness, is clearly displayed. The transducers resolve the separate mode nature of the received signal for good qualitative confirmation of the theory. The major point of the research, that no modes exist for $1 < V/V_R < 1.08$, is borne out by the experimental results of the inset photograph.

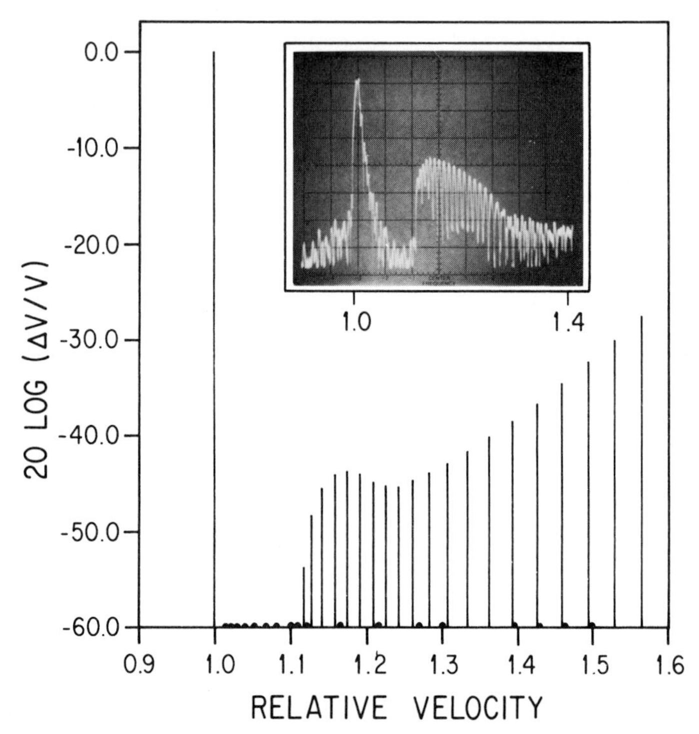

FIG. 11. Experimental and theoretical results for a (zy) $LiNbO_3$ delay line. The calculations were for $\beta t = 60$, while the actual plate tested was much thicker than that. The enhanced dots on the abscissa indicate a family of modes with no electric potential (after Wagers, 1974).

Another illustration of the free plate modes of $LiNbO_3$ is given in Fig. 11. In this case a (zy) $LiNbO_3$ plate is considered. Again, the plate thickness used for the theory differed from that of the experiment. The first electrically coupled plate mode for (zy) $LiNbO_3$ is shown at $V/V_R \simeq 1.12$ about 3% higher in relative velocity than that for (zx) $LiNbO_3$. Confirmation of this fact is provided by the results in the inset photograph. However, the experimental results also show a widening of the Rayleigh wave response (compared to the Fig. 10 Rayleigh wave response). This comes about because of the branch of Fig. 3 marked $\perp$, the SH modes, which are supposed to be piezoelectrically inactive.

The SH modes of (zy) $LiNbO_3$ are shown in Fig. 11 by enhanced dots on the abscissa to indicate their velocities. Theoretically, they have no electric potential. However, even the smallest deviation from coincidence of the sagittal plane and the (zy) $LiNbO_3$ plane adds an electric potential to these modes, thus coupling them to the transducer. In fact, even having a finite width to the transducers will cause excitation of the modes because the transducer's electrostatic potential has a spectrum of k vectors lying in the plane of the crystal's top surface (Wagers, 1975).

A significant difference between the theoretical prediction and the experimental results of Fig. 11 is that the experimental plate mode spectrum decays away monotonically above $V/V_R \simeq 1.15$, whereas the theory predicts the opposite. This occurred because the measured crystals were waxed down to a test fixture that absorbed energy from the plate modes. The effect is evident in Fig. 10 also, but is more striking in Fig. 11.

Reference to Fig. 3 helps illustrate why some modes in Fig. 11 are more affected by the test fixture than others. The coupled family shown in Fig. 11 is associated with the shear wave branch of Fig. 3 labeled $\parallel$. One can see that at low velocities the normal to this curve (the group velocity direction) is at a shallow angle to the surface of the crystal. Thus, a partial wave may travel from one transducer to under the other transducer before reaching the bottom of the crystal. However, when V/V_R is increased to 1.25, the two partial waves associated with the $\parallel$ shear wave branch of Fig. 3 are traveling at $\sim 12^\circ$ and $\sim 35^\circ$ to the surface of the crystal. The higher velocity modes thus have the conditions for greater interaction with the energy absorbing bottom surface of the crystal.

A final example of plate mode excitation in $LiNbO_3$ is offered in Fig. 12. For this crystal, a 46° (zyw) $LiNbO_3$ plate, propagation is along a line between the y and z axes, 46° up from the y axis in Fig. 3. Again the transducers resolve the individual modes of the plate. Although the Rayleigh response and the response in the region $1 < V/V_R < 1.1$ of the inset photograph is rather ragged, the prediction of no excited modes within 10% of the Rayleigh mode is verified by the experimental results. The rough serrated response of the Rayleigh mode occurred because the transducers were designed for application on low-coupling quartz; whereas the 46° (zyw) $LiNbO_3$ cut has $\Delta V/V = 0.013$, some 22 times greater than that of ST-cut quartz. The larger $\Delta V/V$ leads to phase errors in the positions of the electrodes when the transducers are used on $LiNbO_3$.

A dual trace experimental response is shown in Fig. 12. One trace was obtained with the top surface of the plate free; for the other trace, wax was added between the transducers on the top surface until the Rayleigh mode

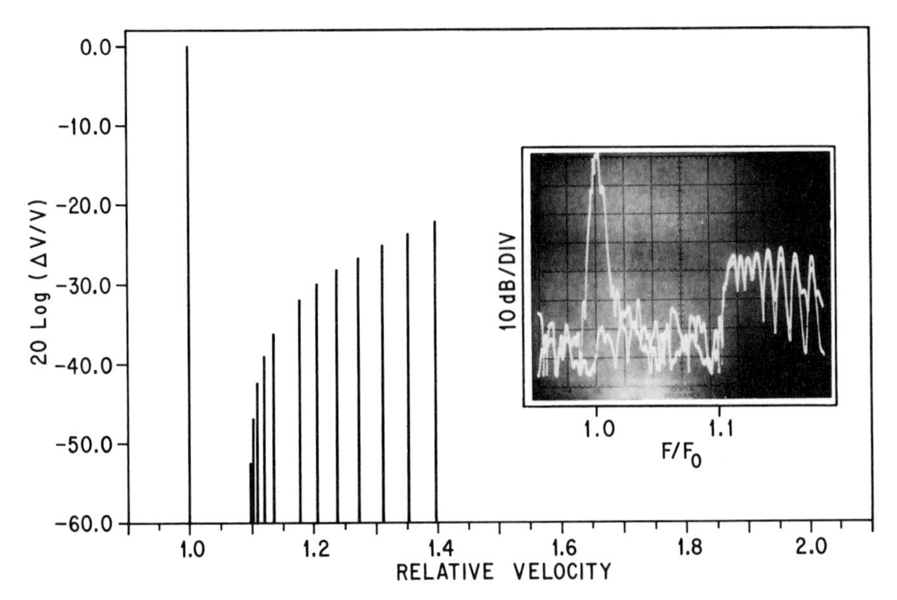

FIG. 12. Experimental and theoretical results for a 46° (zyw) $LiNbO_3$ delay line. The calculations were for $\beta t = 60$, while the plate tested was much thicker than that (after Wagers, 1976a).

was absorbed. Note that in the plate mode regions the traces nearly repeat. The small amount of wax required to completely remove the Rayleigh mode has a negligible effect on the first plate modes; on the eighth plate mode, the last one shown in the inset photograph, 3-dB extra attenuation was added.

D. Suppression of Plate Modes in Surface Acoustic Wave Delay Lines

Because of the relative insensitivity of the slower plate modes to boundary perturbations, they are a subject of commercial interest. For a Rayleigh wave filter to be free of spurious responses, these modes need to be removed from the terminal characteristics of the filter. This does not mean eliminating their excitation necessarily; however, if their excitation is accepted as unavoidable, then steps need to be taken to prevent their detection at the output. One of the methods most investigated is to alter the plate geometry in such a way as to destroy the spatial phase coherence of the plate modes. Since the interior of the crystal is inaccessible and because the upper surface is constrained to be planar, free, and commonly of minimum area, only the bottom surface of the crystal can be altered. Figure 12 shows that the faster plate modes are substantially affected by the back-surface-bonding method, but the slowest modes are relatively unaffected. Consequently, their removal may not be possible for an arbitrary transducer configuration on any substrate orientation.

The 46° (*zyw*) $LiNbO_3$ orientation is one in which the slowest plate modes are relatively sensitive to bottom surface treatments. In particular, if channels are sandblasted into the bottom surface (Wagers *et al.*, 1975) before the crystal is bonded to the test fixture, all but the very slowest modes are completely eliminated, and the slowest modes are substantially reduced.

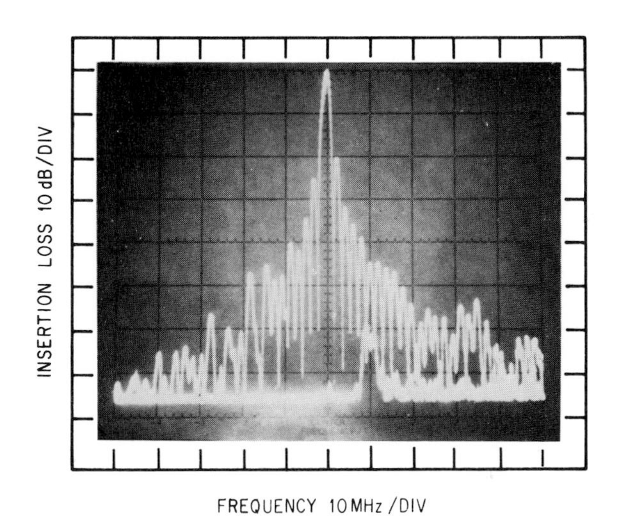

FIG. 13. Frequency response characteristics for a 46° (*zyw*) $LiNbO_3$ plate that has had the bottom surface contoured by sandblasting. The plate was 14.6λ thick and had $N = 40$ transducers on 197λ centers. A dual trace response is shown (a) with the top surface of the plate free, and (b) after damping the Rayleigh response by the addition of wax between the transducers. The aggregate plate responses, which are unaffected by the addition of wax between the transducers, appear centered 10 MHz above the Rayleigh response with a relative strength of −60 dB (after Wagers, 1976b).

Figure 13 illustrates how effective the procedure can be. The substrate was 14.6λ thick at 75 MHz and had two 40λ transducers spaced 197λ apart. A dual trace photograph was used to illustrate the relative strength of the modes in the terminal response. At the Rayleigh response a full 80 dB of separation to any other contribution to the terminal signal is apparent. 8 MHz up from the Rayleigh mode, the aggregate spurious signals reach their maximum −60 dB at the onset of the slowest plate modes. However, the spurious level falls back off rapidly and remains at −70 dB for the remainder of the 66% bandwidth shown above the Rayleigh signal.

Acknowledgments

The author is grateful to B. A. Auld and W. R. Shreve for valuable criticism of the manuscript.

References

Auld, B. A. (1973). "Acoustic Fields and Waves in Solids," Vol. II. Wiley (Interscience), New York.

Guillemin, E. A. (1957). "Synthesis of Passive Networks," pp. 296–308. Wiley, New York.

Kino, G. S., and Reeder, T. (1971). *IEEE Trans. Electron Devices* **18**, 909.

Lagasse, P. E. (1972). *Proc. 1972 Ultrason. Symp.*, IEEE Cat. No. 72 CHO 708-8SU, p. 436.

Meeker, T. R., and Meitzler, A. H. (1964). *In* "Physical Acoustics: Principles and Methods" (W. P. Mason, ed.), Vol. 1, Part A, pp. 111–167. Academic Press, New York.

Mitchell, R. F., and Reed, E. (1975). *IEEE Trans. Sonics Ultrason.* **22**, 264.

Morse, P. M., and Feshbach, H. (1953). "Methods of Theoretical Physics," Vol. 1, pp. 736–739. McGraw-Hill, New York.

Solie, L. (1971). Ph.D. Dissertation, Stanford University, Stanford, California (unpublished).

Wagers, R. S. (1973). *Proc. 1973 Ultrason. Symp.* IEEE Cat. No. 72 CHO 807-8SU, p. 372.

Wagers, R. S. (1974). *Proc. 1974 Ultrason. Symp.* IEEE Cat. No. 74 CHO 896-1SU, p. 337.

Wagers, R. S. (1975). *IEEE Trans. Sonics Ultrason.* **22**, 375.

Wagers, R. S. (1976a). *IEEE Trans. Sonics Ultrason.* **23**, 113.

Wagers, R. S. (1976b). *Proc. IEEE* **64**, 699.

Wagers, R. S., Birch, M., Weirauch, D., and Hartmann, C. (1975). U.S. Patent 3,887,887.

—4—

Anisotropic Surface Acoustic Wave Diffraction

THOMAS L. SZABO

Deputy for Electronic Technology
Hanscom AFB, Massachusetts 01731

I. Introduction

Surface acoustic waves are being used in piezoelectric signal processing devices, for the nondestructive evaluation of materials, and as acoustooptic analogues at frequencies from a megahertz to above a gigahertz. Acoustic diffraction is an unavoidable, natural phenomenon common to all surface wave transducers since they are a few wavelengths or more in width. This chapter will explain why diffraction effects can be significant in the areas mentioned.

A common erroneous assumption is that a surface wave interdigital transducer produces a time harmonic beam of constant width. Figure 1,

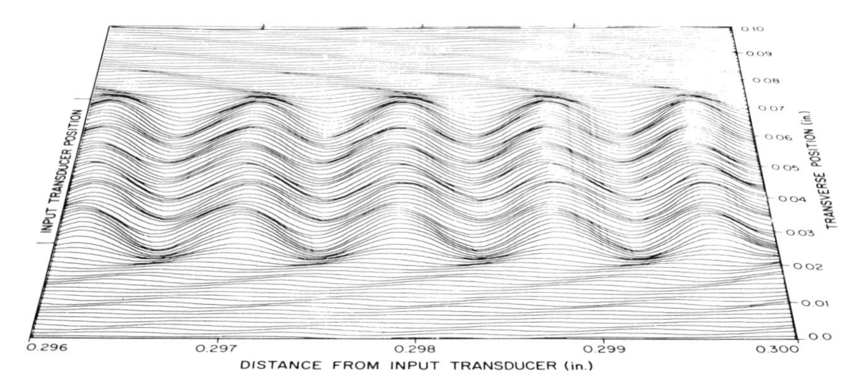

FIG. 1. Electrostatic probe measurements of a 170-MHz surface wave propagating along the z axis of Y-cut lithium niobate. The scale is expanded along z. A wavelength λ is 20 μm and the aperture $L = 59\lambda$ (after Williamson, 1973).

however, depicts the form of an actual beam on the surface of a lithium niobate crystal. This remarkable depiction of surface waves was produced by Williamson (1973) from his electrostatic measurements of near field wave amplitude and phase. Here diffraction causes a nonuniform transverse variation along the crests and troughs; also, beyond the region delineated by the width of the transducer, diffraction creates a backwash of energy that spills away from the central region.

If the diffraction field intensity over a much greater distance could be seen, it would appear as shown in the computer simulation in Fig. 2 with the dimension perpendicular to the transducer greatly compressed. This figure shows that the characteristic evolution of diffraction from a rectangular source consists of rapid fluctuations near the transducer, a gradual peaking, and, finally, significant continuous broadening as the beam moves outward.

Section II will demonstrate that for anisotropic materials usually used for surface wave devices, the rate of diffraction depends on the degree of anisotropy and beam steering occurs. Ways of predicting diffraction will be discussed. As apparent from Fig. 2, a transducer placed in the diffraction field will have a diffraction loss and phase distortion dependent on its size and location; this problem will be analyzed in Section III. The complications caused by diffraction and beam steering have been remedied by very

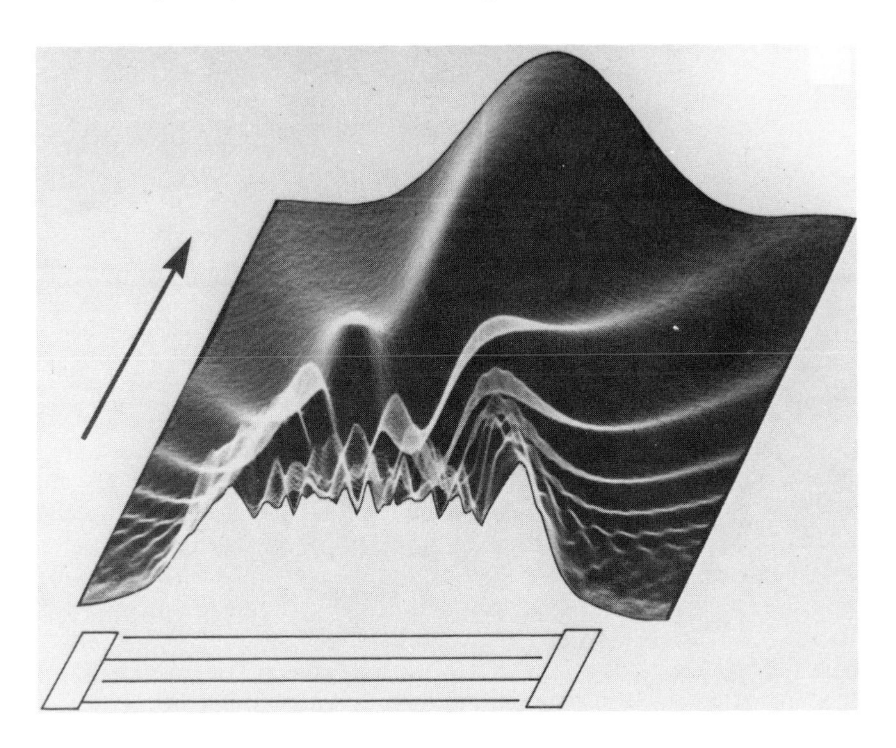

FIG. 2. Three-dimensional view of diffraction from an interdigital transducer. The acoustic intensity is simulated for a 40-wavelength aperture. Note that the scale perpendicular to the transducer is greatly compressed: 1920 wavelengths of the beam are shown along this direction.

different methods, such as acoustic beam shaping (Section III) and material selection (Section IV) including crystals with minimal diffraction properties (Section V) and alterations in the transducer structure (Section IV). A unique class of tiny planar acoustic devices that mimic optical components such as lenses, mirrors, and resonator cavities is covered in Section VI. Special diffraction effects caused by defects and propagation on long time delay lines are described in Section VII.

This chapter will show that all these topics are aspects of the same basic phenomena. The analysis presented is for two-dimensional time-harmonic diffraction. This representation is both simple and sufficiently accurate for most of the problems that will be described. The treatment of diffraction will apply to all kinds of surface wave transducers (White, 1970) with emphasis on interdigital transducers and anisotropic materials. Because of their similarity in structure, the information for interdigital transducers applies equally well to surface wave electromagnetic transducers (Szabo and Frost, 1976).

II. Theory

A. Anisotropy

Nearly all surface acoustic wave devices are fabricated on anisotropic piezoelectric single crystalline materials. Anisotropy considerably complicates the diffraction process and causes beam steering (Szabo and Slobodnik, 1973a). In Fig. 3 a pair of interdigital transducers lie on a substrate at an angle θ with respect to a crystalline axis. In contrast to the isotropic case, the directions of the phase and group velocity vectors coincide only for specific values of θ defined as pure mode axes (θ_0). The acoustic beam in Fig. 3 veers off at the power flow angle ϕ with respect to the center line between the two transducers. For practical devices θ is usually chosen so that $\phi = 0$; however, there is always some unintentional misalignment which results in beam steering loss. A direct measure of the seriousness of this effect is $\partial\phi/\partial\theta$, the slope of the power flow angle. Not only is the severity of beam steering peculiar to transducer alignment and the material chosen, but unlike the isotropic case, the rate of diffraction is unique to the particular anisotropy of the material.

Fortunately, methods for computing anisotropy, power flow, and surface wave data have been developed (Campbell and Jones, 1968; Farnell, 1970). In addition, this information has been calculated and made available for many crystals (Slobodnik *et al.*, 1973).

A typical calculation of anisotropy and power flow is shown in Fig. 4 for 111-cut bismuth germanium oxide ($Bi_{12}GeO_{20}$). Here the $\theta = 0$ direction is the (110) axis. It is important to note that extrema of the velocity curve correspond to pure mode axes and that regions where the velocity is linear correspond to inflection points of the power flow curve. For small deviations near θ_0,

$$\phi \approx \frac{1}{V_0}\frac{\partial V}{\partial \theta}, \tag{1}$$

where V_0 is the pure mode velocity.

Knowledge of material anisotropy such as the velocity surface of Fig. 4 or, alternatively, the propagation wave vector surface $k(\theta)$ ($k = \omega/V$ where ω is angular frequency) is prerequisite to diffraction prediction.

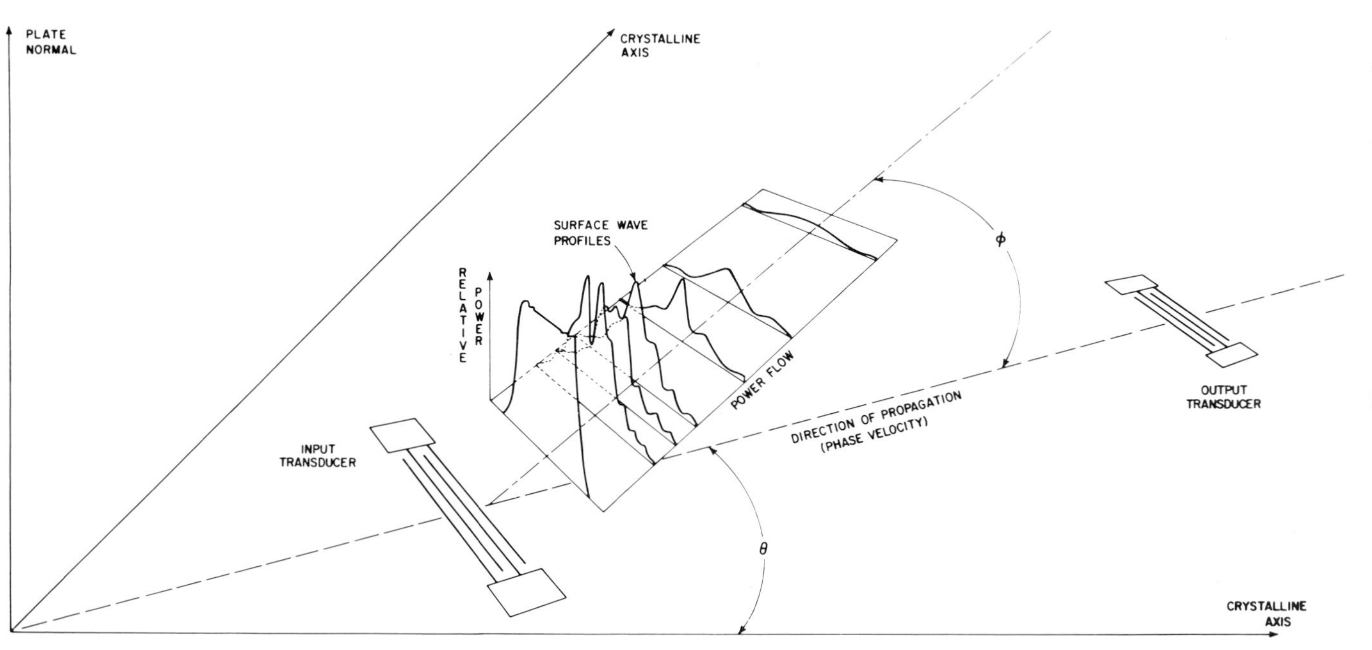

FIG. 3. Schematic representation of the launching and propagation of an acoustic surface wave on an anisotropic substrate. Note diffraction effects and how the phase velocity and power flow directions are separated by the angle ϕ. θ defines the direction of propagation, while ϕ is the power flow angle (after Szabo and Slobodnik, 1973a).

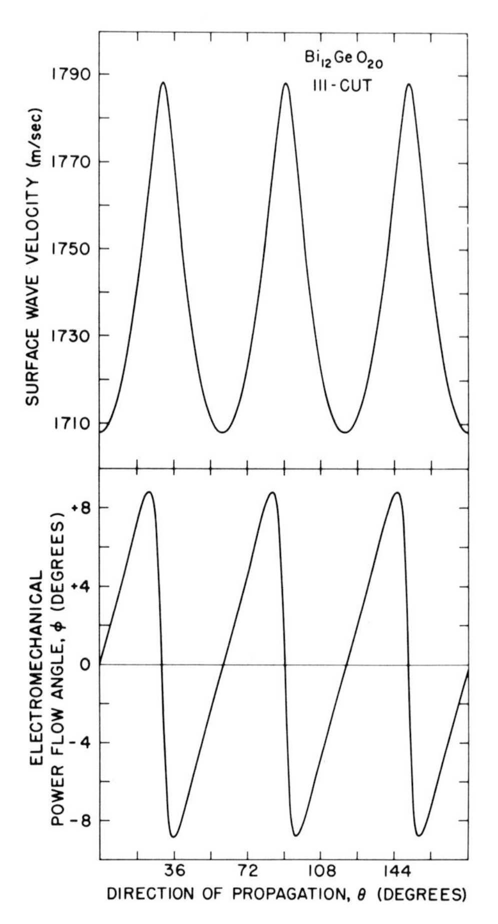

FIG. 4. Surface wave velocity and electromechanical power flow angle as a function of angle from the 110 axis on 111-cut bismuth germanium oxide (after Szabo and Slobodnik, 1973a).

For many cases of interest, a parabolic approximation to the velocity surface is sufficient. (This will be explained in detail later.) The basic assumption for this approximation is that the velocity surface can be well described near θ_0 by

$$\frac{V(\theta)}{V_0} = 1 + \tfrac{1}{2}\gamma(\theta - \theta_0)^2, \tag{2}$$

and, consequently, from Eq. (1),

$$\phi = \gamma(\theta - \theta_0). \tag{3}$$

At the $\theta_0 = 0$ axis in Fig. 4 the velocity surface has the shape of an upward concave parabola. Near this axis, ϕ is positive and linear with a slope of $\partial\phi/\partial\theta = 0.366$. Later it will be shown that a single anisotropy parameter, $\gamma = \partial\phi/\partial\theta$, can be used to predict both the rate of diffraction and the sensitivity of beam steering to misalignment.

B. Methods of Diffraction Prediction

Once the anisotropy of the material has been determined, it is necessary to decide which of the various diffraction theories to use. (Guidelines for this will be discussed in Section II,E.) These theories can be roughly divided into two types: exact and approximate. Only one theory of each type—the angular spectrum of waves (ASW) theory and generalized parabolic approximation approach—can be dealt with in depth in this chapter. Other methods will be described briefly however.

Both the ASW (Kharusi and Farnell, 1970, 1971, 1972) and anisotropic Green's function (Crabb *et al.*, 1971; Ogg, 1971) theories are very powerful exact approaches and require numerical integration. The ASW theory will be described here because it is more widely known, but in principle, the Green's function method should produce equivalent results. The ASW approach can accommodate $k(\theta)$ functions of almost any real shape including highly nonparabolic surfaces and nonpure mode axis orientations. Any transducer source distribution (rectangular or otherwise) can be used, provided it has a Fourier transform. In addition, there is no restriction on calculating fields very close to a transducer.

Other approaches usually employ a parabolic approximation of $k(\theta)$ (Papadakis, 1975). The first of these (Mason, 1971, 1973; Mason and Ash, 1971) is actually an ASW theory in which a parabolic fit to $k(\theta)$ is made in rectangular coordinates. Important results obtained from this relatively simple approach are: a description of diffraction in terms of gaussian modes; rigorous evidence that, for parabolic anisotropy, the rate of diffraction can be described by anisotropic scaling parameters; and analytic proof that, for beam steering, the beam profile retains its pure mode shape and is shifted laterally at the beam steering angle. Although any general source function can, in principle, be treated by this method by resolving it into an infinite set of Gaussian–Hermite modes, it is usually inconvenient to do so, and the method is most suitable for gaussian-shaped sources.

The second approximate approach, also related to ASW theory in its origin, is a parabolic theory proposed by Cohen (1967) for bulk waves and then adapted to the surface wave case by others (Weglein *et al.*, 1970; Kharusi and Farnell, 1971; Szabo and Slobodnik, 1973a). In this instance the parabolic fit to $V(\theta)$ is made in polar coordinates [Eq. (2)]. Because a Fresnel approximation is used in its derivation, the theory is generally valid for distances greater than an aperture's width from the input transducer. Cohen (1967) obtained a diffraction integral identical in form to previous isotropic results, but with the perpendicular distance from the aperture (z) replaced by an effective scaled distance $z' = z|1+\gamma|$ (Szabo and Slobodnik, 1973a). This fundamentally important result shows that, depending on the anisotropy parameter γ, anisotropic diffraction is accelerated or retarded in comparison to the isotropic rate. The parabolic approximation has been generalized to a Fourier transform formulation for arbitrary source distributions (Szabo, 1975, 1977).

Other methods based on an approximation to the velocity surface include a modified Rayleigh integral and parabolic velocity surface approach (Papadakis, 1975); another Green's function approach (Mitchell and Stevens, 1975); a cubic velocity surface theory (Pirio, 1975); and a linear velocity surface theory (Cho *et al.*, 1970).

C. Generalized Fourier Transform Theory for Parabolic Velocity Surfaces

The development of this analysis begins with the general integral obtained by Cohen (1967). For an aperture centered in an x-z coordinate system, having an acoustic source distribution $\tilde{A}(x_0, 0)$ defined over the source coordinate x_0 (parallel to the x axis) and $\tilde{A}(x_0, 0) = 0$ outside of the aperture, the following integral for the field point amplitude[1] can be obtained with some manipulation (Szabo, 1977):

$$A(x, z) = e^{j\pi/4}(\beta_0)^{1/2} \exp(-j\pi\beta_0 x^2) \int_{-\infty}^{\infty} \tilde{A}(x_0, 0) \times \exp(-j\pi\beta_0 x_0^2) \exp(j2\pi\beta_0 x_0 x)\, dx_0, \tag{4}$$

and $\beta_0 = k_0/2\pi z|1+\gamma|$. This form is similar to that obtained for the analogous situation in optics (Papoulis, 1968; Goodman, 1968). The form can be recognized as that of an inverse Fourier transform (Bracewell, 1965), defined as

$$A(x) = \mathscr{F}^{-1}[\tilde{A}(s)] = \int_{-\infty}^{\infty} \tilde{A}(s) e^{-j2\pi sx}\, ds. \tag{5}$$

[1] By "acoustic amplitude" no physically real variable is meant, but rather an amplitude defined by the real acoustic power $P = AA^*/2$. Also, as defined, time harmonic variation $\exp j(k_0 z - \omega t)$ is assumed for amplitude and has been factored out.

By means of the parameter $s = \beta_0 x_0$, Eq. (4) becomes

$$A(x, z) = \frac{e^{j\pi/4}}{(\beta_0)^{1/2}} e^{-j\pi\beta_0 x^2} \mathscr{F}^{-1}[\tilde{A}(s/\beta_0) \exp(-j\pi s^2/\beta_0)], \tag{6}$$

where $\tilde{A}(s/\beta_0) = \tilde{A}(x_0, 0)$. Similarly, the following expression for the far field amplitude (Papoulis, 1968; Szabo, 1977) is obtained:

$$A(x, z) = e^{j\pi/4} \mathscr{F}^{-1}[\tilde{A}(s/\beta_0)]. \tag{7}$$

From these equations simple analytic expressions can be obtained for many source distributions of interest. The most common source shape is rectangular, $\tilde{A}(x_0, 0) = \Pi(x_0/L)$ where the rectangle function (Bracewell, 1965) is defined as

$$\Pi(x_0/L) = \begin{cases} 1, & |x_0| < L/2, \\ \frac{1}{2}, & |x_0| = L/2, \\ 0, & |x_0| > L/2. \end{cases} \tag{8}$$

With some manipulation, Eq. (6) provides the near (as well as far) field amplitude (Szabo, 1977)

$$A(\hat{x}, \hat{z}) = \frac{e^{j\pi/4}}{\sqrt{2}} \left[F\left(\frac{\hat{x} + \hat{L}/2}{\hat{z}|1 + \gamma|/2} \right) - F\left(\frac{\hat{x} - \hat{L}/2}{\hat{z}|1 + \gamma|/2} \right) \right], \tag{9}$$

in which F denotes the Fresnel integral of negative exponential argument (Szabo and Slobodnik, 1973a, 1974), and for convenience, wavelength (λ) scaled parameters are denoted by hats: $\hat{L} = L/\lambda$, $\hat{x} = x/\lambda$, etc. Equation (9) is one of the most important equations in this chapter. It was used for many of the calculations of beam profiles and diffraction loss presented here. The far field equation [which is, of course, only a special case of Eq. (9)] can be obtained directly from Eq. (7):

$$A(x, z) = \frac{e^{j\pi/4} \hat{L}}{\hat{z}|1 + \gamma|} \operatorname{sinc}\left(\frac{\hat{L}\hat{x}}{\hat{z}|1 + \gamma|} \right), \tag{10}$$

where $\operatorname{sinc}(a) = \sin(\pi a)/\pi a$ (Bracewell, 1965).

In order to demonstrate the application of these results to a specific case, a parabolic fit was made to computed velocity values shown in Fig. 4 within a range of $\pm 5^\circ$ of $\theta_0 = 0$. In this case the velocity surface was found to be an excellent parabola with $\gamma = 0.366$. To confirm the accuracy of the parabolic theory in this case, a transducer pair operating at 510 MHz was fabricated along this orientation on $Bi_{12}GeO_{20}$, and beam profiles were measured by the laser probe technique (Szabo and Slobodnik, 1973a). In Fig. 5 these measurements (column 4) are compared with isotropic (column 1), parabolic (column 2), and ASW (column 3) theories. Good agreement is obtained with the parabolic and ASW theories, which give equivalent results for this case. As predicted by the parabolic theory, the rate of diffraction here is accelerated by a factor of $|1+\gamma| = 1.366$ when compared to the isotropic diffraction rate.

D. Angular Spectrum of Waves Theory

In contrast to the $Bi_{12}GeO_{20}$ case, there are situations that cannot be predicted accurately by approximate theories, and a more general method, such as the ASW theory must be used. Kharusi and Farnell (1970, 1971, 1972) presented the ASW theory as the integral

$$A(x, z) = \int_{-\infty}^{\infty} F(k_1) \exp[j(k_1 x + k_3(k_1)|z|)]\, dk_1, \tag{11}$$

where k_3 and k_1 are the projections of the wave vector $k(\theta)$ along the z and x axes, respectively, or, in general, along directions perpendicular and parallel

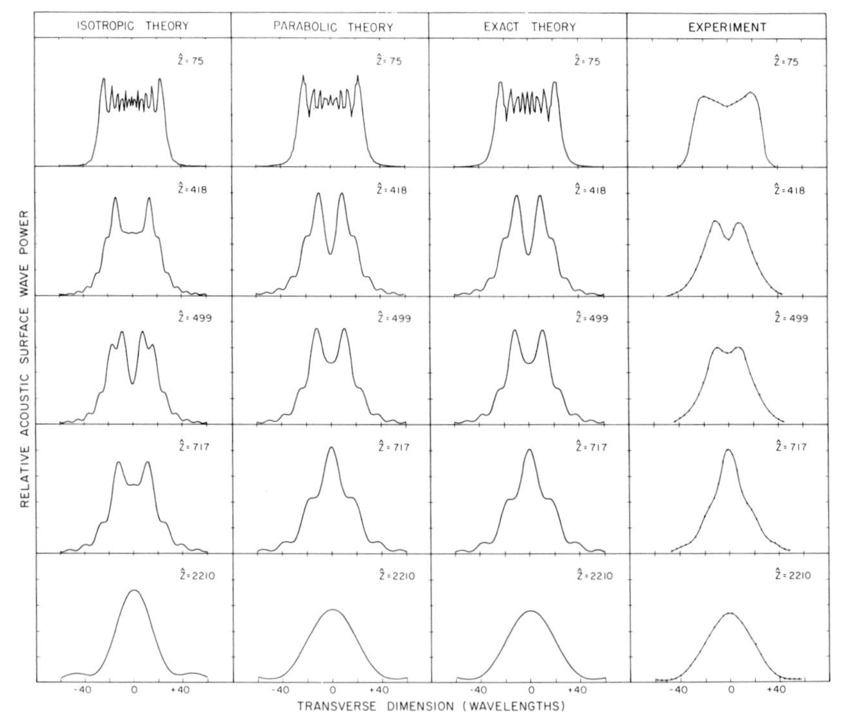

FIG. 5. Theoretical and experimental surface wave profiles illustrating diffraction along the 110 axis of 111-cut bismuth germanium oxide. $\hat{z}$ indicates the distance in wavelengths from the input transducer. Transducers were 61 wavelengths wide at a frequency of 510 MHz (after Szabo and Slobodnik, 1973a).

to the aperture. $F(k_1)$ is the Fourier transform of the source distribution, and for a rectangular source it is

$$F(k_1) = \frac{\sin(k_1 L/2)}{\pi k_1}. \tag{12}$$

The effect of a laser probe of diameter P on the profile measurements can be accounted for by multiplying $F(k_1)$ by $\text{sinc}(k_1 P/2\pi)$ in the integrand (Szabo and Slobodnik, 1973a).

To demonstrate the ability of the ASW theory to predict the fine structure of a diffraction beam, including profile asymmetry and beam steering, for a material having a highly nonparabolic velocity surface, an interdigital transducer was fabricated to launch waves close to the 111 axis of a 211-cut of gallium arsenide at a frequency of 280 MHz (Szabo and Slobodnik, 1973a). For this orientation, $V(\theta)$ is extremely nonparabolic. In Fig. 6 the

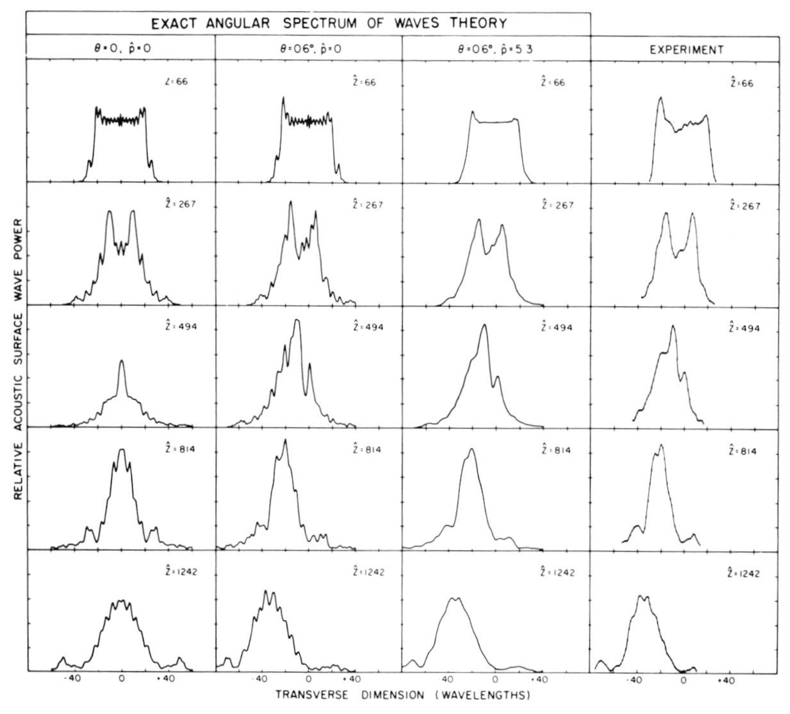

FIG. 6. Theoretical and experimental surface wave profiles illustrating diffraction near the 111 axis of 211-cut gallium arsenide. $\hat{z}$ indicates the distance for propagation in wavelengths from the input transducer. θ gives the misorientation from the 111 axis, and $\hat{P}$ is the laser probe diameter in acoustic wavelengths. Transducers were 51 wavelengths wide at a frequency of 280 MHz (after Szabo and Slobodnik, 1973a).

measured profiles (column 4) are compared to ASW-predicted profiles for the following conditions: no beam steering $\theta = 0$, and no probe $\hat{P} = 0$ (column 1); beam steering $\theta = 0.6^\circ$ and no probe $\hat{P} = 0$ (column 2); and both effects, $\theta = 0.6^\circ$ and $\hat{P} = 5.3$ (column 3). The agreement between the full theory (column 3) and the experiment is excellent.

The off-axis profiles of Fig. 6 clearly show asymmetry characteristic of misalignments near nonparabolic velocity surfaces (Kharusi and Farnell, 1971). Fortunately, this is not the case for misalignments on materials having parabolic velocity surfaces; these beam shapes remain essentially symmetric even for misorientations of a few degrees (Cohen, 1967; Kharusi and Farnell, 1971; Szabo and Slobodnik, 1973b).

E. Practical Implementation of Theories

To obtain accurate results from the ASW theory more programming and computation are needed than necessary when using the parabolic theory (Szabo and Slobodnik, 1973a,b), even though computation schemes (Kharusi and Farnell, 1972) and the use of fast Fourier transforms (J. Field, personal communication, 1976; Wickramasignhe and Ash, 1975a) have streamlined the process considerably. Use of the parabolic theory requires only an evaluation of readily available functions such as Eq. (9) or, for the general source function case, use of efficient fast Fourier transform algorithms; this approach is much faster than equivalent ASW computations (Lakin and Fedotowsky, 1976).

The following discussion will offer guidelines for choosing the simplest appropriate theory with confidence. Certain situations in which neither theory can be used will also be explained.

First, one must determine the degree to which a given velocity surface is parabolic. Quantitative criteria have been established by fitting parabolas to calculated velocity surfaces and noting the percentage of maximum deviation which, when multiplied by 1000, was called $|\delta_m|$ (Szabo and Slobodnik, 1973a). An extensive study (Szabo and Slobodnik, 1973b) of different types of velocity surfaces and their diffraction properties led to the conclusion that "parabolic" means satisfaction of the condition $0 \leq |\delta_m| \leq 2.0$.

Values of $|\delta_m|$ for several materials are listed in Table I. In addition, it was demonstrated that the best agreement between the ASW and parabolic theories was obtained when $\gamma = \partial\phi/\partial\theta$.

When the ASW method was used, surprising results were obtained for two orientations of lithium niobate, the Y cut Z propagating, and $16\frac{1}{2}$ double-rotated cut (Slobodnik and Szabo, 1971). These are shown in Figs. 7 and 8 (Szabo and Slobodnik, 1973b). The measured profiles were not in agreement with the theory, but agreement was obtained for other orientations of lithium niobate. The asymmetry of the measured profiles also in-

TABLE I

SUMMARY OF DATA FOR USE WITH PARABOLIC THEORY[a]

	Material orientation	Recommended γ for use with parabolic theory $(\partial\phi/\partial\theta)$	γ Derived from least squares fit to parabola	Deviation factor $\lvert\delta_m\rvert$
$LiNbO_3$	Y, Z	-1.083	-0.906	7.87
	$16\frac{1}{2}°$ double rotated	-1.087	-1.00	3.99
	$41\frac{1}{2}°$, X	-0.445	-0.454	0.57
	Z, X	$+0.192$	$+0.193$	0.05
	Y, 21.8° cut	$+0.393$	$+0.390$	0.5
$Bi_{12}GeO_{20}$	001, 110	-0.304	-0.304	0.14
	111, 110	$+0.366$	$+0.366$	0.08
	110, 001	$+0.236$	$+0.276$	1.93
$LiTaO_3$	Z, Y	-1.241	-1.13	5.04
	Y, Z	-0.211	-0.209	0.14
	22°, X	$+0.765$	$+0.753$	0.600
Quartz	Y, X	$+0.653$	$+0.645$	0.359
	ST, X	$+0.378$	$+0.373$	0.205
GaAs	110, X	-0.537	-0.536	0.097
	211, 111	-2.58	-2.11	20.1

[a] Parabolic theory can be used with full confidence for $0 \leq \lvert\delta_m\rvert \lesssim 0.2$, it can be used with care for $0.2 \lesssim \lvert\delta\rvert \lesssim 2.0$ and cannot be used for $2.0 \lesssim \lvert\delta_m\rvert < \infty$.

dicated nonparabolic velocity surfaces. Table I shows that both velocity surfaces were nonparabolic with $\partial\phi/\partial\theta \approx -1.08$. The ASW theory, however, gave excellent results for the most nonparabolic velocity surface listed, that of gallium arsenide (discussed previously in this chapter). A search for the cause of these discrepancies revealed that the theory was not at fault, but that the velocity surfaces had not been calculated with sufficient accuracy (Szabo and Slobodnik, 1973a,b). The study indicated that for nonparabolic and parabolic materials with $\partial\theta/\partial\phi \approx -1$, diffraction patterns are exceedingly sensitive to the shape of the velocity surface. The material constants for lithium niobate have not been determined accurately enough to calculate the correct velocity surfaces required for diffraction prediction for these two cuts. In Section V evidence that this situation is also true for lithium tantalate will be given.

Newly proposed methods of determining the shape of the velocity surface experimentally suggest that the solution to these problems will soon be forthcoming. The basic approach (Wickramasinghe and Ash, 1975a,b; Pirio

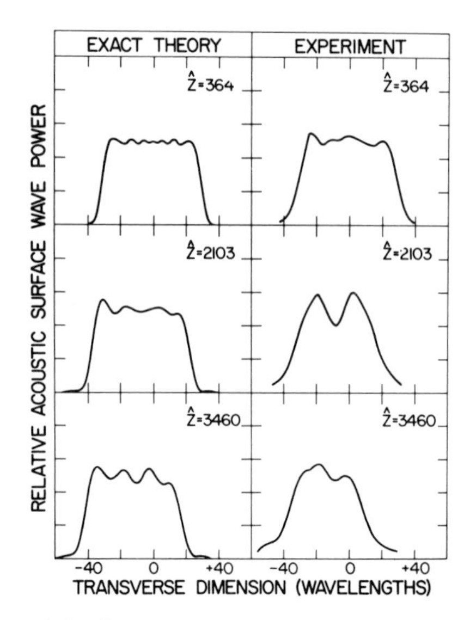

FIG. 7. Illustration of the disagreement between experimental acoustic beam profiles for YZ $LiNbO_3$ and the ASW theory using the calculated velocity surface. Here $f_0 = 900$ MHz, $\hat{L} = 64.5$, $\hat{P} = 10.3$ and $\theta = -0.2$ (after Szabo and Slobodnik, 1973a).

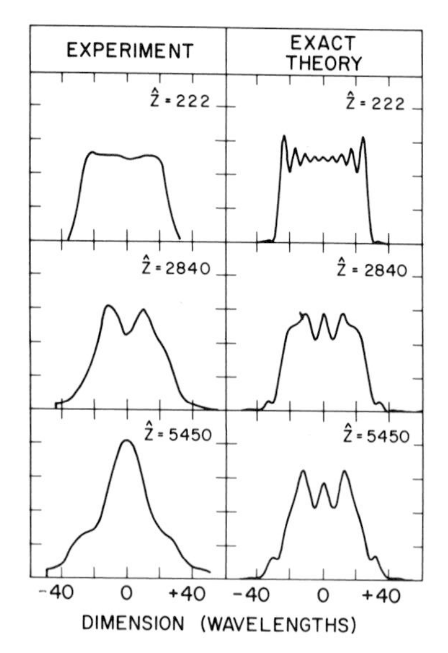

FIG. 8. Illustration of the disagreement between experimental acoustic beam profiles for the $16\frac{1}{2}°$ double-rotated cut of $LiNbO_3$ and the ASW theory using the calculated velocity surface. Here $f_0 = 995$ MHz and $\hat{L} = 56.8$.

and Sinou, 1975) is to obtain $k_3(k_1)$ from an ASW Fourier transform inversion of measured downstream beam profiles

$$F(k_1)e^{jk_3(k_1)z} = \int_{-\infty}^{\infty} A(x, z)e^{-jk_1x}\, dx, \tag{13}$$

from which $F(k_1)$ is eliminated. The velocity surfaces for YZ lithium niobate obtained by this technique differ substantially from the calculated surfaces. Efforts are under way to obtain the more precise measurements necessary for accurate prediction of diffraction in these cases.

III. Diffraction Loss and Phase Effects

A. Theory of Beam Steering and Diffraction Loss

As soon as an acoustic beam is launched from a transducer (as illustrated in Figs. 1 and 2) the energy begins to evolve into a complicated pattern so that the amount of energy intercepted by a receiving transducer depends both on its size and on its placement. The way a receiving transducer of width L_N responds to a complex field distribution is described by the following definition of diffraction loss (Waldron, 1972):

$$E = -10 \log_{10} \left\{ \frac{\left| \int_{-L_N/2}^{L_N/2} A(x, z)\, dx \right|^2}{L \int_{-L_N/2}^{L_N/2} A^2(x, 0)\, dx} \right\}. \tag{14}$$

If $A(x, z)$ is computed from an ASW theory, beam steering loss is automatically included in the definition. Because the generalized parabolic theory does not include beam steering, it must be accounted for separately. Experimental and analytic evidence (described earlier) shows that for small beam steering angles the beam shape remains the same and it is laterally shifted by a distance $x_s = z \tan \phi$. Beam steering can therefore be included in Eq. (14) for the parabolic theory by changing the limits of integration from $\pm L_N/2$ to $x_s \pm L_N/2$ (Szabo and Slobodnik, 1974).

Transducers are usually assumed to have rectangular source and receiving distributions. For this case, a closed-form expression has been obtained from Eq. (14) for combined beam steering diffraction loss for the parabolic theory (Szabo and Slobodnik, 1974). This expression confirms the principle of reciprocity; two transducers have the same beam steering diffraction loss regardless of which one is the receiver. When there is no beam steering, the analysis shows that diffraction loss for equal apertures is a function of $R = \hat{z}\,|1 + \gamma|/\hat{L}^2$, a single scaled parameter. This is shown in the universal loss curve in Fig. 9. In addition, if the apertures are unequal (and there is no beam steering), diffraction loss will be the same for any aperture combina-

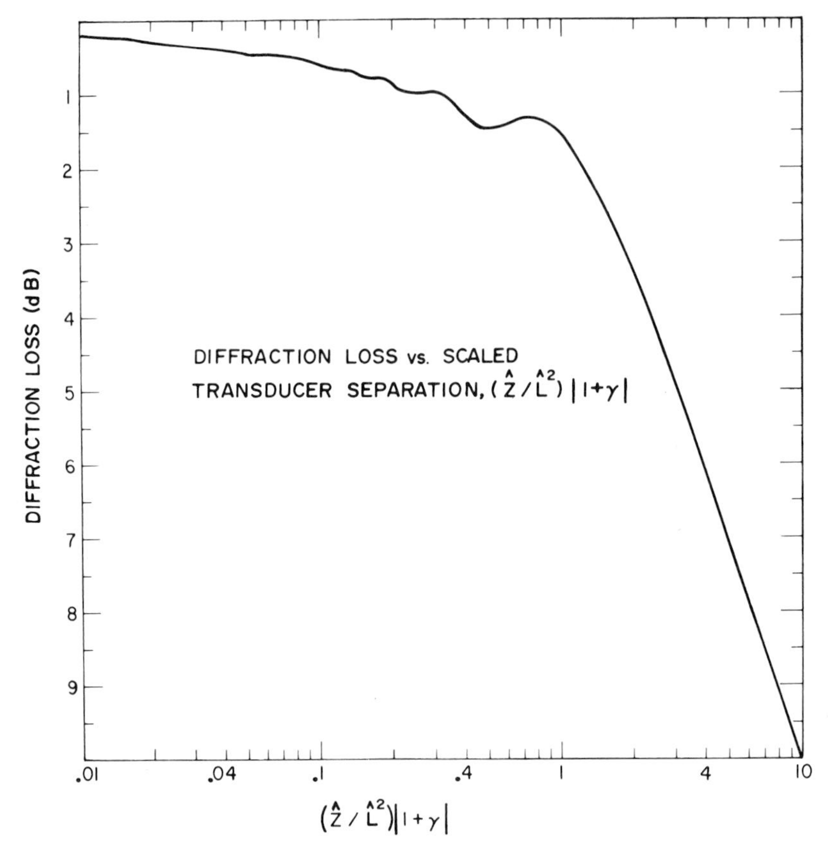

FIG. 9. Universal diffraction loss curve for all parabolic materials as a function of $(\hat{Z}/\hat{L}^2)|1 + \gamma|$ (from Szabo and Slobodnik, 1973a).

tion in the ratio of L_N/L. Application of these results will be dealt with in the sections to follow. It is worth noting at this point that in the absence of diffraction, if an acoustic beam misses only $1/\sqrt{2}$ of the output transducer ($L_N = L$), 3 dB of beam steering loss results.

Finally, diffraction introduces an apparent phase advance that may be of consequence in precise surface wave applications. In Eqs. (7) and (10), diffraction phase reaches 45° in the far field. This phase is computed from the arctangent of the imaginary and real parts of an integration of the field distribution over the limits of the receiving aperture L_N.

B. Diffraction from Tapered Source Functions

As is evident thus far, a rectangular source distribution produces an extremely complicated fluctuating diffraction field. In contrast, tapered source functions are capable of producing very smooth, slowly varying fields (Papadakis, 1975). The ideal source distribution is the gaussian shape which produces a beam that retains its form (Mason, 1973; Szabo, 1975). Tapering for interdigital or electromagnetic surface acoustic wave transducers is easily achieved by apodization, or, conversely, an apodized transducer may be regarded as a tapered source. Well-behaved beams are an advantage in nondestructive evaluation and physical measurement because they make diffraction loss and phase advance corrections more straightforward and less tedious. In particular, when the examination of small defects in the near field of the transducer is unavoidable, a nearly constant-shaped beam is a considerable advantage. In acoustooptics and acoustic analogues of optical elements, gaussianlike beams are conveniently accounted for. For surface wave devices and convolvers, tapering offers a way of minimizing crosstalk (Mason and Ash, 1971) and distortion (Baiocchi and Mason, 1975) usually resulting from diffraction from rectangular sources.

Even though the ASW theory can be used to evaluate the diffraction characteristics of tapered functions (Kharusi and Farnell, 1972), it has not been used for this purpose. The generalized parabolic theory has, however, been employed to obtain relatively simple analytic field expressions for several source distributions (Szabo, 1975, 1977). These expressions are comparable in complexity to that of the rectangular function. In addition, diffraction from a source distribution of general shape can be found by using Eq. (6) and fast Fourier transform algorithms. Three specific source distributions will be discussed here: the ideal gaussian and two truncated functions. The gaussian is defined as

$$\tilde{A}(x_0, 0) = \exp(-\pi x_0^2/\sigma^2). \tag{15}$$

Unfortunately, an ideal gaussian shape cannot be achieved in practice because all transducers have a finite width. For this reason the truncated gaussian is a practical alternative:

$$\tilde{A}(x_0, 0) = \Pi(x_0/L) \exp(-\pi x_0^2/\sigma^2). \tag{16}$$

Another useful source distribution is the truncated cosine,

$$\tilde{A}(x_0, 0) = \Pi(x_0/L) \cos(\pi x_0/M). \tag{17}$$

The diffraction characteristics of these functions were investigated (Szabo, 1975, 1977), and the parameters $\hat{M}$ and $\hat{\sigma}$ were varied to obtain well-behaved beams. Diffraction loss and phase advance were calculated [Eq. (14)] for a tapered source having a width $\hat{L}$ and a rectangular receiving transducer distribution of width $\hat{L}_N$. The results are plotted against a scaled parameter in Figs. 10 and 11. In comparison to the curves for the rectangular function which show a rippled loss slope and phase fluctuations in the

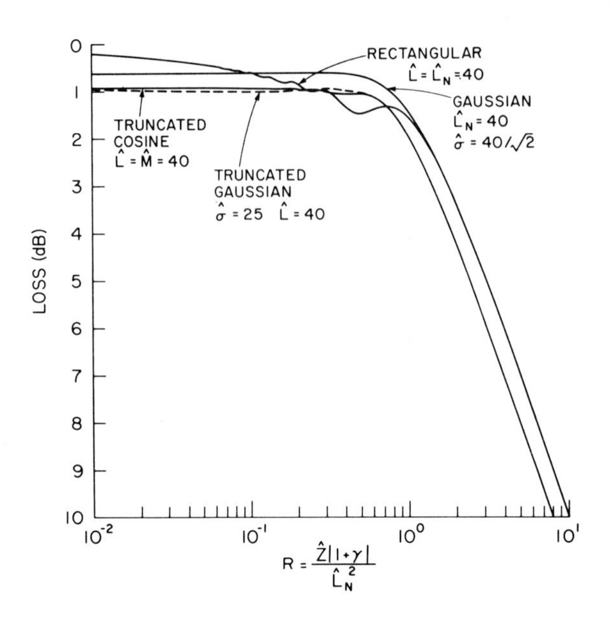

Fig. 10. Diffraction loss for rectangular, gaussian, truncated gaussian, and cosine shaped sources. Here $\hat{L}_N = \hat{L}$ (after Szabo, 1977).

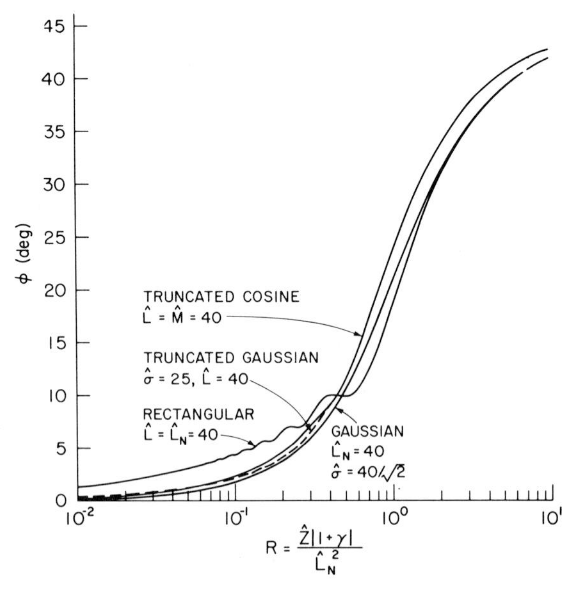

Fig. 11. Diffraction phase for rectangular, gaussian, truncated gaussian, and cosine shaped sources. Here $\hat{L}_N = \hat{L}$ (after Szabo, 1977).

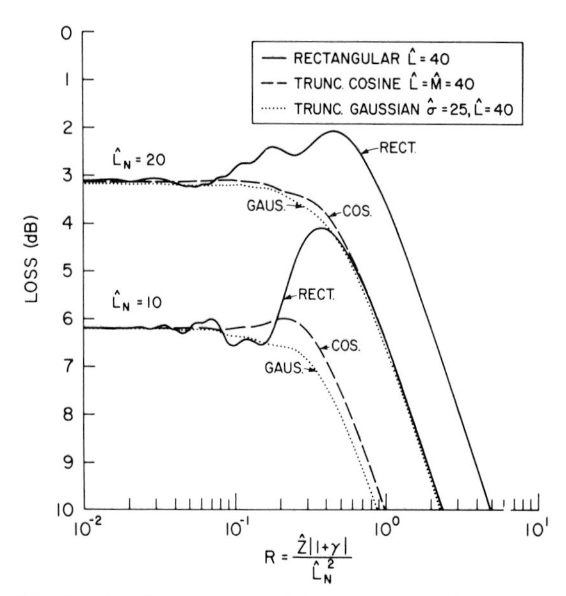

FIG. 12. Diffraction loss for unequal receiving and transmitting apertures in the ratios of $\hat{L}_N/\hat{L} = 0.25$ or 0.5. Note the truncated cosine and gaussian curves are much smoother than that of the rectangular source (after Szabo, 1977).

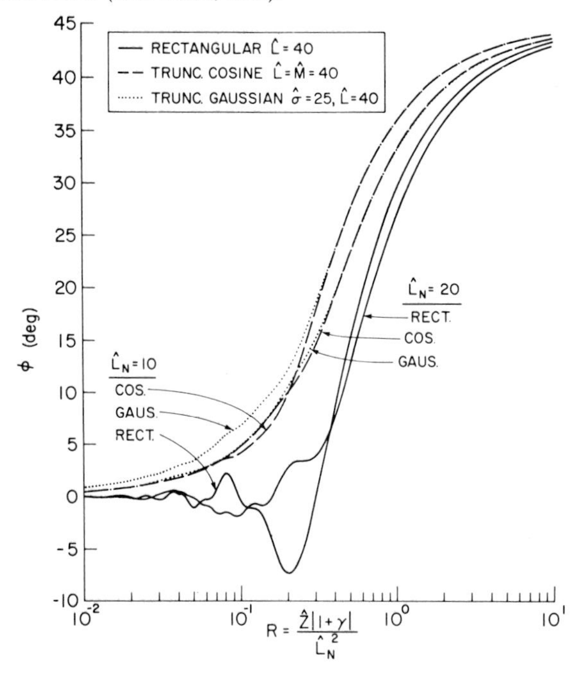

FIG. 13. Diffraction phase for unequal receiving and transmitting apertures in the ratios of $\hat{L}_N/\hat{L} = 0.25$ or 0.5. The curves for the rectangular sources differ from the others. All curves approach 45° in the far field (after Szabo, 1977).

near field, the curves for the gaussian distribution have a virtually constant near field loss and a substantially reduced phase advance in this region. In addition, a similar improvement in near field characteristics is achieved by using truncated gaussian and cosine shapes.

Diffraction loss and phase advance curves were also computed for these functions with $\hat{L}_N \neq \hat{L}$, and they are shown in Figs. 12 and 13 (Szabo, 1977). For a rectangular source, a smaller receiving transducer has the effect of magnifying field variations near the center of the beam where the most violent fluctuations occur. On the other hand, the truncated gaussian and cosine tapers provide a dramatic improvement.

Tapered sources are useful for examining defects in the near field because they provide a more constant illumination than that obtained with a rectangular source. Figures 12 and 13 may be regarded as the average intensity and phase advance on an illuminated defect of width $\hat{L}_N$. It is worth noting that in Fig. 12 the near field loss is approximately $E = -10 \log L_N/L$, and the far field loss slope is the same in all cases, 10 dB/decade.

IV. Effect of Diffraction on Device Design

A. Material Selection

The design of surface acoustic wave delay lines and signal processing devices, especially at microwave frequencies, requires a consideration of the properties of materials and the effects of physical factors (Szabo and Slobodnik, 1973a; Slobodnik, 1973, 1976). The naturally occurring combinations of these parameters for a given material may not be suitable for a specific application. Compromises must sometimes be made in the choice of the following: piezoelectric coupling $\Delta V/V$, temperature coefficient of delay, surface wave velocity, thermal scattering loss, air loading loss, and diffraction and beam steering loss. Design data are listed in Table II for popular surface wave materials (Szabo and Slobodnik, 1973a). Beam steering and diffraction losses, for example, always involve a tradeoff. Because some unintentional misalignment of transducers always occurs, values of $\partial\theta/\partial\phi \approx 0$ are preferable to minimize beam steering loss. Although for this value of the slope of the power flow angle, diffraction loss proceeds at the isotropic rate, it can be considerably reduced by choosing a material with $-1 < \partial\theta/\partial\phi < 0$.

For any specific situation, criteria can be developed for the selection of a material. For example, a figure of merit was used to compare the materials in Table II for a 1-GHz uniform transducer pair with $\hat{L} = 40$, and $L = 200$ μm for a misalignment angle of 0.2°. Values at which each loss mechanism contributed 3 dB of loss were used in the figures of merit shown in the bottom row of Table II.

B. Optimal Design Including Diffraction and Beam Steering

Under some circumstances, a simple, separate accounting of diffraction and beam steering of the sort described in Section IV,A is not adequate. These situations include optimal transducer design at microwave frequencies, the design of long time delay lines (Coldren, 1973), and the assessment of conditions under which these losses become significant for a given set of design constraints and tolerances (Slobodnik and Szabo, 1974).

Beam steering and diffraction are actually interrelated sources of loss, as is evident from Eq. (14). Diffraction is, of course, an everpresent, unavoidable source of loss which depends only on geometric factors ($\hat{L}$, $\hat{L}_N$, $\hat{z}$) for a given material. On the other hand, the contribution of beam steering to the combined loss can be controlled by precise X-ray alignment although this involves increased cost.

The interdependence of these sources of loss as a function of the anisotropy parameter is shown in Fig. 14 for the specific case of uniform transducers with $\hat{L}_N = \hat{L} = 80$, $\hat{z} = 5000$ and a misalignment of 0.1° (Slobodnik

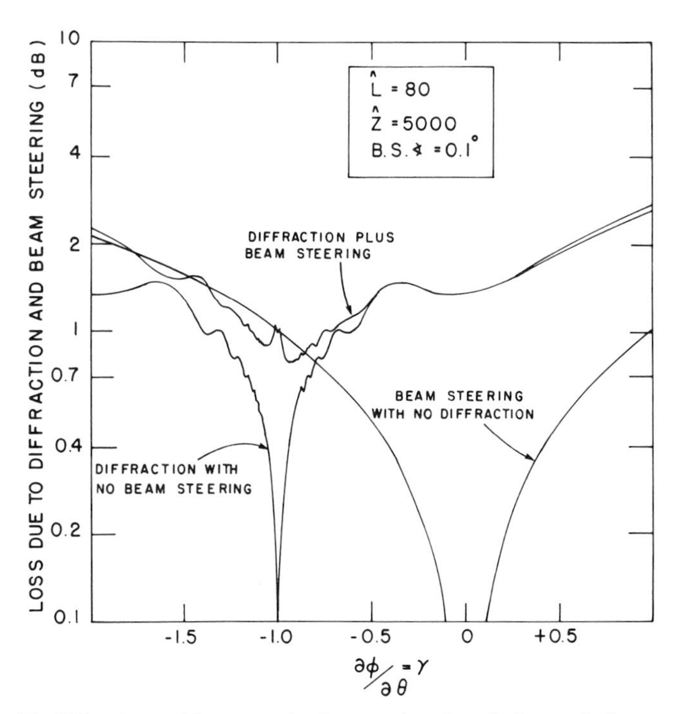

FIG. 14. Diffraction and beam steering loss as a function of γ for parabolic materials for a specific geometry. BS ∡ is the beam steering angle (after Slobodnik and Szabo, 1974).

TABLE II

Material	$LiNbO_3$	$LiNbO_3$	$LiNbO_3$	$LiNbO_3$	$LiNbO_3$	$Bi_{12}GeO_{20}$	$Bi_{12}GeO_{20}$
Orientation	Y cut Z prop	$16\frac{1}{2}^\circ$ double rotated cut	$41\frac{1}{2}^\circ$ rotated cut X prop	Z cut X prop	X cut Z prop	001 cut 110 prop	111 cut 110 prop
Surface wave velocity V_∞ (m/sec)	3488	3503	4000	3798	3483	1681	1708
Estimate of electromagnetic to acoustic coupling $\Delta V/V_\infty$	0.0241	0.0268	0.0277	0.0026	0.0252	0.0068	0.0082
Power flow angle ϕ (deg) (electromechanical)	0	0	0	0	−1.726	0	0
Temperature coefficient of delay $\frac{1}{\tau}\frac{\partial\tau}{\partial T}$ (ppm)	94	96	72	77	93	—	—
SW attn in air at 1 GHz (dB/μ sec)	1.07	1.15	1.05	0.93	—	1.64	1.64
Air loading at 1 GHz (dB/μ sec)	0.19	0.21	0.3	0.24	—	0.19	0.19
Vac attn at 1 GHz (dB/μ sec)	0.88	0.94	0.75	0.69	—	1.45	1.45
3 dB Air prop loss time delay at 1 GHz A (μ sec)	2.8	2.6	2.9	3.2	—	1.8	1.8
Slope of electromechanical power flow curve $\partial\phi/\partial\theta$	−1.083	−1.087	−0.445	+0.192	−0.610	−0.304	+0.366
Slope of electromechanical power flow curve $\partial\phi/\partial\mu$	−0.117	+0.151	0	0	0	0	0
3 dB Beam steering loss time delay B (μ sec)	7.4	7.0	18.9	46.0	15.8	65.7	53.7
3 dB Diffraction loss time delay at 1 GHz C (μ sec)	29.1	29.3	5.1	2.4	7.6	4.1	2.1
Material figure of merit $F_M = ABC\left(\frac{\Delta V}{V_\infty}\right)^2$	0.350	0.383	0.214	0.002	—	0.022	0.014
Time-BW Product figure of merit $F_{TB} = \frac{4\times 10^3 ABC}{V_\infty}\left(\frac{\Delta V}{V_\infty}\right)^2$	0.402	0.437	0.214	0.003	—	0.053	0.032

$_{12}GeO_{20}$	$LiTaO_3$	$LiTaO_3$	$LiTaO_3$	Quartz	Quartz	GaAs	GaAs	$Ba_2NaNb_5O_{15}$
10 cut 1 prop.	Z cut Y prop.	Y cut Z prop.	22^0 Rotated cut X prop.	Y cut X prop.	ST cut X prop.	110 cut X prop.	211 cut 111 prop.	Y cut Z prop.
1624	3329	3230	3302	3159	3158	2822	2621	3177
0.0037	0.0059	0.0033	0.0027	0.0009	0.00058	0.00008	0.00012	0.0005
0	0	0	0	0	0	0	0	0
—	69	35	68	−24	0	—	—	—
—	1.0	1.14	—	2.6	3.09	4.22	3.62	~3.7
—	0.23	0.20	—	0.45	0.47	0.40	0.27	—
—	0.77	0.94	—	2.15	2.62	3.82	3.35	—
—	3.0	2.6	—	1.2	0.97	0.71	0.83	~0.8
).236	−1.241	−0.211	+0.764	+0.653	+0.378	−0.537	−2.58	+0.071
0	+0.556	−0.229	0	0	0	0	0	0
7.6	4.4	15.6	13.3	16.3	28.1	22.2	5.0	149.0
2.3	11.3	3.6	1.6	1.7	2.1	6.1	1.50	2.65
—	0.005	0.002	—	0.00003	0.00002	6×10^{-7}	1×10^{-7}	—
—	0.006	0.002	—	0.00003	0.00002	9×10^{-7}	1×10^{-7}	—

and Szabo, 1974). Diffraction loss alone is symmetric about $\gamma = -1$ where autocollimation occurs. Beam steering loss, by itself, is symmetric about $\gamma = 0$, the condition of isotropy. The complicated combined loss curve is no longer symmetric and is not simply the sum of the two losses considered separately. Thus beam steering diffraction loss is unique to a specific set of conditions. Typically, a set of curves is computed for a chosen parameter, such as beam steering angle, frequency, γ, $\hat{L}$, or $\hat{z}$, in order to ascertain the impact of the total loss on a particular design.

Standard optimization procedures for identical uniform periodic interdigital transducers have been extended to include beam steering diffraction loss as well as electrical losses. This is illustrated in Fig. 15 (Slobodnik and Szabo, 1974). Here, a delay line on ST quartz operating at a center frequency $f_0 = 660$ MHz with a time delay t of 10 μs ($\hat{z} = tf_0$) was designed for minimal insertion loss by varying the aperture width $\hat{L}$. Curve 4 shows transducer insertion loss when only the usual matched equivalent circuit is used. Curve 3 includes parasitic capacitive and resistive effects. Propagation loss is added to insertion loss in curve 2. The final insertion loss curve including beam steering diffraction loss is substantially different from all the others. It allows a determination of the optimal aperture $\hat{L} = 100$.

C. Diffraction Compensation for Apodized Transducers

Because surface acoustic wave interdigital transducers can closely approximate an ideal transversal filter, straightforward Fourier transform pair and digital filter procedures can be used to synthesize many different filter responses. A typical periodic filter, shown in Fig. 16, consists of a uniform transducer and an apodized transducer. The uniform transducer usually has a wide bandwidth so that the overall filter characteristic is determined by the narrower frequency response of the second transducer, which has a Fourier transform relation to its apodization function. Ideally, the usual design procedure assumes that any finger pair having an overlap $\hat{L}_N$ at a position $\hat{z}_N$ away from the center of the launching transducer intercepts a uniform beam of width $\hat{L}_0$. This situation is the same as that encountered in the discussion of diffraction for unequal apertures (Section III,B). Likewise, in this case diffraction is a complicated function of $\hat{L}_N$, $\hat{L}$, and $\hat{z}$ (as described in Figs. 12 and 13), and as a result the ideal Fourier transform frequency response apodization relationship is destroyed in actual practice. In order to recover this relationship a direct synthesis method which modifies the original apodization will be described (Szabo and Slobodnik, 1974).

For the situation just described, the signal amplitude transferred to an electrical load from an acoustic aperture L_N, irradiated by an acoustic beam of amplitude $A(x, z)$, can be shown to be

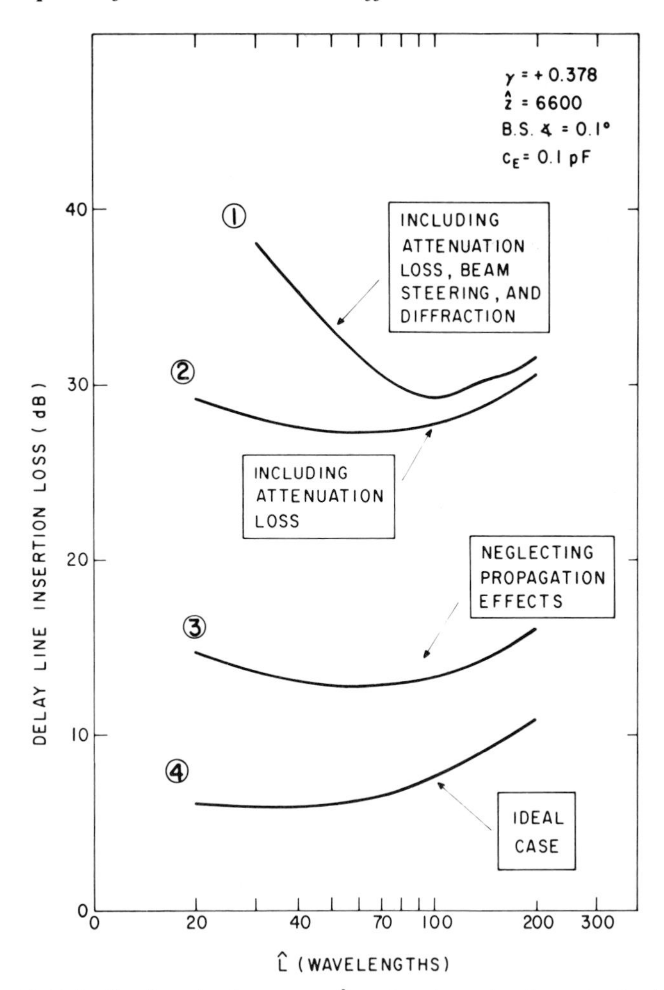

FIG. 15. Delay line insertion loss versus $\hat{L}$ used to determine the optimal value of $\hat{L}$ for minimal insertion loss. Curve 1 is the final curve including real transducer effects, attenuation loss, beam steering, and diffraction. Curve 2 includes real transducer effects and attenuation loss. Curve 3 includes real transducer effects. Curve 4 includes only ideal transducer characteristic without propagation, diffraction, or beam steering losses (after Slobodnik and Szabo, 1974).

$$S = C \int_{-L_N/2}^{L_N/2} A(x, z_N)\, dx, \tag{18}$$

in which the cancellation of electrical and acoustical factors results in a constant C independent of L_N. Each finger pair in Fig. 16 is treated separately, and superposition is used to obtain the total result. In the

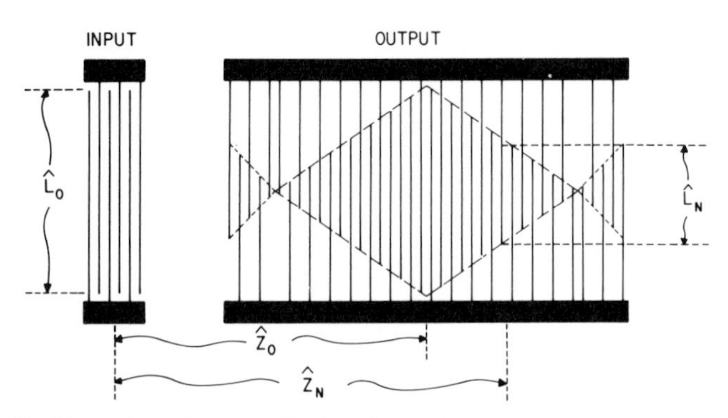

FIG. 16. Illustration of an apodized and a uniform transducer pair geometry. Both transducers are $\hat{L}_0$ wide at their center separation $\hat{Z}_0$. Each Nth aperture of $\hat{L}_N$ at $\hat{Z}_N$ would become $\hat{L}'_N$ at $\hat{Z}'_N$ after correction (after Szabo and Slobodnik, 1974).

absence of diffraction, the response S for the Nth finger pair of width L_N with respect to that of the widest finger pair (width L_0) is

$$\frac{\int_{-L_N/2}^{L_N/2} A(0, z_N)\, dx}{\int_{-L_0/2}^{L_0/2} A(0, z_0)\, dx} = \frac{L_N}{L_0}. \tag{19}$$

In order to achieve this ratio in the presence of diffraction, the equation

$$\frac{\int_{-L_N'/2}^{L_N'/2} A(x, z'_N)\, dx}{\int_{-L_0/2}^{L_0/2} A(x, z_0)\, dx} = \frac{L_N}{L_0}, \tag{20}$$

must be solved for an unknown aperture L'_N and its position z'_N. This is accomplished by setting $z'_N = z_N$ and solving for L'_N with the use of a computer iterative method.

Once amplitude correction (or the new apodization function) is obtained, phase correction is accomplished where necessary (Hagon and Lakin, 1974) by analyzing the new Nth finger pair response with L'_N placed at z_N to determine the *relative* phase ξ_N of each finger pair. By setting

$$\hat{z}'_N = \hat{z}_N + (\xi_N/2\pi), \tag{21}$$

a good approximation to phase correction results.

Typical results for amplitude correction are shown in Fig. 17. In the majority of cases the correction ratio $\hat{L}'_N/\hat{L}_N$ is less than one so that the original aperture is reduced. It is interesting to note that no correction ($\hat{L}'_N/\hat{L}_N = 1$) is required if all finger pairs lie either in the extreme near or far fields. This result is in consonance with earlier observations (Section III,B) that very near field diffraction loss for unequal apertures is constant and that the far field loss slopes are identical.

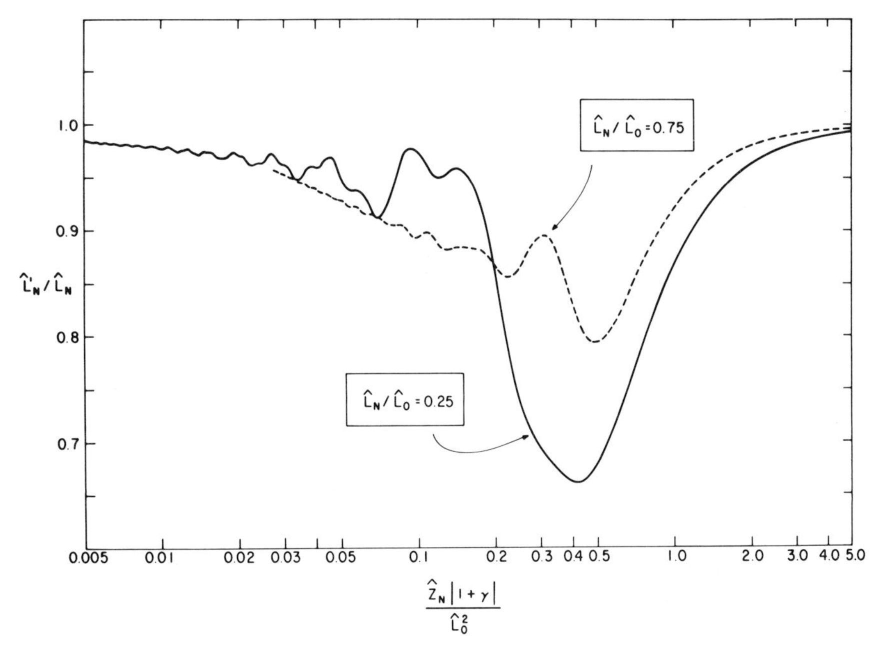

FIG. 17. Diffraction correction as a function of the scaled acoustic aperture separation factor. The original unequal aperture ratio $\hat{L}_N/\hat{L}$ is the parameter. Here $\hat{Z}_0 = \hat{Z}_N$ (after Szabo and Slobodnik, 1974).

An assumption central to the preceding development is that the uniform transducer which consists of several finger pairs can be represented by a single aperture. Kharusi and Farnell (1971, 1972) confirmed this assumption when they determined that the diffracted fields from these two types of sources do not differ significantly. Kharusi *et al.* (1972) also used the ASW theory to calculate the beam profiles from a nonperiodic apodized transducer. Maines *et al.* (1972) presented a diffraction correction scheme for nonperiodic apodized transducers. For narrow band periodic transducers, corrections are usually made at the center frequency only ($\lambda = V/f_0$). Wagers (1976) has shown that in certain cases such as a pair of long, narrow, closely spaced withdrawal weighted transducers, this is not a good assumption.

V. Minimal Diffraction Cuts

A surprising consequence of the parabolic theory is that for the value $\gamma = -1$, ideally there is no diffraction corresponding to a condition of autocollimation. Searches for crystalline orientations having this property will now be discussed (Slobodnik and Szabo, 1973; Slobodnik *et al.*, 1975).

Because of the previously described difficulty with diffraction prediction on materials having nonparabolic velocity surfaces with $\partial\theta/\partial\phi \approx -1$ (Section II,E), important criteria were established to avoid these problems. These conditions were that the velocity surfaces had to be parabolic satisfying the condition $|\delta_m| \lesssim 2$ from Section II,E, and also that the material constants had to be extremely accurately known for the theoretical predictions to translate into practice. In addition, the computer search required a continuum of possible cuts having $\phi = 0$ and parabolic surfaces in the vicinity of the search.

Because the material constants of $Bi_{12}GeO_{20}$ had undergone extensive measurement, this crystal was chosen as a minimal diffraction cut (MDC) candidate. The best cut meeting the criteria was the 45°, 40.04°, 90° orienta-

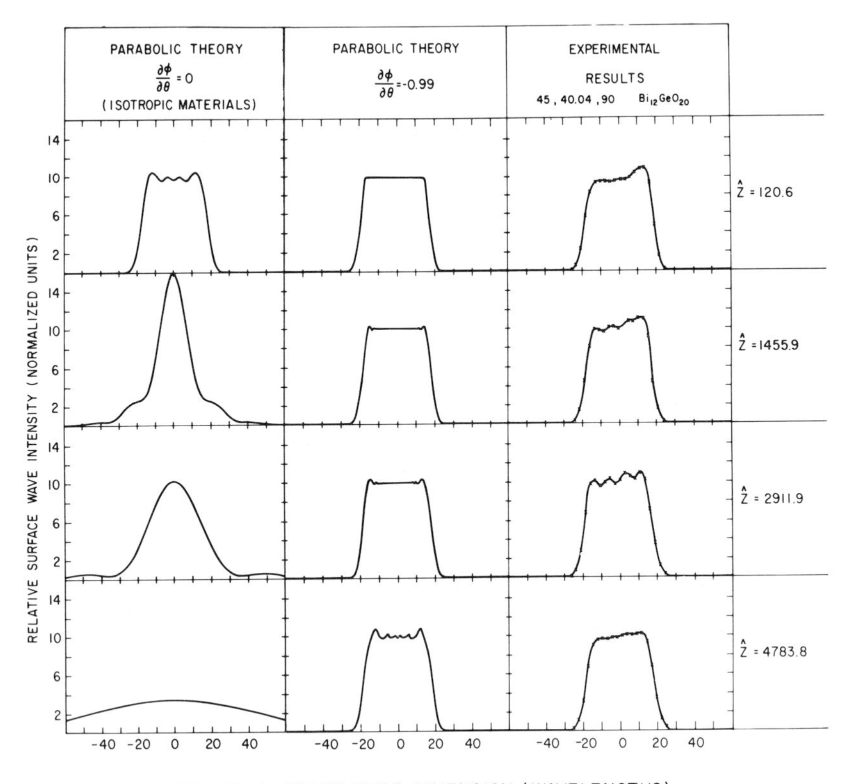

FIG. 18. Experimental profiles for the 40.04 $Bi_{12}GeO_{20}$ minimal diffraction cut illustrating diffraction suppression by more than two orders of magnitude compared to the isotropic case. Here $\hat{L} = 40.56$ and $f_0 = 380$ MHz (after Slobodnik and Szabo, 1973).

tion for which the velocity surface is sufficiently parabolic with $\gamma = -1.000$ and $|\delta_m| = 1.53 < 2$ (Slobodnik and Szabo, 1973). The piezoelectric coupling parameter is $\Delta V/V = 0.0031$, comparable to other cuts in use. Experimental confirmation of the MDC properties is demonstrated in Fig. 18. Here experimental profiles from a 380-MHz transducer are compared to the parabolic theory with $\gamma = -0.99$ and to isotropic theory. Experimentally observed diffraction is suppressed by a factor of over 100 compared to the isotropic case.

A search for an MDC on lithium tantalate was also initiated (Slobodnik *et al.*, 1975). The same procedure was followed until the theoretically predicted orientations were found to disagree with experiment. In this case, as with lithium niobate, the material constants are not accurate enough for a successful theoretical prediction of a velocity surface having $\gamma \approx -1$. Finally, an extensive experimental search was necessary to find an MDC. The cut discovered has a diffraction suppression of 20 compared to isotropic materials, relatively high piezoelectric coupling, and unusually good spurious mode rejection.

These MDCs are allowing a new class of highly apodized filters and long time delay lines to be realized.

VI. Diffraction for Acoustic Analogues of Optical Components

At the beginning of the surface wave device boom, it was recognized that accessibility of the substrate surface could be exploited to create a new generation of acoustic counterparts to optical devices. Conventional photolithographic techniques could be used to fabricate these compact planar devices that are smaller than optical components by three orders of magnitude. Early enthusiasm was later tempered by a respect for the unique problems that anisotropy presented, which had not been considered in the analogous optical situations. Workers in this area discovered that anisotropy provided degrees of latitude not available in optics. Ironically, the new theories developed for anisotropic acoustic problems were in advance of current optical approaches.

The development of these acoustic devices has a fascinating history in which many optical concepts were exploited even though their realization often required completely different physical principles and considerable ingenuity. These acoustic analogues include the following: lenses, Fresnel lenses, and focusing mirrors (curved transducers) (Engan, 1969; Van Duzer, 1970; Mason and Ash, 1971; Mason, 1973; Kharusi and Farnell, 1972; Bridoux *et al.*, 1974; Hartemann and Ménager, 1973; Gedeon, 1973); prisms, ribbon waveguides, and prism couplers (wedge transducers) (Oliner, 1975); mirrors (multistrip couplers) (Marshall *et al.*, 1973); grating filters and Fabry–Perot cavity resonators (Bell and Li, 1976); transducer arrays and acoustooptic modulators and deflectors (Tsai and Nguyen, 1975; Schmidt, 1976); and holography (Wickramasinghe and Ash, 1975a). Even though it is not possible to describe these devices here, the problem of anisotropic focusing will be discussed as an example.

Diffraction plays a role in the operation of these devices because their dimensions are on the order of wavelengths. Even though the ASW and parabolic theories have been used to examine certain specific cases, a more general approach for these analogues has not yet been proposed. The complexity of the physical situation often does not permit diffraction modeling. For some cases, however, when the parabolic and paraxial approximations are valid, the author suggests that the generalized Fourier transform theory, like the powerful methods employed in Fourier optics (Papoulis, 1968; Goodman, 1968), can be extended to the problems of acoustic analogue devices. As applications of this approach, curved transducers and lenses, represented schematically in Fig. 19, will now be discussed.

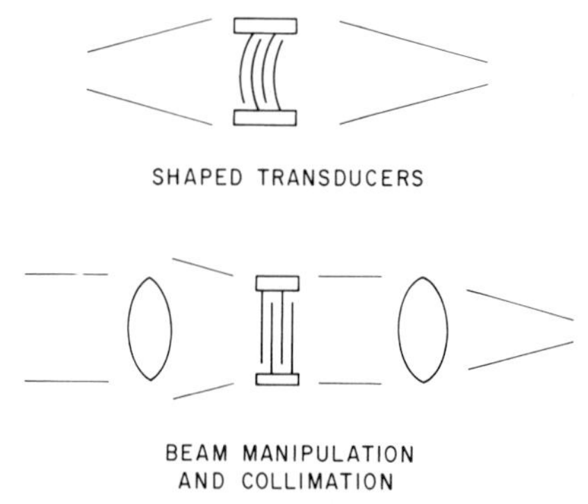

FIG. 19. Schematic illustration of beam focusing possible with curved transducers and lenses on anisotropic materials.

For the first example, consider the curved transducer, the analogue of the focusing mirror. For a transducer of curvature R and aperture width L,

$$\tilde{A}(x_0, 0) = \Pi(x_0/L) \exp\left(j \frac{k_0 x_0^2}{2R}\right) \tag{22}$$

can be substituted in either Eq. (4) or (6) to obtain expressions for the field amplitude. By inspection, this substitution reveals that focusing occurs at a distance $z = R/|1 + \gamma|$. At the focus, the amplitude distribution is the Fourier transform of the rectangle function as given in Eq. (10) and is in agreement with the observations of Cohen (1967). Kharusi and Farnell (1972) made the interesting discovery that by shaping the transducer in the form of the group velocity surface, better focusing is obtained at the geometric focus $z = R$. The beam width of the intensity at focus for a parabolic phase velocity surface ($\gamma < 1$) appears narrower by a factor of $1/|1 + \gamma|$. Unfortunately, group velocity surfaces have shapes difficult to realize in practice.

In the second example, a convex lens of aperture L has a uniformly illuminated object next to it. Equation (22) can be used with the focal length f replacing the radius R. Similarly, at the focus $z = f/|1 + \gamma|$, there is a sinc amplitude distribution. Mason and Ash (1971) have explained the remarkable range of effects that can be produced with lenses (and curved transducers) through a combination of curvature and anisotropy. (Either $|1 + \gamma|$ or f can be positive or negative.) In addition, Mason's (1973) design method for anisotropic lens aberration correction is demonstrated in Fig. 20 for a gold condenser lens on Y cut, X propagating quartz. Gedeon (1973) has

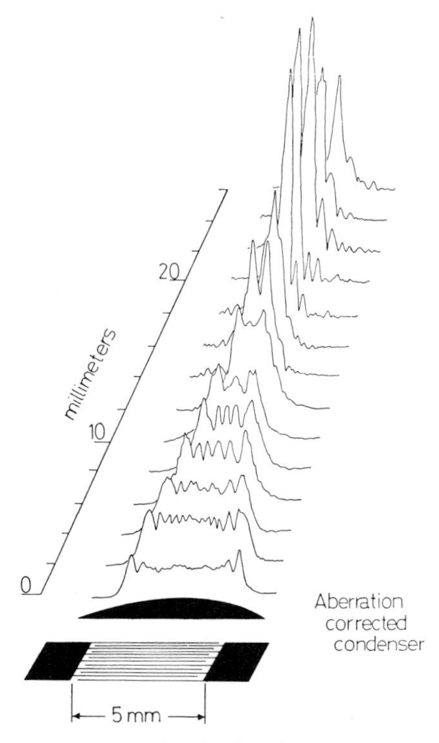

FIG. 20. Mason's measurements for the focusing pattern of his aberration corrected anisotropic condenser lens. The lens was made of 4800-Å thick gold with an effective index of 1.09 on Y-cut, X-propagating quartz having a $\gamma = 0.64$. Here $f_0 = 60$ MHz, $L = 5$ mm, and $f = 20$ mm (after Mason, 1973).

extended isotropic Fresnel (phase-reversal zone plate) lens theory for surface waves with nonparabolic velocity surfaces.

VII. Miscellaneous Effects

A. Defects

Methods for solving the "inverse diffraction" problem provide a means of calculating what the (two-dimensional) diffraction field is at points other than an observation scan line. For example, a measurement of complex field amplitude along a line parallel to the x axis can be used to determine the source distribution $\tilde{A}(x_0, 0)$. This is essentially the same problem encountered in Section II,E for obtaining $k_3(k_1)$ from $A(x, z)$. The basic idea behind this ASW approach (Wickramasinghe and Ash, 1975a) is stated in Eq. (13) in which a Fourier transform of the source distribution $F(k_1)$ is obtained. In this case, however, $k_3(k_1)$ is either known or eliminated so that the source function $A(x_0, 0)$ can be recovered through a second transform of $F(k_1)$. This process can be used to evaluate a transducer for radiation uniformity and defective regions. In a similar way, the generalized Fourier transform theory has also been used to "inspect" transducer sources for the isotropic case (Lakin and Fedotowsky, 1976).

By this method Wickramasinghe and Ash (1975a) have been able to reconstruct the entire diffraction field from scan measurements at two different z's. As a result, surface defects, situated away from the measurement scan lines, were located and resolved.

B. Diffraction on Anisotropic Cylinders and Disks

Long time delay lines (Slobodnik, 1973; Slobodnik *et al.*, 1974; Coldren, 1973; Coldren and Shaw, 1976; Papadofrangakis *et al.*, 1973), employing surface wave propagation on cylinders and on plates or disks with rounded edges, have diffraction problems that warrant special consideration. The anisotropy of these unusually shaped crystals changes along the path of propagation and, in some cases, it may actually guide the acoustic energy.

On a helical delay line, for example, the actual acoustic path is not a simple spiral but wanders according to a complicated trajectory determined by the anisotropy and transducer cant angle (Slobodnik, 1973). In addition, as the energy travels, there is a continual change in anisotropy along the curved surface.

Papadofangakis and Mason (1973) have proposed that, within the limits of the parabolic approximation, a wave on a curved surface encounters an effective anisotropy factor, an integration along a path of length z_T,

$$\gamma_{\mathrm{EFF}} = \frac{1}{z_{\mathrm{T}}} \int_{0}^{z_{\mathrm{T}}} \gamma(z)\, dz. \tag{23}$$

They realized that propagation around a rounded (or sharp) corner can act as a geodesic lens to refocus the acoustic energy. They also discovered that by using gaussian-weighted transducers to launch waves on an anisotropic disk, unusually long time delays could be achieved. Certain trajectories, dictated by the disk anisotropy, resulted in a refocusing of acoustic energy caused by the rounded corners acting as geodesic edge lenses (Mason *et al.*, 1975).

More research is needed to fully account for these unusual diffraction effects on curved surfaces.

VIII. Conclusion

This review has covered anisotropic surface wave diffraction in three areas of research and application. Despite the range of subjects treated, all are aspects of the same fundamental phenomena which can be well described by the two-dimensional scalar diffraction theories presented. Surface wave diffraction on anisotropic materials is more interesting than that for isotropic materials because the anisotropy of a substrate can influence transducer design as well as offer degrees of freedom for new applications not possible otherwise. Future research on this subject will continue to turn up some surprises.

Acknowledgments

I wish to express my appreciation to Andrew J. Slobodnik, Jr., for his contribution to the work we did together on diffraction, and I would also like to thank Martha Bartlett and Mary Szabo for their many helpful suggestions and their enthusiasm in the preparation of the manuscript. Much of the research described here was conducted at the Deputy for Electronic Technology, formerly Air Force Cambridge Research Laboratories.

References

Baiocchi, O. R., and Mason, I. M. (1975). *IEEE Trans. Sonics Ultrason.* **22**, 347.
Bell, D. T., Jr., and Li, R. C. M. (1976). *Proc. IEEE* **64**, 711.
Bracewell, R. (1965). "The Fourier Transform and its Applications." McGraw-Hill, New York.
Bridoux, E., Rouvaen, J. M., Moriamez, M., Torquet, R., and Hartemann, P. (1974). *J. Appl. Phys.* **45**, 5156.
Campbell, J. J., and Jones, W. R. (1968). *IEEE Trans. Sonics Ultrason.* **15**, 1223.
Cho, F. V., Lawson, R. L., and Hunsinger, B. J. (1970). *IEEE Trans. Sonics Ultrason.* **17**, 199.
Cohen, M. G. (1967). *J. Appl. Phys.* **38**, 3821.
Coldren, L. A. (1973). *IEEE Trans. Sonics Ultrason.* **20**, 17.
Coldren, L. A., and Shaw, H. J. (1976). *Proc. IEEE* **64**, 598.
Crabb, J. C., Maines, J. D., and Ogg, N. R. (1971). *Electron. Lett.* **7**, 253.

Engan, H. (1969). *Electron. Res. Lab. Norw. Inst. Tech. Rep.* No. TE-128 (unpublished).
Farnell, G. W. (1970). *In* "Physical Acoustics: Principles and Methods" (W. P. Mason and R. N. Thurston, eds.), Vol. 6, pp. 109–166. Academic Press, New York.
Gedeon, A. (1973). *Appl. Phys.* **2**, 15.
Goodman, J. W. (1968). "Introduction to Fourier Optics." McGraw-Hill, New York.
Hagon, P. J., and Lakin, K. M. (1974). *Proc. 1974 Ultrason. Symp.* IEEE Cat. No. 74 CHO 896-1SU, p. 341.
Hartemann, P., and Ménager, D. (1973). *Proc. 1973 Ultrason. Symp.* IEEE Cat. No. 73 CHO 807-8SU, p. 378.
Kharusi, M. S., and Farnell, G. W. (1970). *J. Acoust. Soc. Am.* **48**, 665.
Kharusi, M. S., and Farnell, G. W. (1971). *IEEE Trans. Sonics Ultrason.* **18**, 35.
Kharusi, M. S., and Farnell, G. W. (1972). *Proc. IEEE* **60**, 945.
Kharusi, M. S., Tancrell, R. H., and Williamson, R. C. (1972). *Electron. Lett.* **8**, 238.
Lakin, K. M., and Fedotowsky, A. (1976). *IEEE Trans. Sonics Ultrason.* **23**, 317.
Maines, J. D., Moule, G. L., and Ogg, N. R. (1972). *Electron Lett.* **8**, 431.
Marshall, F. G., Newton, C. D., and Paige, E. G. S. (1973). *IEEE Trans. Sonics Ultrason.* **20**, 134.
Mason, I. M. (1971). *Electron. Lett.* **7**, 344.
Mason, I. M. (1973). *J. Acoust. Soc. Am.* **53**, 1123.
Mason, I. M., and Ash, E. A. (1971). *J. Appl. Phys.* **42**, 5343.
Mason, I. M., Papadofrangakis, E., and Chambers, J. (1975). *Electron. Lett.* **11**, 348.
Mitchell, R. F., and Stevens, R. (1975). *Wave Electron.* **1**, 201.
Ogg, N. R. (1971). *J. Phys. A: Gen. Phys.* **4**, 382.
Oliner, A. A. (1975). *In* "Optical and Acoustical Micro-electronics" (J. Fox, ed.), Vol. 23, pp. 1–17. Polytechnic Press, New York.
Papadakis, E. P. (1975). *In* "Physical Acoustics: Principles and Methods" (W. P. Mason and R. N. Thurston, eds.), Vol. 11, pp. 151–211). Academic Press, New York.
Papadofrangakis, E., and Mason, I. M. (1973). *Electron. Lett.* **9**, 304.
Papadofrangakis, E., Mason, I. M., and Chambers, J. (1973). *Proc. 1973 Ultrason. Symp.* IEEE Cat. No. 73 CHO 807-8SU, p. 131.
Papoulis, A. (1968). "Systems and Transforms with Applications in Optics." McGraw-Hill, New York.
Pirio, F. (1975). *In* "Optical and Acoustical Micro-Electronics" (J. Fox, ed.), Vol. 23, pp. 311–318. Polytechnic Press, New York.
Pirio, F., and Sinou, P. (1975). *Proc. 1975 Ultrason. Symp.* IEEE Cat. No. 75 CHO 994-4SU, p. 492.
Schmidt, R. V. (1976). *IEEE Trans. Sonics Ultrason.* **23**, 22.
Slobodnik, A. J., Jr. (1973). *IEEE Trans. Sonics Ultrason.* **20**, 315.
Slobodnik, A. J., Jr. (1976). *Proc. IEEE* **64**, 581.
Slobodnik, A. J., Jr., and Szabo, T. L. (1971). *Electron. Lett.* **7**, 257.
Slobodnik, A. J., Jr., and Szabo, T. L. (1973). *J. Appl. Phys.* **44**, 2937.
Slobodnik, A. J., Jr., and Szabo, T. L. (1974). *IEEE Trans. Microwave Theory Tech.* **22**, 458.
Slobodnik, A. J., Jr., Conway, E. D., and Delmonico, R. T. (1973). "Microwave Acoustics Handbook," Vol. 1A, TR-73-0597 (unpublished).
Slobodnik, A. J., Jr., Kearns, W. J., Silva, J. H., and Fenstermacher, T. E. (1974). *Proc. 1974 Ultrason. Symp.* IEEE Cat. No. 74 CHO 896-1SU, p. 185.
Slobodnik, A. J., Jr., Fenstermacher, T. E., Kearns, W. J., Roberts, G. A., and Silva, J. H. (1975). *Proc. 1975 Ultrason. Symp.* IEEE Cat. No. 75 CHO 994-4SU, p. 405.
Szabo, T. L. (1975). *Proc. 1975 Ultrason. Symp.* IEEE Cat. No. 75 CHO 994-4SU, p. 116.
Szabo, T. L. (1977). *J. Acoust. Soc. Am.* To be published.
Szabo, T. L., and Frost, H. M. (1976). *IEEE Trans. Sonics Ultrason.* **23**, 323.

Szabo, T. L., and Slobodnik, A. J., Jr. (1973a). *IEEE Trans. Sonics Ultrason.* **20**, 240.
Szabo, T. L., and Slobodnik, A. J., Jr. (1973b). "Acoustic Surface Wave Diffraction and Beam Steering," AFCRL-TR-73-0302 (unpublished).
Szabo, T. L., and Slobodnik, A. J., Jr. (1974). *IEEE Trans. Sonics Ultrason.* **21**, 114.
Tsai, C. S., and Nguyen, L. T. (1975). *In* "Optical and Acoustical Micro-electronics" (J. Fox, ed.), Vol. 23, pp. 583–597. Polytechnic Press, New York.
Van Duzer, T. (1970). *Proc. IEEE* **58**, 1230.
Wagers, R. S. (1976). *IEEE Trans. Sonics Ultrason.* **23**, 249.
Waldron, R. A. (1972). *IEEE Trans. Sonics Ultrason.* **19**, 448.
Weglein, R. D., Pedinoff, M. E., and Winston, H. (1970). *Electron. Lett.* **6**, 654.
White, R. M. (1970). *Proc. IEEE* **58**, 1238.
Wickramasignhe, H. K., and Ash, E. A. (1975a). *In* "Optical and Acoustical Micro-Electronics" (J. Fox, ed.), Vol. 23, pp. 413–431. Polytechnic Press, New York.
Wickramasinghe, H. K., and Ash, E. A. (1975b). *Proc. 1975 Ultrason. Symp.* IEEE Cat. No. 75 CHO 994-4SU, p. 496.
Williamson, R. C. (1973). *IEEE Trans. Sonics Ultrason.* **19**, 436.

—5—

Doubly Rotated Thickness Mode Plate Vibrators

ARTHUR BALLATO

U.S. Army Electronics Technology & Devices Laboratory
Fort Monmouth, New Jersey 07703

I. Introduction

This chapter deals with the properties of piezoelectric plate vibrators having laterally unbounded, parallel faces that have arbitrary inclinations to the natural axes of the crystals from which they are fashioned. The plate motions under discussion are the simple thickness modes; these do not depend on lateral coordinates, making the situation inherently one-dimensional. With the thickness coordinate the only direction of interest, plate orientation is specified uniquely by two angles, so the crystal cuts are spoken of as being "doubly rotated."

The use of such cuts is not new, nor is any practical problem, strictly speaking, one-dimensional. Why, then, an article on the topic at this time? There are two major reasons. Recent discovery of compensation of nonlinear effects in quartz offers the promise of new advances in the area of high precision frequency control, while the advent of new, high coupling, zero temperature coefficient materials is seen as providing exciting prospects for bulk wave filter and transducer technology as well as important applications in surface acoustic wave devices. In both cases, doubly rotated cuts are expected to be widely utilized in the future. Moreover, several aspects of the problems associated with this more general type of vibrator, such as the simultaneous excitation of two or three modes and their interaction have meanwhile been clarified. These have made possible a coherent account of the properties of these cuts, better approximations, and an indication of areas for further investigation and improvement, so that a survey of the situation now is particularly pertinent. As for the limitation to thickness modes, this case has the estimable virtue of possessing an exact solution, even for plates devoid of crystal symmetry, while retaining those features most relevant to the characterization of practical devices.

Singly and doubly rotated plates are shown in Fig. 1 in relation to the crystallographic axes X, Y, Z. In standard notation (Anonymous, 1949) the

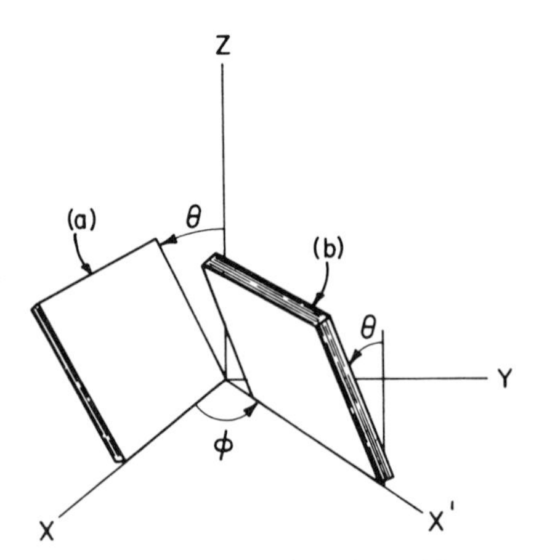

FIG. 1. Conventions for specifying plate orientations with respect to crystal axes X, Y, Z. (a) Singly rotated plate. (b) Doubly rotated plate.

singly rotated Y cut shown is described as $(YXl)\theta$; the AT and BT cuts of quartz are representative (Mason, 1964). Doubly rotated quartz cuts $(YXwl)\varphi/\theta$ were investigated by Bokovoy and Baldwin (1935) starting from the singly rotated X cut, and this led to their discovery of the V-cut family. Improvements in frequency–temperature behavior motivated this early work on quartz plates, and remains a continued concern for most modern applications. Having two orientation angles available permits satisfaction of additional requirements simultaneously with temperature coefficient constraints, as will become apparent in Sections IV and V. We shall lead up to these considerations by describing first the propagation of acoustic plane waves in piezoelectric crystals and the solution to the general unbounded plate problem.

II. Waves and Vibrations in Piezoelectric Media

One hundred years ago, Elwin Christoffel (1877) gave a formalism for treating acoustic plane wave propagation in arbitrarily anisotropic, linear media. Lawson (1941) extended it to include piezoelectricity, finding correct expressions for the three stiffened phase velocities. Subsequent applications of the theory to thickness vibrations of plates proved rigorous only for particular cases involving excitation of a single mode until Tiersten (1963) produced the exact solution that showed the three modes to be piezoelectrically coupled by the plate surfaces. The exact analysis also disclosed in retrospect that various quantities required for discussing plate vibrations, such as the correct piezoelectric coupling factor driving each mode, could be obtained easily without considering any boundaries.

This section outlines the Christoffel method and its application to simple thickness modes of plates. In keeping with the convention seen in Fig. 1, a right-handed orthogonal coordinate system is taken to undergo two successive rotations from the crystallographic axes X_i $(= X, Y, Z)$: a first rotation about X_3 by angle φ, followed by a rotation about X'_1 by angle θ. Plane waves are then taken to propagate along X''_2, which will subsequently become the thickness axis of our plate.

A. Plane Acoustic Waves in Piezoelectric Crystals

In the following a subscripted index preceded by a comma denotes differentiation with respect to the coordinate having that subscript, and twice-repeated indices denote summation. A time harmonic dependence $\exp(+j\omega t)$ is assumed. The pertinent sets of equations, referred to the crystallographic axes X_i, that are to be solved consist of the stress equations of motion

$$T_{ij,i} = -\rho\omega^2 u_j; \tag{1}$$

the definition of mechanical strain

$$S_{kl} = \tfrac{1}{2}(u_{k,l} + u_{l,k}); \tag{2}$$

the Maxwell divergence relation

$$D_{i,i} = 0; \tag{3}$$

the quasi-static electric field–potential equation

$$E_n = -\Phi_{,n}; \tag{4}$$

and the linear piezoelectric constitutive relations characterizing the medium

$$T_{ij} = c^{\mathrm{E}}_{ijkl} S_{kl} - e_{nij} E_n, \tag{5}$$

$$D_i = e_{ikl} S_{kl} + \varepsilon^{\mathrm{S}}_{in} E_n. \tag{6}$$

In these equations T_{ij}, u_j, S_{kl}, D_i, and E_n are the components of mechanical stress, displacement and strain, electric displacement and field, respectively; ρ and Φ are the mass density and electric potential. The material parameters c^{E}_{ijkl}, e_{nij}, and $\varepsilon^{\mathrm{S}}_{in}$ are the elastic stiffnesses at constant electric field, the piezoelectric stress constants, and the dielectric permittivities at constant strain. All indices have the range 1, 2, 3.

Equations (2) and (4) are used in Eqs. (5) and (6) to eliminate S_{kl} and E_n. The resulting relations are substituted into Eqs. (1) and (3) to yield four equations in the displacements u_j and the potential. These are referred to the direction of propagation X''_2 by making use of the transformation

$$X''_i = \alpha_{ij} X_j. \tag{7}$$

With the abbreviations ζ for X''_2 and α_i for α_{2i} the resulting equations reduce to

$$\Gamma^{\mathrm{E}}_{jk} u_{k,\zeta\zeta} + \Xi_j \Phi_{,\zeta\zeta} = -\rho\omega^2 u_j \tag{8}$$

$$\Xi_k u_{k,\zeta\zeta} - \varepsilon^{\mathrm{S}}_{\zeta} \Phi_{,\zeta\zeta} = 0, \tag{9}$$

where

$$\Gamma^{\mathrm{E}}_{jk} = c^{\mathrm{E}}_{ijkl} \alpha_i \alpha_l \tag{10}$$

$$\Xi_k = e_{nkl} \alpha_n \alpha_l \tag{11}$$

$$\varepsilon^{\mathrm{S}}_{\zeta} = \varepsilon^{\mathrm{S}}_{in} \alpha_i \alpha_n. \tag{12}$$

Equations (8) and (9) yield

$$\bar{\Gamma}_{jk} u_{k,\zeta\zeta} = -\rho\omega^2 u_j \tag{13}$$

with

$$\bar{\Gamma}_{jk} = \bar{c}_{ijkl} \alpha_i \alpha_l = \Gamma^{\mathrm{E}}_{jk} + \Xi_j \Xi_k / \varepsilon^{\mathrm{S}}_{\zeta}, \tag{14}$$

and an equation in the potential that integrates to

$$\Phi = \Xi_k u_k/\varepsilon_\zeta^S + \mathscr{A}\zeta + \mathscr{B}. \qquad (15)$$

In order that Eq. (13) be compatible with the wave equation

$$\bar{c}u_{j,\zeta\zeta} + \rho\omega^2 u_j = 0 \qquad (16)$$

requires that

$$(\bar{\Gamma}_{jk} - \bar{c}\delta_{jk})u_{k,\zeta\zeta} = 0. \qquad (17)$$

The solution to the eigenvalue equation yields three real roots $\bar{c}_m$, ordered according to the relations

$$\bar{c}_a > \bar{c}_b \geq \bar{c}_c, \qquad (18)$$

which determine the piezoelectrically stiffened phase velocities v_m:

$$v_m = (\bar{c}_m/\rho)^{1/2}. \qquad (19)$$

Corresponding to the roots $\bar{c}_m$ are the orthonormal eigenvectors $\bar{\gamma}_m$ which determine the direction cosines of particle displacement for each mode m with respect to the crystal axes X_i.

One may include the phenomenological effects of small loss by considering the viscosity as the imaginary part of the elastic stiffness (Lamb and Richter, 1966):

$$\hat{c}_{ijkl} = c_{ijkl}^E + j\omega\eta_{ijkl}. \qquad (20)$$

As long as $\omega\eta_{ijkl} \ll c_{ijkl}^E$ one may avoid the complex eigenvalue problem and compute the effective viscosity for each mode as

$$\eta_m = H_{jk}\bar{\gamma}_{mj}\bar{\gamma}_{mk} \qquad (21)$$

using the lossless eigenvectors $\bar{\gamma}_m$ and the analog of Eq. (10)

$$H_{jk} = \eta_{ijkl}\alpha_i\alpha_l. \qquad (22)$$

For quartz and other high purity dielectric crystals at room temperature, $\eta_m/\bar{c}_m$ is typically 10^{-14} sec, so that Eq. (21) is an excellent approximation into the microwave region.

Applications involving the use of crystal plates as acoustic transducers or as devices for beam steering require a knowledge of the direction of energy flux for an assumed direction of phase progression. The power flow associated with each mode m consists of two parts: a mechanical portion with components

$$P_{iM}^{(m)} = +j\omega T_{ij}^{(m)}u_j^{(m)*} \qquad (23)$$

and corresponding electrical portion

$$P_{iE}^{(m)} = -j\omega\Phi^{(m)}D_i^{(m)*}. \qquad (24)$$

An asterisk denotes complex conjugation. A forward-traveling wave propagating in direction ζ produces a displacement field with components proportional to $\bar{\gamma}_{mk}$. Straightforward substitution of this plane wave solution into the pertinent foregoing equations yields the following expressions for the time average power flows due to mode m:

$$P_{iM}^{(m)} = P_0 \omega^2 (\tilde{c}_{ijkl} \bar{\gamma}_{mj} \alpha_k \bar{\gamma}_{ml}) / v_m \quad \text{(no } m \text{ sum)}, \tag{25}$$

where

$$\tilde{c}_{ijkl} = c_{ijkl}^{E} + e_{kij} \Xi_l / \varepsilon_\zeta^{S}, \tag{26}$$

$$P_{iE}^{(m)} = P_0 \omega^2 (\tilde{e}_{ikl} \bar{\gamma}_{mk} \alpha_l (\Xi_j \bar{\gamma}_{mj}) / \varepsilon_\zeta^{S}) / v_m \quad \text{(no } m \text{ sum)} \tag{27}$$

with

$$\tilde{e}_{ikl} = e_{ikl} - \Xi_k \varepsilon_{il}^{S} / \varepsilon_\zeta^{S}. \tag{28}$$

The quantity P_0 is an arbitrary amplitude factor. From the sum of Eqs. (25) and (27) the power flow direction is easily obtained.

B. Simple Thickness Modes of Plates

The simple thickness modes of plates are those for which all variations depend solely on the thickness coordinate—in our instance ζ. We take the plate to be traction-free and driven by potentials $\pm\Phi_0$ on the surfaces $\zeta = \pm h$. These boundary conditions lead to the vanishing of the transformed stresses at the surfaces

$$\alpha_{ij}(\bar{\Gamma}_{jk} u_{k,\zeta} + \Xi_j \mathscr{A}) = 0 \qquad \text{at} \quad \zeta = \pm h. \tag{29}$$

Together with the potential boundary condition, Eq. (29) determines the constants in Eq. (15) as

$$\mathscr{A} = (\Phi_0 / h) \Bigg/ \left[1 - \sum_m k_m^2 \tan X^{(m)} / X^{(m)} \right], \tag{30}$$

$$\mathscr{B} = 0, \tag{31}$$

where

$$X^{(m)} = \omega h / v_m \tag{32}$$

and

$$k_m^2 = (\bar{\gamma}_{mj} \Xi_j)^2 / \bar{c}_m \varepsilon_\zeta^{S} \qquad \text{(no } m \text{ sum)}. \tag{33}$$

Transformation of Eq. (6) leads to

$$D_\zeta = \Xi_k u_{k,\zeta} - \varepsilon_\zeta^{S} \Phi_{,\zeta}. \tag{34}$$

With the help of Eq. (15) this reduces to

$$D_\zeta = -\varepsilon_\zeta^S \mathscr{A}; \tag{35}$$

the current density is then just

$$J_\zeta = -j\omega D_\zeta = +j\omega\varepsilon_\zeta^S \mathscr{A}. \tag{36}$$

For any patch of area A, the total current passing through the plate is AJ_ζ, so the input admittance is obtained for this portion by taking the quotient of total current to applied voltage $2\Phi_0$. The static capacitance of the plate section is

$$C_0 = \varepsilon_\zeta^S A/2h, \tag{37}$$

and the input admittance is found to be (Yamada and Niizeki, 1970)

$$Y_{\text{in}} = j\omega C_0 \bigg/ \left[1 - \sum_m k_m^2 \frac{\tan X^{(m)}}{X^{(m)}}\right]. \tag{38}$$

The k_m are the piezoelectric coupling factors responsible for excitation of the modes m. They are defined in terms of effective piezoelectric constants e_m such that comparison with Eq. (33) leads to the identification

$$e_m = \bar{\gamma}_{mj}\Xi_j. \tag{39}$$

Therefore, the effective piezoelectric constants and coupling factors are simply obtained from the solution for the infinite medium. In the presence of small loss, Eq. (38) is used with $X^{(m)}$ from Eq. (32) replaced by

$$\hat{X}^{(m)} \simeq X^{(m)}(1 - j\omega\eta_m/2\bar{c}_m). \tag{40}$$

III. Critical Plate Frequencies

The piezoelectric plate vibrator has critical frequencies defined by the zeros and poles of the input admittance, Eq. (38). The zeros determine three harmonically related sequences of open-circuit antiresonance frequencies $f_{mA}^{(M)}$. M is an odd integer, the harmonic number, and $N_m = v_m/2$ is the frequency constant for mode m:

$$f_{mA}^{(M)} = MN_m/2h. \tag{41}$$

In the antiresonance case the three thickness modes uncouple from each other. The short-circuit resonance frequencies, obtained from the poles of the admittance, are coupled so that the frequencies cannot be ascribed to any mode, except in a loose manner of speaking. They are denoted f_R and found from the solution of

$$\sum_m k_m^2 \frac{\tan X^{(m)}}{X^{(m)}} = 1, \tag{42}$$

where

$$X^{(m)} = (\pi/2)(f_R/f_{mA}^{(1)}). \tag{43}$$

Operation of the resonator in an air gap or in series with an external capacitor will have no effect upon the antiresonance frequencies; the resonance frequencies, however, will all be shifted to higher values. If the capacitance value inserted in series is C_L, then the modified resonance frequencies are found from Eq. (42) using

$$\hat{k}_m^2 = k_m^2/(1 + C_0/C_L). \tag{44}$$

A. Two and Three Modes Excited

Doubly rotated crystal plates generally will have all three k_m finite; yet it is still possible in many instances to treat the resonance frequencies as if they were uncoupled. This will be addressed in Section III,B. Here we consider the situation where such a simplification is not justified, and the coupling must be taken into account. The three mode case follows the discussion for two modes, so we start with the two mode case. A practically important instance is the rotated-Y-cut family of cuts in crystal class $3m$, which includes lithium tantalate and lithium niobate, discussed in Section V.

The interplay between two excited modes is seen in the dispersion diagram sketched in Fig. 2. The heavy lines trace branches of the solution of Eq. (42); plotted along the abscissa is the antiresonance frequency ratio

$$\beta = f_{2A}^{(1)}/f_{1A}^{(1)} = v_2/v_1, \tag{45}$$

and the ordinate is normalized frequency

$$\Omega = f/f_{1A}^{(1)}. \tag{46}$$

Bounds are readily established, as will be described below; these permit one to construct any portion of the spectrum rapidly, and from the construction to infer the behavior of the input admittance with driving frequency.

Construction of the terrace plot for two coupled modes proceeds by first finding the lines corresponding to the antiresonances for both modes. For mode one,

$$\Omega_{1A}^{(M)} = M \qquad (M = 1, 3, 5, \ldots). \tag{47}$$

These are a sequence of horizontal lines in the diagram. For mode two,

$$\Omega_{2A}^{(N)} = N\beta \qquad (N = 1, 3, 5, \ldots), \tag{48}$$

giving lines of slope N. Next, Eq. (42) is solved twice: once with k_1 set to

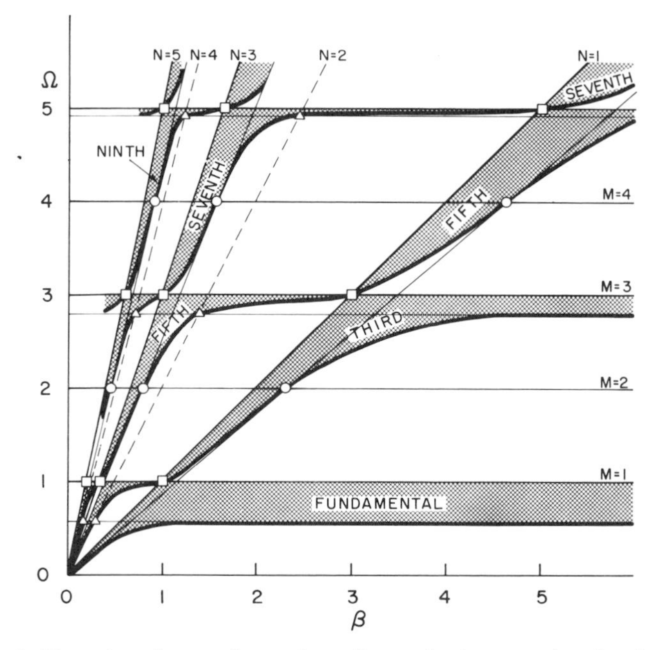

FIG. 2. Dispersion diagram for a plate vibrator having two piezoelectrically excited modes. Normalized frequency Ω is plotted against β, the ratio of fundamental antiresonance frequencies. Regions between succeeding resonances and antiresonances, for fixed β, are crosshatched.

zero, and then with $k_2 = 0$. (Single mode solutions are discussed by graphical construction in Section III,B.) When k_2 is zero, the solution yields for $\Omega_{1R}^{(M)}$ a series of horizontal lines, each below the antiresonance line having the same M, but approaching successively closer to the $\Omega_{1A}^{(M)}$ lines with increasing M. With k_1 set to zero, the solution set consists of lines radiating from the origin. The line for $\Omega_{2R}^{(N)}$ has slope less than N, but approaching N for N large.

The next stage in the construction consists in finding where the horizontal $\Omega_{1R}^{(M)}$ lines intersect the solutions of Eq. (42); this is easily shown to occur where the $\Omega_{1R}^{(M)}$ lines cross radial lines of slope N, with N even. Similarly, the radial lines for $\Omega_{2R}^{(N)}$ cross the horizontal lines with M even at positions satisfying Eq. (42) also. Solutions of Eq. (42) for M even and N odd are marked in Fig. 2 as circles; those with M odd and N even are marked as triangles.

What happens when one of the two tangent functions in Eq. (42) approaches a pole? In order to balance the equation, the other tangent must obviously also approach a pole in such a fashion as to cancel the effect. This line of thought may be carried out to show that the resonance frequency loci in the coupled mode case pass through the intersections with M and N odd; these are shown as squares.

Using the circles, triangles, and squares as a lattice of points through which the spectrum must pass, and with the horizontal and radial lines acting as boundaries, the solution branches to Eq. (42) may be found easily. In Fig. 2 the heavy solid lines have been drawn to show Ω_R for relatively large coupling factors. The shaded areas are those for which the input reactance is positive; i.e., the resonator appears inductive.

For a given crystal cut, the ratio β will be a fixed number, usually between 1 and 2. A vertical line on the diagram at this value immediately gives the mode spectrograph sequence of resonances, the pole–zero separation of each resonance, and their location in frequency. In addition, one clearly sees how the resonances move with respect to each other when β is changed and how the separations are affected.

When the full set of three k_m is allowed, a three-dimensional construction is required, consisting of two β axes

$$\beta_2 = f_{2A}^{(1)}/f_{1A}^{(1)}, \qquad \beta_3 = f_{3A}^{(1)}/f_{1A}^{(1)}, \tag{49}$$

and the Ω axis of Eq. (46). The result consists of the same sequence of horizontal resonance and antiresonance lines in both β-Ω planes as in Fig. 2. The radial resonance lines differ in each plane if $k_2 \neq k_3$. A terraced set of surfaces then describes the complete solution, the construction of which follows in general the procedure outlined here.

B. Single Mode Excited

Equation (42) reduces to the single mode equation

$$k^2 \tan X = X \tag{50}$$

when two k_m vanish, as occurs for certain combinations of crystal symmetry and plate orientation. An example is the excitation of rotated-Y-cut members in crystal class 32, which includes the AT and BT cuts of quartz. In many instances Eq. (50) may be used as an adequate approximation to Eq. (42) even when two or three k_m are nonzero. The conditions permitting the approximation may be seen for the two mode case from Fig. 2. In those regions of the diagram where the curves lie near the lines $\Omega_{1R}^{(M)}$ or $\Omega_{2R}^{(N)}$, Eq. (50) may be used, including specifically the points marked by circles and triangles, where it is exact. As the coupling factors become smaller, the extent of the applicable regions increases, and conversely. Always to be avoided are the areas where, for N and M odd,

$$M \simeq N\beta; \tag{51}$$

these positions are those marked in Fig. 2 with squares. Here Eq. (50) cannot be used.

A graphical solution to Eq. (50) is shown in Fig. 3. Intersections of the tangent function branches with a straight line of slope k^{-2} define the frequencies of resonance $f_{\mathrm{RO}}^{(M)}$; the poles of the tangent function correspond to the antiresonance frequencies $f_{\mathrm{AO}}^{(M)}$. The zero subscripts denote the

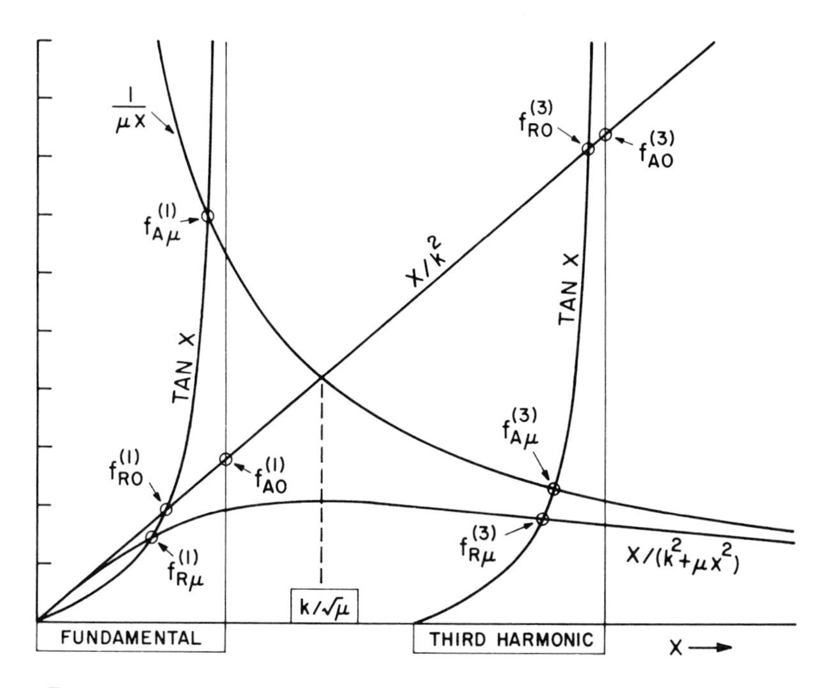

FIG. 3. Graphical solutions to the single mode frequency equations for resonance and antiresonance. The mass loaded case is distinguished by the subscript μ.

absence of mass loading due to the electrode, which is considered in the next subsection. From the construction one sees that the solution consists of a progression of roots successively drawing closer to the harmonically related antiresonances. As the piezoelectric coupling factor k increases, the resonance–antiresonance separation increases at each harmonic. Using Eq. (38), the effective input capacitance of the resonator in the limit of zero frequency is

$$C_{\mathrm{in}} = C_0 \Big/ \left(1 - \sum_m k_m^2\right). \tag{52}$$

Since the stored energy of a capacitor must be positive, the limiting value for the sum of the three k_m^2 is unity. If only one k_m is finite, unity becomes its limiting value; such a value would reduce $f_{\mathrm{RO}}^{(1)}$ to zero frequency—a clearly unphysical result. Piezoelectric materials, however, are available today with coupling factors in excess of 70%, which means that the fundamental resonance frequency is depressed by the large coupling to values less than three-quarters of the antiresonance frequency. Because of the anharmonic ratios of the frequencies $f_{\mathrm{RO}}^{(M)}$ arising from Eq. (50), it is more convenient to characterize many of the plate properties in terms of the antiresonance frequencies $f_{\mathrm{AO}}^{(M)}$, and to use $f_{\mathrm{AO}}^{(1)}$ as the basis of normalizations.

C. Effects of Mass Loading on Frequencies

It is current practice to plate metallic electrodes directly on the major surfaces of thickness mode vibrators in order to apply the driving voltage. The electrode coating is assumed here to be devoid of elastic stiffness and to be negligibly thick so that it is represented as a lumped mass $\bar{m}$ per unit area on each plate surface. Replacing Eq. (29) is the stress boundary condition

$$(\bar{\Gamma}_{jk} u_{k,\zeta} + \Xi_j \mathscr{A}) = \pm\omega^2 \bar{m} u_j \qquad \text{at} \quad \zeta = \pm h. \tag{53}$$

Reduced mass loading is defined as

$$\mu = \bar{m}/\rho h, \tag{54}$$

and in terms of this quantity Eq. (53) leads to a modification of Eq. (38) in the presence of electrode coatings:

$$Y_{\mathrm{in}} = j\omega C_0 \Bigg/ \left\{ 1 - \sum_m \left[\left(\frac{k_m^2}{1 - \mu X^{(m)} \tan X^{(m)}} \right) \frac{\tan X^{(m)}}{X^{(m)}} \right] \right\}. \tag{55}$$

As in the traction-free situation, the open-circuit frequencies are uncoupled, but now each is determined from the roots of

$$\mu X^{(m)} \tan X^{(m)} = 1, \tag{56}$$

so they are no longer harmonically related. The resonance frequencies, on the other hand, remain coupled and are found from

$$\sum_m \left[\left(\frac{k_m^2}{1 - \mu X^{(m)} \tan X^{(m)}} \right) \frac{\tan X^{(m)}}{X^{(m)}} \right] = 1. \tag{57}$$

The discussion of Fig. 2 can be extended to cover mass loading, with two or three k_m nonzero. Bounds are readily constructed from which the depressed spectra are obtained in the general case; we show here the effect on a single mode, which is applicable whenever the conditions of Section III,B hold.

For one mode, Eq. (57) becomes

$$\tan X = X/(k^2 + \mu X^2) \tag{58}$$

with associated graphical solution shown in Fig. 3. Roots of Eq. (58) are designated $f_{R\mu}^{(M)}$; those of Eq. (56), $f_{A\mu}^{(M)}$. In each case μ has the effect of lowering the critical frequency; but for harmonics where $X < k/\mu^{1/2}$, the piezoelectric contribution to the lowering predominates and the resonance frequencies with and without μ cluster while the antiresonance frequencies form a separate group. Mass-loading effects predominate for harmonics for which the inequality is reversed, with the mass-loaded frequencies occurring together and the unloaded frequencies forming a separate group.

A more convenient measure of frequency shift due either to mass loading or piezoelectric coupling is frequency displacement, defined by Ballato and Lukaszek (1974) as

$$\delta = M - 2X/\pi. \tag{59}$$

In terms of antiresonance and resonance displacements, Eqs. (56) and (58) respectively become

$$\tan(\delta_A^{(M)}\pi/2) = \mu(M - \delta_A^{(M)})\pi/2, \tag{60}$$

and

$$\tan(\delta_R^{(M)}\pi/2) = \mu(M - \delta_R^{(M)})\pi/2 + 2k^2/\pi(M - \delta_R^{(M)}). \tag{61}$$

Figure 4 shows how $\delta_R^{(M)}$ varies with M and k as a function of μ; for $k = 0$, $\delta_R^{(M)}$ becomes $\delta_A^{(M)}$.

The foregoing has considered some effects of mass loading on plate frequencies when both surfaces are equally loaded. A fact often overlooked in manufacturing is the influence of mass imbalances on the frequency spectrum. When an imbalance exists, even harmonics become excited as well as those that are odd. Although the effect is usually small as far as the magnitudes of the even resonances are concerned, nonlinear coupling via third order elastic constants can produce undesired activity dips at the operating resonance. In addition, the families of anharmonic overtones associated with each thickness mode by virtue of couplings to lateral modes in finite plates all couple to the even modes and their families. The result can be a very complex mode spectrum under the best of circumstances; but when doubly rotated cuts are used with their attendant complexities due to lack of orientational symmetry, any further geometrical asymmetry such as electrode imbalance further worsens the spectrum and decreases the utility of these elements in oscillators and filters.

D. Approximations for Critical Frequencies

Apart from the antiresonances of traction-free plates, the critical frequencies are determined as roots of transcendental equations. Approxi-

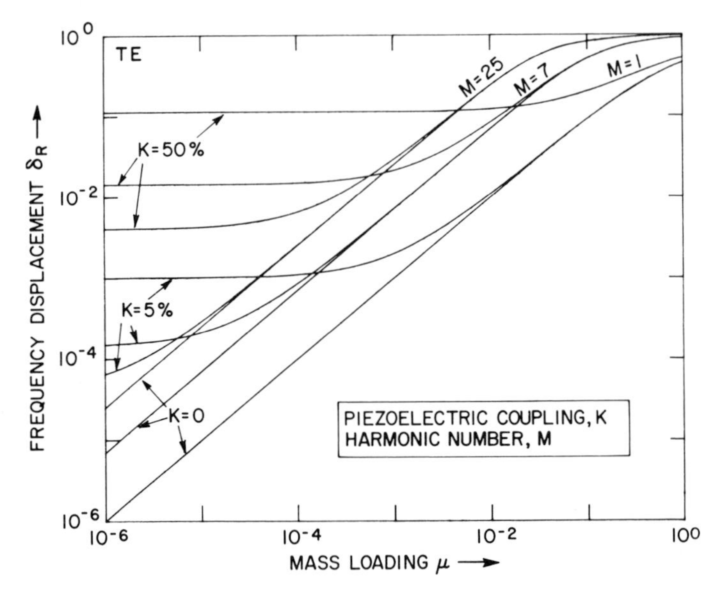

FIG. 4. Single mode frequency displacement versus mass loading for various piezoelectric coupling and harmonic values. Frequency displacement is a normalized measure of departure from antiresonance frequency with zero mass loading.

mate solutions to these equations, adequate for most applications, take the form of simple algebraic functions. Some of these are given here.

In the absence of mass loading, the resonance displacements are approximated by

$$\delta_{\mathrm{R0}}^{(M)} \simeq 4k^2/\pi^2 M(1 + 4k^2/\pi^2 M^2) \simeq 4k^2/\pi^2 M. \tag{62}$$

The last form is accurate for all k when $M > 1$, and up to $k = 30\%$ when $M = 1$; the first form is within 1% at the fundamental up to $k = 50\%$.

With mass loading the displacements are found from

$$\delta_{\mathrm{R}\mu}^{(M)} \simeq (\delta_{\mathrm{R0}}^{(M)} + M\mu)/(1 + M\mu) \tag{63}$$

with $\delta_{\mathrm{R0}}^{(M)}$ from Eq. (62). Setting $k = 0$ in Eq. (62) produces the result for $\delta_{\mathrm{A}\mu}^{(M)}$ using Eq. (63).

Differences between resonance and antiresonance displacements are found to vary in μ dependence according to whether $M = 1$ or not. The expressions for the approximations are

$$\delta_{\mathrm{R}\mu}^{(1)} - \delta_{\mathrm{A}\mu}^{(1)} \simeq (2/\pi)(1 - (1 - k^2)^{1/2})\delta_{\mathrm{R0}}^{(1)}/[(2/\pi)(1 - (1 - k^2)^{1/2}) + \delta_{\mathrm{R0}}^{(1)}\mu^{1/2}] \tag{64}$$

when $M = 1$, and for $M > 1$,

$$\delta_{\mathrm{R}\mu}^{(M)} - \delta_{\mathrm{A}\mu}^{(M)} \simeq (2/\pi)^4 k^2 \delta_{\mathrm{R0}}^{(M)}/[(2/\pi)^4 k^2 + (M - 1)^3 \delta_{\mathrm{R0}}^{(M)} \mu^2]. \tag{65}$$

Operation of the vibrator with series load capacitance modifies the effective coupling factor as shown in Eq. (44). The modified value for k^2 introduced into Eqs. (62)–(65) produces the approximations pertinent to the use of $\delta_{L0}^{(M)}$ and $\delta_{L\mu}^{(M)}$.

IV. Static Frequency–Temperature Behavior

One of the most important practical considerations for frequency control and selection is that of temperature behavior. The AT cut has maintained its dominant position among the thickness mode resonators for so many years mainly because of its superior characteristics in this regard. This section outlines how the temperature influence on the various critical frequencies is treated phenomenologically under the assumption that thermal changes take place slowly enough so that the vibrator is constantly at thermal equilibrium. The static behavior is described for single and multiple modes along with associated approximations.

A. First Order Temperature Coefficients

Bechmann (1956) expanded the static frequency–temperature function as a power series and found that three terms were normally sufficient for an adequate description of AT- and BT-cut quartz resonators. This formalism has subsequently been applied to general, doubly rotated cuts of quartz and to other materials with satisfactory results.

If at reference temperature T_0 the critical frequency of interest is f_0, the expansion is

$$(f - f_0)/f_0 = \Delta f/f_0 = \sum_n^3 T_f^{(n)}(T - T_0)^n \tag{66}$$

with

$$T_f^{(n)} = [(\partial^n f/\partial T^n)/n!\, f_0]_{T=T_0}. \tag{67}$$

The higher order terms with $n = 2$ and 3 are dealt with in Section IV,B; here the linear term is considered. As with the critical frequencies themselves, the temperature coefficients may be coupled or uncoupled. Simplest of all to describe are the first order temperature coefficients of the antiresonance frequencies. These are found by taking the logarithmic derivative of Eq. (41) with substitution of Eq. (19) to obtain

$$T_{f_{mA}}^{(1)} = \tfrac{1}{2}T_{\bar{c}_m}^{(1)} - \tfrac{1}{2}T_{\rho}^{(1)} - T_h^{(1)}. \tag{68}$$

$T_\rho^{(1)}$ and $T_h^{(1)}$ are obtained from the thermoelastic constants for the material, while $T^{(1)}_{\bar{c}m}$ requires differentiation of the cubic that arises from Eq. (17).[1] Because the $\bar{\Gamma}_{jk}$ involve the elastic, piezoelectric, and dielectric material constants plus the plate orientation angles, the expression for $T^{(1)}_{\bar{c}m}$ is lengthy in general, but is nevertheless straightforward. In Eq. (68) the superscript relating to the harmonic (M) has been omitted; the antiresonance result is independent of M.

Using Eq. (68), the locus of $T^{(1)}_{f_{mA}} = 0$ was mapped in detail by Bechmann *et al.* (1962) for quartz, with the result shown in Fig. 5. The angles are

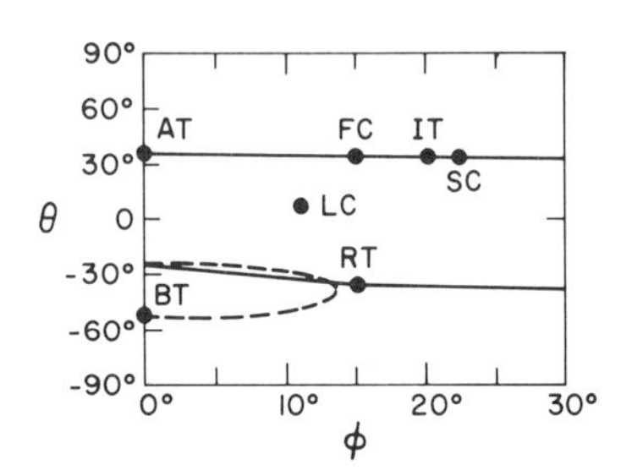

FIG. 5. Loci of zeros of first order temperature coefficient of antiresonance frequency in quartz. The angles are those shown in Fig. 1. Solid lines denote the slower quasi-shear mode c; the dashed loop indicates the faster quasi-shear mode b. Locations of a number of useful cuts are shown.

those described in connection with Fig. 1. In Fig. 5 the locations of the AT and BT cuts are seen to belong to different branches. Shown dashed is the zero temperature coefficient locus for the b mode, while the solid lines indicate the locus for the c mode. Also indicated are a number of additional cuts: the IT cut (Bottom and Ives, 1951); the RT cut (Bechmann, 1961); the LC cut (Hammond and Benjaminson, 1965); the FC cut (Lagasse *et al.*, 1972); and the SC cut (Holland, 1974; EerNisse, 1975). These cuts will be discussed more extensively in Section V,A in connection with properties of quartz. It is important to notice here that the presence of a continuous locus of zero

[1] In defining the temperature derivative of the elastic constants, ρ, h, and $\bar{c}_m$ are allowed to vary. However, more fundamental equations could have been derived in terms of reference (material) coordinates, and ρ and h would have been constant. The result would have been an equally valid, but numerically different, set of effective elastic constant temperature coefficients. In either case the temperature dependence of the effective elastic constants depends explicitly on temperature and strain, including the strain dependence on temperature.

Very recently (Tiersten, 1975a), an analysis has been made using the third-order elastic stiffnesses; the portions of dependence of the effective elastic constants on the third-order stiffnesses and on the change in strain with temperature have been calculated. From these results the fundamental dependence of the effective elastic constants on temperature can be obtained (H. F. Tiersten, private communication, 1976).

The alternate procedure, used here, was that followed by Bechmann *et al.* (1962); the third-order elastic constants only became available at a later date (Thurston *et al.*, 1966).

temperature coefficient orientations affords the possibility of locating a cut having some other optimal property simultaneously with good static thermal characteristics.

Figure 5, plotted for the thickness modes, also furnishes the approximate loci for other modes, e.g., contour modes (Baldwin and Bokovoy, 1936), and surface acoustic wave (SAW) modes (Schulz *et al.*, 1970). When $\bar{c}_m$ and h are defined properly, Eq. (68) is just the negative of the temperature coefficient of delay employed in SAW analyses; to characterize these elements requires the addition to Fig. 5 of a third axis representing propagation angle so the zero locus is a surface in angle space.

Resonance frequencies differ from the antiresonances in depending upon the coupling factor. For the single mode case, Onoe (1969) showed that the difference between the resonance and antiresonance temperature coefficients is proportional to the temperature coefficient of coupling:

$$T_{f_R}^{(1)} - T_{f_A}^{(1)} = T_X^{(1)} = -G_0 T_k^{(1)}, \tag{69}$$

where

$$G_0 = 2k^2/(X^2 + k^2(k^2 - 1)), \tag{70}$$

and where X is the Mth root of Eq. (50). In the limit of small k, Eq. (70) becomes

$$G_0 \simeq 8k^2/\pi^2 M^2. \tag{71}$$

This approximation is suitable for many quartz applications, and on the scale of Fig. 5 no sensible difference between $T_{f_R}^{(1)}$ and $T_{f_A}^{(1)}$ would be manifest. For high precision applications, as well as for high coupling materials, Eq. (70) must be used since the difference in the temperature coefficients can be appreciable. An example is given in Section V,C. Equations (69)–(70) contain the influence of harmonic on $T_{f_R}^{(1)}$; the additional effect of mass loading has been given recently (Ballato, 1976). Insertion of a series load capacitor C_L also shifts the temperature behavior, its influence acting like a shift to a higher harmonic with the exception that the harmonic shift is quantized. The load frequency temperature coefficient is approximately

$$T_{f_L}^{(1)} \simeq T_{f_A}^{(1)} - G_0 T_k^{(1)}/(1 + C_0/C_L). \tag{72}$$

When considering the multimode case, Eq. (42) must be differentiated to find the temperature coefficient of resonance frequency. The result is

$$T_{f_R}^{(1)} = \sum_p^3 k_p^2 (A_p T_{f_{pA}}^{(1)} + 2B_p T_{k_p}^{(1)}) \Big/ \sum_q^3 k_q^2 A_q, \tag{73}$$

where

$$B_m = \tan X^{(m)}/X^{(m)}, \qquad A_m = B_m - \sec^2 X^{(m)}, \tag{74}$$

and where $X^{(m)}$ is a root of Eq. (42).

Alternatively, one may find $T_{f_R}^{(1)}$ numerically by computing the effective values of the elastic, piezoelectric, and dielectric constants at two temperatures using their temperature coefficients in the analog of Eq. (66) and then finding f_R at those temperatures in the manner described earlier. From these values and the temperature interval, $T_{f_R}^{(1)}$ follows approximately from Eq. (67).

B. Higher Order Temperature Coefficients

Expressions analogous to Eq. (68) for $n = 2$ and 3 were derived in Bechmann *et al.* (1962). These are more succinctly written as

$$\hat{T}_{f_{mA}}^{(n)} = \tfrac{1}{2}\hat{T}_{\bar{c}_m}^{(n)} - \tfrac{1}{2}\hat{T}_{\rho}^{(n)} - \hat{T}_{h}^{(n)}, \tag{75}$$

where

$$\hat{T}_Y^{(n)} = \begin{cases} T_Y^{(1)} & \text{for} \quad n = 1, \quad (76) \\ T_Y^{(2)} - \tfrac{1}{2}(T_Y^{(1)})^2 & \text{for} \quad n = 2, \quad (77) \\ T_Y^{(3)} - T_Y^{(2)} \cdot T_Y^{(1)} + \tfrac{1}{3}(T_Y^{(1)})^3 & \text{for} \quad n = 3, \quad (78) \end{cases}$$

and where Y is f_{mA}, $\bar{c}_m$, ρ, or h. Altitude charts for all three orders of the c mode in quartz are given in Hafner (1974). Only one orientation exists where $T_{f_{mA}}^{(n)} = 0$ simultaneously for $n = 2$ and 3; this occurs for the c mode, and defines the LC cut (Hammond and Benjaminson, 1965). At this orientation the frequency–temperature curve is ultralinear and the cuts are well suited for thermometric application.

The higher order temperature coefficients are listed in Table I for selected cuts along the upper zero locus of Fig. 5. The inflection temperature T_i, defined as the temperature for which the second derivative of Eq. (66) vanishes, is obtained from

$$T_i - T_0 = -T_{f_A}^{(2)}/3T_{f_A}^{(3)}; \tag{79}$$

it is seen from Table I to increase monotonically with angle φ.

TABLE I

HIGHER ORDER TEMPERATURE COEFFICIENTS OF QUARTZ CUTS (MODE c)

Cut	$T_{f_A}^{(2)}$ $(10^{-9}/\mathrm{K}^2)$	$T_{f_A}^{(3)}$ $(10^{-12}/\mathrm{K}^3)$
AT	−0.45	108.6
5° V	−1.77	104.7
10° V	−2.65	96.2
13.9° V	−4.37	85.6
FC	−5.55	82.2
IT	−10.1	68.4
SC	−12.3	58.2
25° V	−13.8	47.4
30° V	−12.4	31.4

Resonance frequency temperature coefficients for the single mode case are determined from the relations

$$T_X^{(2)} = T_{f_R}^{(2)} - T_{f_A}^{(2)} - T_X^{(1)} T_{f_A}^{(1)}, \tag{80}$$

$$T_X^{(3)} = T_{f_R}^{(3)} - T_{f_A}^{(3)} - T_X^{(2)} T_{f_A}^{(1)} - T_X^{(1)} T_{f_A}^{(2)} \tag{81}$$

with $T_X^{(1)}$ from Eq. (69), and $T_X^{(2)}$ and $T_X^{(3)}$ determined approximately from

$$\hat{T}_X^{(n)} \simeq -G_0 \hat{T}_k^{(n)} \tag{82}$$

for $n = 2, 3$. G_0 is given in Eq. (70); X and k are to be used for Y from Eqs. (77)–(78) in the expansions of Eq. (82). These equations and those corresponding for $n = 1$ relate $T_{f_R}^{(n)}$ to $T_{f_A}^{(n)}$ and the associated coupling factor coefficients of different orders. $T_{f_A}^{(n)}$ is found from Eqs. (75)–(78) in the manner described for $n = 1$ following Eq. (68); $T_k^{(n)}$ is obtained from differentiation of Eq. (33).

Inclusion of mass-loading effects leads to modification of Eqs. (75), (79)–(82), and is treated in detail by Ballato and Lukaszek (1975). Experimental data relating μ to changes in the higher order coefficients are presently very scarce; T. J. Lukaszek (private communication, 1976) has found a lowering of T_i with μ.

The higher order coefficients for the resonance frequencies in the multimode case may be determined after the manner of Eq. (73), but the results are quite lengthy, and recourse to the numerical method is not to be discouraged.

V. Properties of Doubly Rotated Cuts

This section is devoted to a compilation of data on doubly rotated cuts of four materials: quartz, aluminum phosphate, lithium tantalate, and lithium niobate. Quartz and aluminum phosphate are members of crystal class 32 and have remarkably similar properties. Quartz is treated in considerable detail. Its position as perennial favorite for many frequency control applications certainly merits an expanded account. However, the attention given this familiar material here is largely for a different reason: that a new, and hitherto unsuspected, but most desirable, property has recently come to light. It appears that certain doubly rotated orientations in quartz possess compensating combinations of nonlinear elastic properties that render the plate frequencies insensitive to mechanical and thermal shocks of various types. This fortunate circumstance can be expected to have widespread consequences for a number of high precision uses. Aluminum phosphate, like quartz, has a locus of zero temperature coefficient cuts—a property only recently determined (Chang and Barsch, 1976). It has an appealing potential for wider band resonator, filter, and SAW applications because it combines low loss with piezoelectric coupling factors about 2.5 times that of quartz.

Lithium tantalate and lithium niobate are refractory oxides of crystal class $3m$. The tantalate possesses a locus of zero temperature cuts and moderate to large coupling values, making it suitable for wideband filter and transducer uses. Lithium niobate lacks cuts having zero temperature coefficients, but its large coupling coefficients commend it for a variety of transducer applications.

Although they could not be included here, a number of additional types of materials deserve mention as potential candidates for future work. These include the tungsten bronze structures (crystal class $4mm$) comprising the solid solutions consisting of mixtures of strontium barium niobate with barium lithium niobate; for various ratios these are likely to have good temperature behavior with sizeable piezoelectric coupling. A group of sulfosalts appears to have promise as high coupling, low velocity, and zero temperature coefficient materials. Tl_3VS_4, in class $\bar{4}3m$ is representative (Weinert and Isaacs, 1975). These and other crystals from the 20 piezoelectric classes may be characterized as to their resonator properties by means of the theory outlined in previous sections. Compilations of measured crystal constants may be found in the Landolt–Börnstein tables (1966, 1969).

A. Quartz

In the first part of this section the linear properties of quartz pertinent to vibrators are discussed. The next portion considers those effects relying on nonlinear elastic behavior. Tradeoffs between doubly rotated quartz cuts and the traditional AT cut are given in the last part of the section. All calculations were carried out using as input data the values given by Bechmann (1951, 1958) and Bechmann *et al.* (1962), along with the temperature coefficients of permittivity from Landolt–Börnstein (1966). Alternative sets of constants such as those of Koga *et al.* (1958), McSkimin *et al.* (1965), or Zelenka and Lee (1971) produce results differing from those given here generally by only a small percentage, which is comparable to the error likely to be encountered in using the one-dimensional approximation. Additional results for this material are given in Section VI. An excellent resume of other aspects of quartz crystal resonators, their technology, and applications is given by Hafner (1974) with further references to the literature.

1. *Linear Properties of Quartz Resonators*

A global view of the most important resonator quantities of interest is given in Fig. 6 for orientations $(YXwl)\varphi/\theta$, with $\varphi = 0°(6°)30°$ and $-90° \leq \theta \leq 90°$. The three modes are labeled in accordance with Eq. (18). Frequency constants N_m are obtained from halving the velocities in Eq. (19); coupling factors are found from Eq. (33); and Eq. (68) yields the antiresonance temperature coefficients.

For cuts $(YXw)\varphi$, $|k_a|$ rises nearly linearly from zero at $\varphi = 0°$ to about 9% at $\varphi = 30°$; $|k_b|$ rises in a parabolic fashion from zero at $\varphi = 0°$ to a 7% maximum at $\varphi \simeq 11°$, and falls parabolically to zero at $\varphi = 30°$; $|k_c|$ falls nearly linearly from about 13.5% at $\varphi = 0°$ to zero at $\varphi = 30°$. All modes nearly coincide at 6.5% for $\varphi = 18°$.

Because X_1 is a digonal symmetry axis in quartz, the elastic constants with indices 15, 16, 25, 26, 35, 36, 45, and 46 vanish along with the piezoelectric constants with indices 15, 16, 21, 22, 23, 24, 31, 32, 33, and 34. For rotated-Y-cut plates $(YXl)\theta$, the digonal axis is, moreover, contained in the plane of the plate, and this leads to zero φ derivatives of many quantities. These cuts are therefore less complicated than doubly rotated cuts. For example, only the pure shear mode is piezoelectrically driven. In the sequel we shall give values for various quantities and their angle derivatives for a number of doubly rotated cuts so that the effects of angle variations may be calculated from the relation

$$q(\varphi, \theta) \simeq q(\varphi_0, \theta_0) + (\partial q/\partial \varphi)(\varphi - \varphi_0) + (\partial q/\partial \theta)(\theta - \theta_0), \qquad (83)$$

where q is any quantity. When $\varphi_0 = 0°$, one usually has to extend the expansion to second order. The most important instance of this is the expansion for the temperature coefficient of the AT-cut c mode:

$$T_{fA}^{(1)}(\varphi, \theta) - T_{fA}^{(1)}(0°, \theta_0) \simeq (-5.1 \times 10^{-6}/\mathrm{K}, °\theta)(\theta - \theta_0) + (-9.0 \times 10^{-9}/\mathrm{K}, (°\varphi)^2)(\varphi)^2. \qquad (84)$$

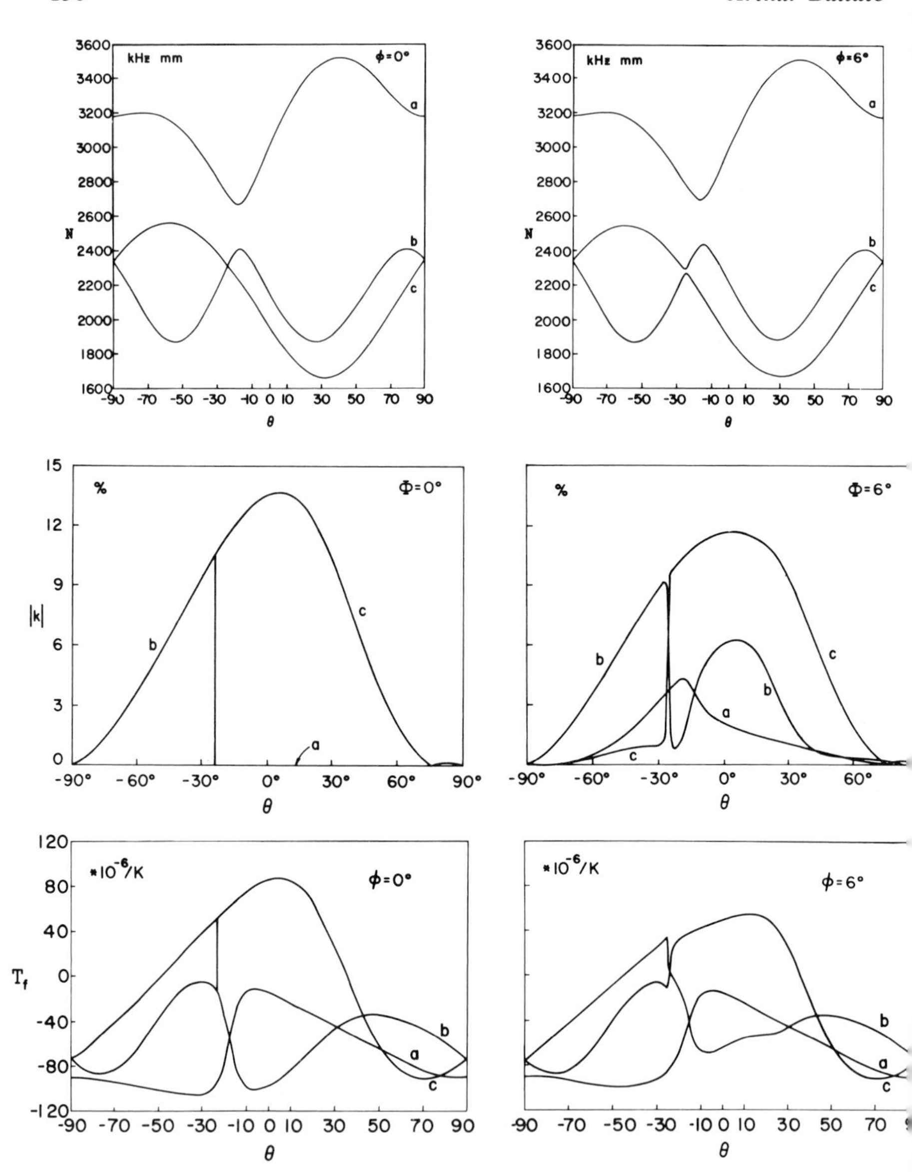

FIG. 6. Thickness mode properties of doubly rotated cuts of quartz. Frequency constants N are in kHz mm, coupling factors $|k|$ are in percent, and temperature coefficients $T_{f_\lambda}^{(1)}$ are in 10^{-6}/K.

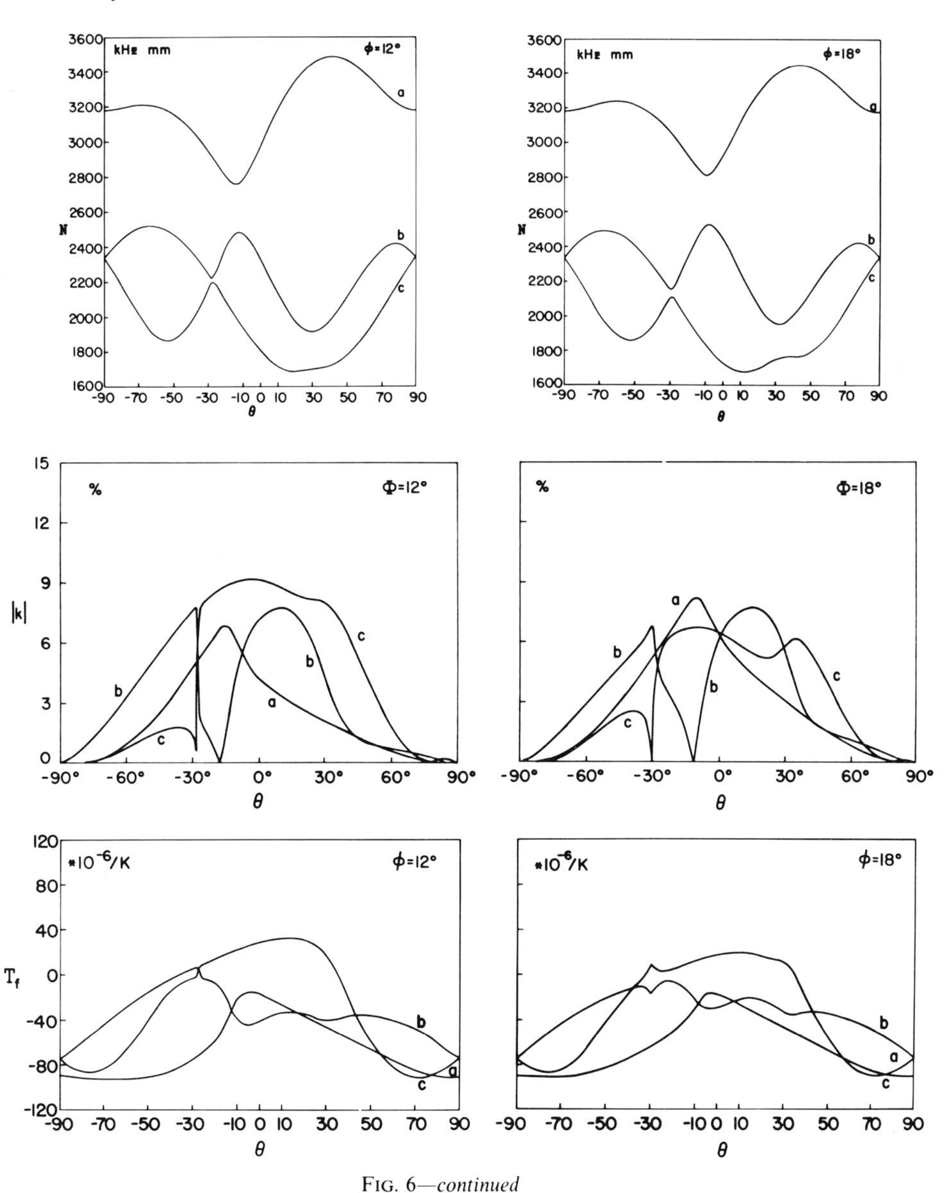

Fig. 6—*continued*

Equations (83)–(84) are extremely useful for determining the influence of manufacturing deviations on yield and the tradeoff between angular errors.

Figure 5 displays the loci of $T_{f_A}^{(1)} = 0$ for modes *b* and *c*. From Fig. 5 one

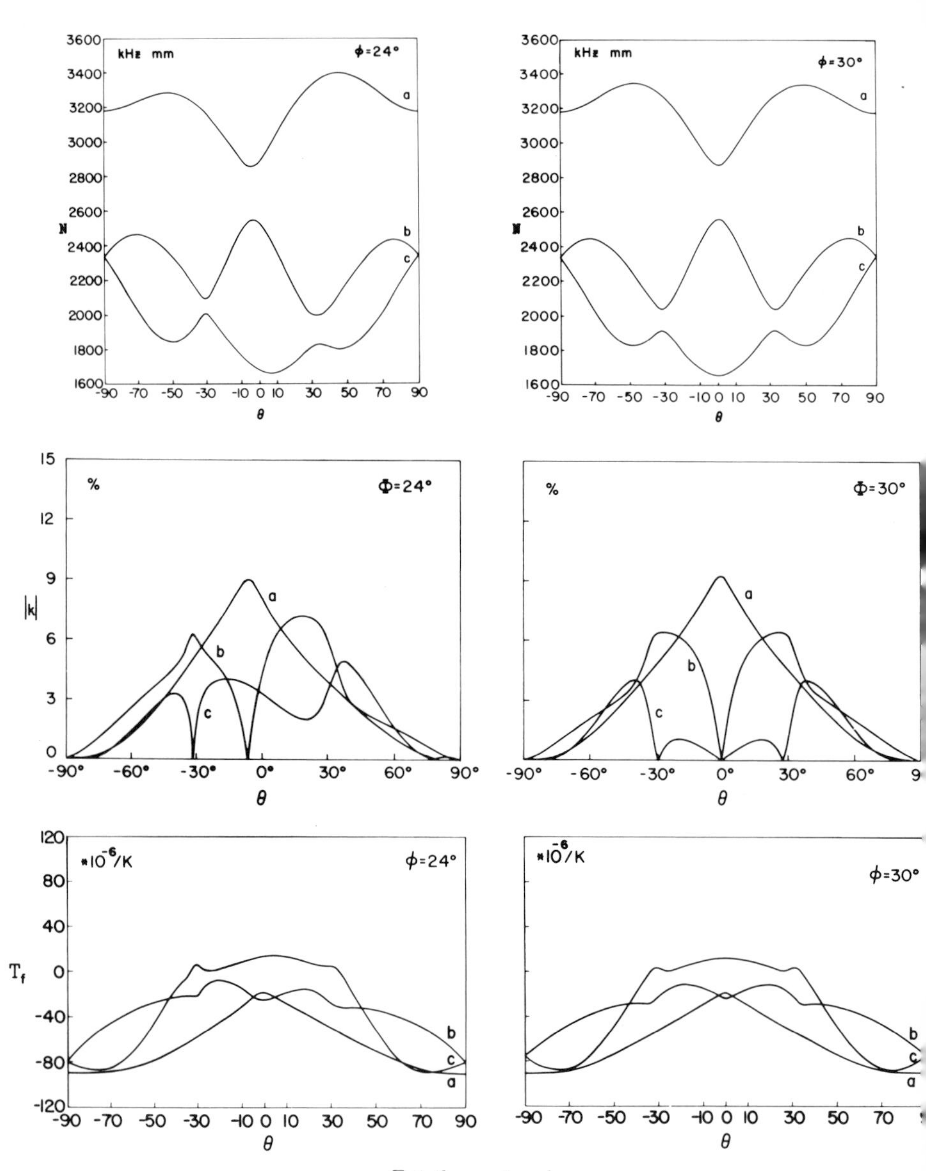

FIG. 6—*continued*

sees that the b- and c-mode curves touch at $\varphi = 0^\circ$, $\theta \simeq -24^\circ$, corresponding to the point in Fig. 6 where $N_b = N_c$. Another point of degeneracy exists in quartz. It was found by Epstein (1973) and occurs at the point in Fig. 5

TABLE II

FREQUENCY CONSTANTS AND ANGLE GRADIENTS. QUARTZ CUTS $(YXwl)\varphi/\theta > 0°$ [a]

Crystal cut	φ	θ	N_a	N_b	N_c	$\partial N_a/\partial\theta$	$\partial N_b/\partial\theta$	$\partial N_c/\partial\theta$	$\partial N_a/\partial\varphi$	$\partial N_b/\partial\varphi$	$\partial N_c/\partial\varphi$
	degrees		kHz mm			kHz mm/deg θ			kHz mm/deg φ		
AT	0	35.25	3504	1900	1661	3.60	6.91	2.09	0	0	0
5 V	5	34.94	3497	1903	1668	3.89	6.46	1.92	−2.45	1.90	3.02
10 V	10	34.64	3477	1915	1690	4.34	5.59	1.87	−4.89	3.70	5.95
13.9 V	13.90	34.40	3454	1931	1717	4.81	4.71	1.83	−6.76	4.96	8.12
FC	15	34.33	3446	1936	1726	4.96	4.45	1.80	−7.28	5.29	8.71
IT	19.10	34.08	3411	1959	1766	5.59	3.56	1.51	−9.17	6.39	10.8
SC	21.93	33.93	3382	1977	1797	6.08	3.15	1.03	−10.4	7.05	12.0
25 V	25	33.72	3346	2000	1836	6.67	3.05	0.06	−11.7	7.76	13.1
30 V	30	33.42	3281	2041	1905	7.76	4.36	−3.13	−13.7	9.14	13.9
LC	11.17	9.39	3165	2140	1727	19.0	−17.9	−7.15	−6.07	17.1	−10.0

[a] $T_{f\Lambda}^{(1)} = 0$ for the c mode of these cuts.

where the b-mode loop intersects the c-mode line at $\varphi \simeq 10.4°$, $\theta \simeq -26.6°$. Near this point the power flow angles exhibit their maximum departures from the propagation direction ζ.

The cuts selected for inclusion in the following tables are those having $T_{f\Lambda}^{(1)} = 0$, plus the LC cut. Of greatest importance is the c-mode locus for

TABLE III

FREQUENCY CONSTANTS AND ANGLE GRADIENTS. QUARTZ CUTS $(YXwl)\varphi/\theta < 0°$ [a]

Crystal cut	φ	θ	N_a	N_b	N_c	$\partial N_a/\partial\theta$	$\partial N_b/\partial\theta$	$\partial N_c/\partial\theta$	$\partial N_a/\partial\varphi$	$\partial N_b/\partial\varphi$	$\partial N_c/\partial\varphi$
	degrees		kHz mm			kHz mm/deg θ			kHz mm/deg φ		
BT	0	−49.20	3089	2536	1884	−10.2	−3.84	5.56	0	0	0
	5	−46.56	3075	2508	1901	−10.6	−5.76	8.44	5.69	−6.32	−0.89
	10	−38.63	3032	2391	1990	−11.8	−10.9	15.9	12.9	−14.0	−2.92
	5	−24.98	2769	2296	2277	−15.5	−6.66	15.2	12.1	−10.8	−3.85
	10	−32.23	2950	2313	2107	−13.8	−13.1	20.5	15.4	−16.3	−3.71
	12.5	−33.33	3004	2286	2075	−12.6	−13.6	19.5	16.1	−17.0	−4.63
RT	15	−34.50	3059	2260	2040	−11.3	−13.9	18.2	16.1	−17.0	−5.46
	17.5	−35.78	3112	2236	2003	−9.96	−14.1	16.5	15.8	−16.6	−6.13
	20	−36.79	3161	2209	1971	−8.72	−14.2	14.9	15.3	−16.0	−6.76

[a] $T_{f\Lambda}^{(1)} = 0$ for the b mode of first three entries; for the c mode of remainder.

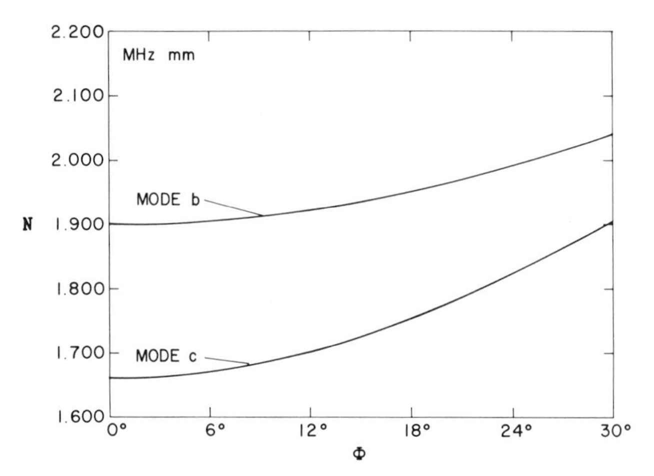

FIG. 7. Antiresonance frequency constants of the b and c modes in quartz along the AT-SC locus of Fig. 5. Mode separation declines from 14% at the AT cut to 10% at the SC cut.

$\theta > 0°$ which is adequately described by a straight line with equation

$$\theta° = +35.25° - (11/180)\varphi°. \tag{85}$$

Its importance derives from the nonlinear behavior discussed in Section V,A,2. Figure 7 shows, for the b and c modes, the antiresonance frequency constants along this locus. Tables II and III list N_m and their angle gradients for $\theta \gtrless 0°$, respectively. Tables IV and V similarly pertain to $|k_m|$, with Fig. 8 showing plots of the coupling factors on the upper zero temperature coefficient locus.

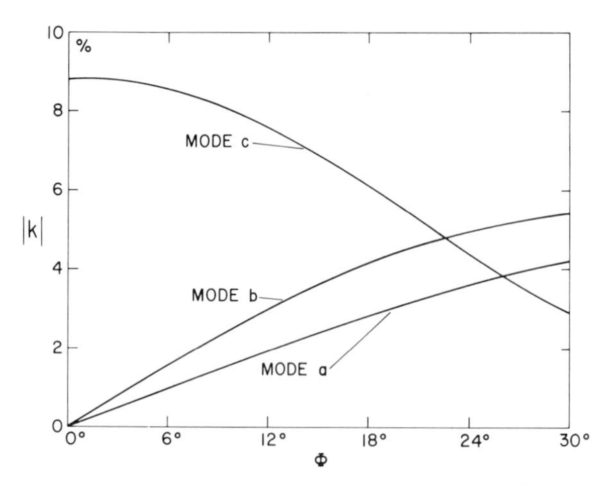

FIG. 8. Piezoelectric coupling factors along the AT–SC locus of Fig. 5. At the SC cut, the b and c mode excitations are approximately equal.

TABLE IV

COUPLING FACTORS AND THEIR ANGLE GRADIENTS. QUARTZ CUTS $(YXwl)\varphi/\theta > 0°$ [a]

Crystal cut	φ	θ	$\lvert k_a \rvert$	$\lvert k_b \rvert$	$\lvert k_c \rvert$	$\frac{\partial \lvert k_a \rvert}{\partial \theta}$	$\frac{\partial \lvert k_b \rvert}{\partial \theta}$	$\frac{\partial \lvert k_c \rvert}{\partial \theta}$	$\frac{\partial \lvert k_a \rvert}{\partial \varphi}$	$\frac{\partial \lvert k_b \rvert}{\partial \varphi}$	$\frac{\partial \lvert k_c \rvert}{\partial \varphi}$
	degrees		percent			10^{-3}/deg θ			10^{-3}/deg φ		
AT	0	35.25	0	0	8.80	0	0	−2.97	1.57	2.57	0
5 V	5	34.94	0.80	1.29	8.63	−0.24	−1.28	−2.69	1.59	2.50	−1.00
10 V	10	34.64	1.60	2.53	7.97	−0.47	−2.45	−1.88	1.54	2.21	−1.90
13.9 V	13.90	34.40	2.20	3.39	7.16	−0.65	−3.13	−0.93	1.47	1.87	−2.42
FC	15	34.33	2.37	3.61	6.89	−0.69	−3.27	−0.61	1.44	1.76	−2.53
IT	19.10	34.08	2.96	4.33	5.79	−0.87	−3.69	0.72	1.33	1.30	−2.78
SC	21.93	33.93	3.33	4.71	4.99	−0.98	−3.81	1.75	1.23	0.98	−2.78
25 V	25	33.72	3.71	5.05	4.11	−1.10	−3.81	3.01	1.09	0.70	−2.58
30 V	30	33.42	4.23	5.43	2.87	−1.28	−3.56	5.19	0.82	0.39	−1.73
LC	11.17	9.39	3.21	7.64	9.21	−0.67	0.03	−0.45	0.28	1.15	−4.63

[a] $T_{f_A}^{(1)} = 0$ for the c mode of these cuts.

TABLE V

COUPLING FACTORS AND THEIR ANGLE GRADIENTS. QUARTZ CUTS $(YXwl)\varphi/\theta < 0°$ [a]

Crystal cut	φ	θ	$\lvert k_a \rvert$	$\lvert k_b \rvert$	$\lvert k_c \rvert$	$\frac{\partial \lvert k_a \rvert}{\partial \theta}$	$\frac{\partial \lvert k_b \rvert}{\partial \theta}$	$\frac{\partial \lvert k_c \rvert}{\partial \theta}$	$\frac{\partial \lvert k_a \rvert}{\partial \varphi}$	$\frac{\partial \lvert k_b \rvert}{\partial \varphi}$	$\frac{\partial \lvert k_c \rvert}{\partial \varphi}$
	degrees		percent			10^{-3}/deg θ			10^{-3}/deg φ		
BT	0	−49.20	0	5.62	0	0	1.95	0	2.34	0	1.11
	5	−46.56	1.16	5.96	0.64	0.71	1.85	0.20	2.08	−0.68	1.27
	10	−38.63	3.02	6.73	1.49	1.36	1.48	0.15	1.78	−1.53	1.43
	5	−24.98	3.24	8.67	4.42	1.08	−31.2	63.1	5.11	−13.0	20.7
	10	−32.23	3.93	7.63	1.49	1.49	1.32	−0.16	2.07	−1.89	1.17
	12.5	−33.33	4.19	7.03	1.80	1.54	1.27	−0.40	1.35	−1.77	1.06
RT	15	−34.50	4.27	6.46	2.12	1.55	1.26	−0.54	0.82	−1.63	1.03
	17.5	−35.78	4.23	5.90	2.45	1.52	1.27	−0.53	0.42	−1.50	1.07
	20	−36.79	4.14	5.41	2.77	1.48	1.30	−0.49	0.12	−1.39	1.10

[a] $T_{f_A}^{(1)} = 0$ for the b mode of first three entries; for the c mode of remainder.

The temperature coefficients of antiresonance frequency and their angle gradients for the selected cuts are given in Tables VI and VII. Corresponding entries for the coupling temperature coefficients appear in Tables VIII and IX. Use of the $T_{k_m}^{(1)}$ values in Eq. (69) discloses that along the locus expressed by Eq. (85), $T_X^{(1)}$ for mode c remains nearly constant at -0.25×10^{-6}/K for $\varphi < 18°$ and then nearly linearly increases to -0.15×10^{-6}/K at $\varphi = 30°$.

Associated with the solution to Eq. (17) are the piezoelectrically

TABLE VI

ANTIRESONANCE FREQUENCY TEMPERATURE COEFFICIENTS AND THEIR ANGLE GRADIENTS. QUARTZ CUTS $(YXwl)\varphi/\theta > 0°$ [a]

Crystal cut	φ	θ	$Tf(a)$	$Tf(b)$	$Tf(c)$	$\partial Tf(a)/\partial\theta$	$\partial Tf(b)/\partial\theta$	$\partial Tf(c)/\partial\theta$	$\partial Tf(a)/\partial\varphi$	$\partial Tf(b)/\partial\varphi$	$\partial Tf(c)/\partial\varphi$
	degrees		10^{-6}/K			10^{-6}/K deg θ			10^{-6}/K deg φ		
AT	0	35.25	−48.9	−31.3	0	−0.86	1.54	−5.08	0	0	0
5° V	5	34.94	−48.8	−31.3	0	−0.87	1.52	−4.99	−0.09	0.16	−0.10
10° V	10	34.64	−49.2	−30.6	0	−0.87	1.40	−4.71	−0.18	0.31	−0.18
13.9° V	13.90	34.40	−49.8	−29.5	0	−0.87	1.26	−4.41	−0.25	0.41	−0.21
FC	15	34.33	−50.1	−29.1	0	−0.87	1.21	−4.32	−0.28	0.43	−0.21
IT	19.10	34.08	−51.2	−27.5	0	−0.87	1.07	−3.98	−0.37	0.50	−0.20
SC	21.93	33.93	−52.1	−26.2	0	−0.88	1.02	−3.78	−0.44	0.54	−0.18
25° V	25	33.72	−53.5	−24.7	0	−0.88	1.05	−3.65	−0.53	0.60	−0.15
30° V	30	33.42	−56.3	−21.5	0	−0.88	1.45	−3.80	−0.71	0.85	−0.22
LC	11.17	9.39	−26.1	−39.7	39.8	−0.93	0.55	0.34	−0.19	2.58	−2.28

[a] $Tf(m)$ stands for $T_{f_A}^{(1)}$ (mode m).

TABLE VII

ANTIRESONANCE FREQUENCY TEMPERATURE COEFFICIENTS AND THEIR ANGLE GRADIENTS. QUART CUTS $(YXwl)\varphi/\theta < 0°$ [a]

Crystal cut	φ	θ	$Tf(a)$	$Tf(b)$	$Tf(c)$	$\partial Tf(a)/\partial\theta$	$\partial Tf(b)/\partial\theta$	$\partial Tf(c)/\partial\theta$	$\partial Tf(a)/\partial\varphi$	$\partial Tf(b)/\partial\varphi$	$\partial Tf(c)/\partial\varphi$
	degrees		10^{-6}/K			10^{-6}/K deg θ			10^{-6}/K deg φ		
BT	0	−49.20	−95.6	0	−30.9	−0.74	2.06	2.84	0	0	0
	5	−46.56	−94.5	0	−24.0	−0.46	1.82	2.61	1.09	−1.05	−0.10
	10	−38.63	−86.6	0	−7.71	0.35	1.18	1.57	2.06	−2.02	−0.05
	5	−24.98	−91.4	39.7	0	1.49	−8.31	8.37	3.19	−4.81	1.77
	10	−32.23	−83.4	11.7	0	0.69	1.06	0.55	2.48	−2.65	0.28
	12.5	−33.33	−78.5	4.52	0	0.74	0.78	0.96	2.07	−2.28	0.32
RT	15	−34.50	−74.6	−1.49	0	0.76	0.56	1.37	1.71	−1.92	0.32
	17.5	−35.78	−71.7	−6.47	0	0.76	0.41	1.75	1.42	−1.58	0.25
	20	−36.79	−69.2	−10.4	0	0.76	0.29	2.09	1.19	−1.31	0.19

[a] $Tf(m)$ stands for $T_{f_A}^{(1)}$ (mode m).

TABLE VIII

TEMPERATURE COEFFICIENTS OF COUPLING AND THEIR ANGLE GRADIENTS. QUARTZ CUTS $(YXwl)\varphi/\theta > 0°$ [a]

Crystal cut	φ	θ	Tk_a	Tk_b	Tk_c	$\partial Tk_a/\partial\theta$	$\partial Tk_b/\partial\theta$	$\partial Tk_c/\partial\theta$	$\partial Tk_a/\partial\varphi$	$\partial Tk_b/\partial\varphi$	$\partial Tk_c/\partial\varphi$
	degrees		$10^{-6}/K$			$10^{-6}/K$ deg θ			$10^{-6}/K$ deg φ		
AT	0	35.25	—	—	88.2	—	—	16.0	0	0	0
5° V	5	34.94	−192	−183	89.1	−1.81	−27.5	15.9	0	−1.17	2.45
10° V	10	34.64	−192	−184	104	−1.81	−27.9	14.0	−0.04	−2.33	5.34
13.9° V	13.90	34.40	−191	−188	127	−1.83	−28.8	10.9	−0.05	−3.34	8.22
FC	15	34.33	−191	−190	136	−1.84	−29.2	9.58	−0.05	−3.65	9.19
IT	19.10	34.08	−191	−200	181	−1.87	−31.5	1.39	−0.06	−5.04	13.6
SC	21.93	33.93	−191	−210	224	−1.90	−33.7	−9.60	−0.05	−6.25	17.5
25° V	25	33.72	−191	−225	290	−1.94	−36.9	−32.6	−0.04	−8.01	22.1
30° V	30	33.42	−190	−262	428	−2.01	−42.9	−12.0	−0.01	−11.8	21.1
LC	11.17	9.39	−157	−93.9	−138	−0.87	3.14	7.85	0.17	−2.81	5.89

[a] $T_{f_A}^{(1)} = 0$ for the c mode of these cuts.

stiffened eigenvectors $\bar{\gamma}_m$ which determine the directions of particle displacement and enter the expressions for power flow, effective coupling constant, etc. For rotated Y cuts, the driven mode is uncoupled and has motion strictly along the digonal axis X_1. In general, the doubly rotated cut has particle motion out of the plane of the plate with consequent increased coupling to mounting supports and to the compressional mode in the ambient fluid. Both of these mechanisms will lead to increased loss. In order to show the displacements belonging to each mode in a simple and symmetrical fashion, we adopt the following conventions defining the displacement angles $\varphi_d^{(m)}$ and $\theta_d^{(m)}$ with respect to the plate (X_i'') axes:

For mode a, which is predominantly thickness extensional, the displacement direction is that corresponding to X_2'' when it undergoes the rotations

$$(X_2'' X_1'' wl)\varphi_d^{(a)}/\theta_d^{(a)}. \tag{86}$$

For mode b, the fast quasi-thickness-shear mode, the displacement direction is that corresponding to X_3'' when it undergoes the rotations

$$(X_3'' X_2'' wl)\varphi_d^{(b)}/\theta_d^{(b)}. \tag{87}$$

For mode c, the slow quasi-thickness-shear mode, the displacement direction is that corresponding to X_1'' when it undergoes the rotations

$$(X_1'' X_3'' wl)\varphi_d^{(c)}/\theta_d^{(c)}. \tag{88}$$

TABLE IX

TEMPERATURE COEFFICIENTS OF COUPLING AND THEIR ANGLE GRADIENTS. QUARTZ CUTS $(YXwl)\varphi/\theta < 0°$ [a]

Crystal cut	φ	θ	Tk_a	Tk_b	Tk_c	$\frac{\partial Tk_a}{\partial\theta}$	$\frac{\partial Tk_b}{\partial\theta}$	$\frac{\partial Tk_c}{\partial\theta}$	$\frac{\partial Tk_a}{\partial\varphi}$	$\frac{\partial Tk_b}{\partial\varphi}$	$\frac{\partial Tk_c}{\partial\varphi}$
	degrees		$10^{-6}/K$			$10^{-6}/K$ deg θ			$10^{-6}/K$ deg φ		
BT	0	−49.20	—	−444	—	—	5.57	—	0	0	0
	5	−46.56	−218	−433	189	3.40	5.26	−13.8	−0.41	−1.26	0.97
	10	−38.63	−197	−402	115	2.81	4.91	−7.25	−0.84	−2.33	3.21
	5	−24.98	−152	−354	−249	2.35	−7.45	−9.00	−2.42	−7.14	−21.2
	10	−32.23	−180	−372	84.0	2.58	4.63	−2.26	−1.27	−1.23	11.2
	12.5	−33.33	−185	−381	112	2.48	5.50	4.14	−0.84	−1.71	12.3
RT	15	−34.50	−190	−393	136	2.42	6.69	7.75	−0.50	−1.95	11.8
	17.5	−35.78	−194	−407	153	2.39	8.07	7.96	−0.26	−2.09	10.1
	20	−36.79	−196	−421	168	2.36	9.92	7.79	−0.09	−1.86	8.61

[a] $T^{(1)}_{f_A} = 0$ for the b mode of first three entries; for the c mode of remainder.

Aside from symmetry, this choice of definitions for $\varphi_d^{(m)}$ and $\theta_d^{(m)}$ has the advantage of showing deviations from pure mode displacements for which both angles vanish. Figure 9 displays $|\varphi_d^{(m)}|$ as solid lines for all modes and $|\theta_d^{(m)}|$ as dashed lines along the locus of Eq. (85). The AT cut ($\varphi = 0°$) c mode is pure, with the a and b modes coupled. As φ increases, the out-of-plane displacements for the desired c mode increase appreciably. Tables X and XI provide displacement angles for the selected cuts.

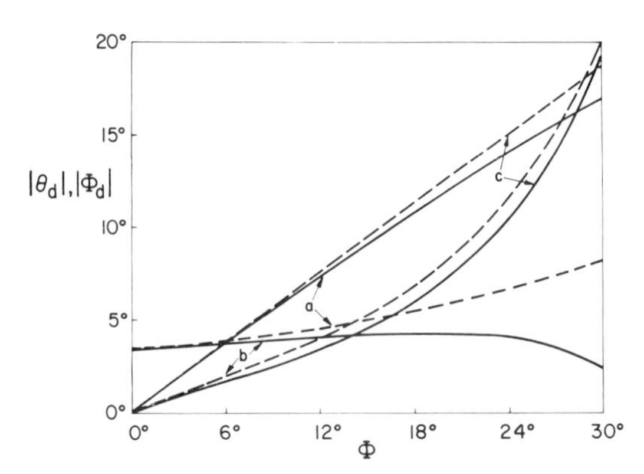

FIG. 9. Angles of particle displacement for quartz plates along the AT–SC locus of Fig. 5. Solid lines are for $|\varphi_d^{(m)}|$, and dashed for $|\theta_d^{(m)}|$.

TABLE X

PARTICLE DISPLACEMENT ANGLES AND MOTIONAL TIME CONSTANTS. QUARTZ CUTS $(YXwl)\varphi/\theta > 0^\circ$ [a]

Crystal cut	φ	θ	$\lvert\varphi_d^{(a)}\rvert$	$\lvert\varphi_d^{(b)}\rvert$	$\lvert\varphi_d^{(c)}\rvert$	$\lvert\theta_d^{(a)}\rvert$	$\lvert\theta_d^{(b)}\rvert$	$\lvert\theta_d^{(c)}\rvert$	$\tau_1^{(a)}$	$\tau_1^{(b)}$	$\tau_1^{(c)}$
	degrees		degrees			degrees			femtoseconds		
AT	0	35.25	0	3.56	0	3.56	0	0	10.2	6.71	11.8
5° V	5	34.94	3.04	3.78	1.46	3.87	1.66	3.14	10.3	6.73	11.8
10° V	10	34.64	6.05	4.04	2.84	4.36	3.27	6.26	10.3	6.69	11.8
13.9° V	13.90	34.40	8.36	4.23	4.12	4.88	4.76	8.69	10.4	6.62	11.8
FC	15	34.33	9.00	4.28	4.55	5.05	5.24	9.37	10.4	6.59	11.8
IT	19.10	34.08	11.3	4.37	6.60	5.76	7.49	11.9	10.5	6.46	11.8
SC	21.93	33.93	12.9	4.28	8.57	6.31	9.57	13.7	10.5	6.35	11.7
25° V	25	33.72	14.5	3.95	11.5	7.01	12.6	15.6	10.6	6.21	11.7
30° V	30	33.42	16.9	2.38	19.3	8.29	20.0	18.6	10.8	5.97	11.6
LC	11.17	9.39	6.33	23.6	27.9	20.4	27.5	4.40	11.6	9.70	10.8

[a] $T_{f_A}^{(1)} = 0$ for the c mode of these cuts.

TABLE XI

PARTICLE DISPLACEMENT ANGLES AND MOTIONAL TIME CONSTANTS. QUARTZ CUTS $(YXwl)\varphi/\theta < 0^\circ$ [a]

Crystal cut	φ	θ	$\lvert\varphi_d^{(a)}\rvert$	$\lvert\varphi_d^{(b)}\rvert$	$\lvert\varphi_d^{(c)}\rvert$	$\lvert\theta_d^{(a)}\rvert$	$\lvert\theta_d^{(b)}\rvert$	$\lvert\theta_d^{(c)}\rvert$	$\tau_1^{(a)}$	$\tau_1^{(b)}$	$\tau_1^{(c)}$
	degrees		degrees			degrees			femtoseconds		
BT	0	−49.20	0	—	90.0	11.8	90.0	11.8	11.8	4.89	8.68
	5	−46.56	7.51	21.7	24.0	12.5	80.6	13.3	11.9	5.06	8.81
	10	−38.63	15.9	24.7	17.2	14.0	70.1	17.6	12.1	5.84	9.06
	5	−24.98	12.9	6.78	13.0	17.8	69.2	13.9	15.2	7.27	6.36
	10	−32.23	18.1	25.5	16.4	16.2	68.2	20.5	12.7	6.37	8.83
	12.5	−33.33	19.3	29.8	12.2	14.6	64.6	20.3	12.3	6.46	9.13
RT	15	−34.50	19.8	25.5	8.02	13.0	61.2	19.9	11.8	6.44	9.41
	17.5	−35.78	19.7	22.0	3.98	11.4	58.0	19.3	11.5	6.42	9.68
	20	−36.79	19.3	18.7	0.17	9.97	54.7	18.7	11.2	6.34	9.92

[a] $T_{f_A}^{(1)} = 0$ for the b mode of first three entries; for the c mode of remainder.

Also given in Tables X and XI are the motional time constants

$$\tau_1^{(m)} = \eta_m/\bar{c}_m, \tag{89}$$

where η_m and $\bar{c}_m$ are obtained from Eqs. (21) and (17), respectively. Viscosity values of Lamb and Richter (1966) for room temperature have been used for the calculations. The time constant is related to the resonator quality factor Q_m by the relation

$$Q_m = 1/\omega_0 \tau_1^{(m)}, \tag{90}$$

with ω_0 the nominal angular frequency of the mode; the relation of $\tau_1^{(m)}$ to the parameters of the equivalent electrical network will be discussed in Section VI,A. The loss represented by η_m in Eq. (89) pertains solely to the bulk wave attenuation due to the crystal viscosity, and Eq. (90) hence gives the intrinsic Q for each mode. To this will be added any other loss mechanisms such as that due to mounting of the plate and mode conversion losses due to surface features. The frequency–temperature curves along the c mode locus of Eq. (85) may be constructed using the data of Tables I and VI plus Eq. (83). This is done in Fig. 10 for the case of the SC cut. From the graph it

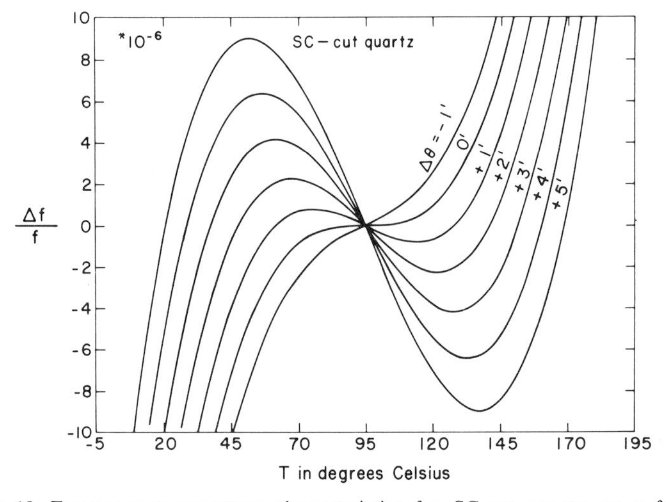

FIG. 10. Frequency–temperature characteristics for SC-cut quartz, as a function of variations in angle θ. The lower turning point is convenient for oven-controlled applications, and is flatter than the corresponding AT cut at its upper turning point.

is apparent that this cut should be excellent for oven-controlled applications due to its high inflection temperature of 95.4°C; comparison with the corresponding AT-cut curves discloses that the SC cut is flatter than the AT cut and less sensitive to changes in θ.

2. *Nonlinear Effects in Quartz Resonators*

A brief and qualitative account is given in this section of a number of effects whose explanations lie in the area of nonlinear elasticity (Thurston, 1964). Some of the effects are small in magnitude; but for high and even medium precision quartz resonators, all constitute nonnegligible error budget entries. Although they are interrelated, the effects are grouped, for convenience, in four categories:

(a) initial stress and acceleration,
(b) nonlinear resonance and intermodulation,
(c) nonlinear mode coupling,
(d) dynamic thermal and film stress.

Recent theoretical predictions by Holland (1974) and EerNisse (1975) of doubly rotated quartz cuts exhibiting greatly reduced category d effects, followed by experimental confirmation (Kusters, 1976), will very probably have far-reaching consequences in opening up for detailed exploration and use multiply rotated cuts of quartz and other materials. Because the nonlinear effects listed above are related, one may minimize one effect by a choice of φ and θ and expect that, by comparison to the standard AT cut, at least some of the other effects would simultaneously be reduced in size. Most applications will dictate a zero temperature coefficient requirement, so that the choice of angles follows the curves of Fig. 5. Experimental considerations indicate that the region of greatest interest is the *c*-mode locus for $\theta > 0^\circ$ containing the FC, IT, and SC cuts, and along this locus is where we concentrate our attention. It is entirely possible that a second SC cut exists on the $\theta < 0^\circ$ locus as well, along with an analogous SAW mode cut.

a. *Initial stress and acceleration effects.* In-plane diametric forces applied to the periphery of vibrating plates produce frequency changes that depend upon the azimuth angle ψ in the plane of the plate. If ψ is measured from the X_1'' axis, then it is found experimentally (Gerber and Miles, 1961) that for the AT cut the effect is zero at ψ values of 60° and 120°. For the IT cut at $\varphi = 19.1^\circ$, Ballato (1960) found the zeros to occur at $\psi = 85^\circ$ and 163° with a maximum value only one-third that of the AT cut. This points to a reduced coefficient at the SC cut as well. Calculations for rotated Y cuts (Lee and Haines, 1974; Lee *et al.*, 1975) show excellent agreement with experiment.

Plate resonators subjected to acceleration fields experience distributed body forces in place of the concentrated edge loadings discussed above, but the effect is similar. Frequency shifts are comparable in both cases (order 10^{-9}), and a factor ten improvement is needed in applications involving operation in shock and vibration environments. Work in this area has been carried out by Valdois *et al.* (1974) as well as by Lee and Wu (1976) with encouraging results.

b. *Nonlinear resonance and intermodulation effects.* The nonlinear resonance, or amplitude–frequency effect (Hammond *et al.*, 1963) pertains to the shape of the resonance amplitude versus frequency curve. Resonance curves for linear systems with small loss will be symmetric; with nonlinearities present, the curve is distorted by leaning toward higher or lower frequencies. The AT cut leans toward higher frequencies, its effective stiffness increasing with drive level, while the BT cut behaves in opposite fashion as a soft spring. Indirect evidence indicates that the curve becomes symmetric in the region between the FC and SC cuts, with a compensation of the nonlinearities taking place.[2] Figure 11 represents a typical surface for a hard

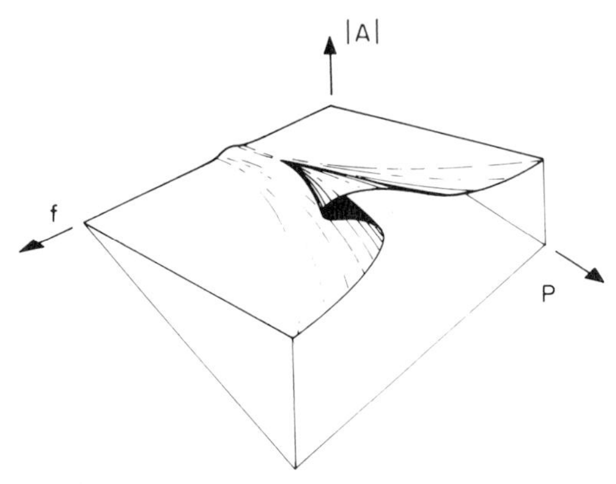

FIG. 11. Pleated surface depicting the frequency (f), input power (P), amplitude $|A|$ behavior of a nonlinear resonator with hard spring. AT cuts are representative; around the SC cut the pleat disappears until much higher power levels are reached.

spring, with amplitude $|A|$ plotted as function of resonator frequency f and input power level P. Beyond a certain power level (normally in the microwatt range) the amplitude–frequency curve becomes distinctly asymmetric; at still higher levels it ceases to become a single-valued function of frequency. Advances in characterizing this important effect have been made by Gagnepain (1972). Tiersten (1976) has made a very accurate analysis of the problem.

Closely related to the nonlinear resonance phenomenon is that of intermodulation, where energy supplied to a resonator at one frequency "spills over" and appears at another frequency. This cross talk occurs with both discrete resonator and monolithic filter (Spencer, 1972) structures. It is important in front-end filters and in adjacent channel filtering with transmitters proximate to receivers. The effect appears due to nonlinearities

[2] This has been experimentally confirmed in cuts of orientation $\varphi = 21.93°$, $\theta = +34.11°$. The resonators were planoconvex, of 1-diopter contour, and diameter 15 mm. They operated at 10 MHz, third harmonic. The lower turning points occurred at $72 \pm 4°C$ (J. A. Kusters, private communication, 1976).

associated with the bulk material as well as to surface features such as microcracks and microtwinning (Smythe, 1974; Tiersten, 1975b).

c. *Nonlinear mode coupling effects.* Anomalies in the frequency or admittance temperature characteristic of a resonator are called "activity dips" or "bandbreaks." They are generally conceded to be caused by various combinations of different modal frequencies coming into coincidence at particular temperatures because of differing temperature coefficients. The presence of activity dips is a persistent problem and necessitates a good deal of costly testing for medium and high precision resonator units. Doubly rotated cuts generally could be expected to have even more problems in this regard than AT cuts since they exhibit less symmetry and have, therefore, a more complicated mode spectrum when lateral boundaries are taken into account.

The modal interference that takes place may be linear or nonlinear. If the impressed voltage can drive the desired thickness mode and at the same time drive a harmonic of a flexural mode, e.g., then the vibrator admittance will reflect this fact as the linear superposition of the separate modal admittances. With temperature changes it is possible for the two resonance frequencies to cross and produce an anomaly. Linear activity dips have been described by Wood and Seed (1967) and by Fukuyo *et al.* (1967).

Nonlinear activity dips are less well understood and perhaps more important. Wood and Seed (1967) found the AT-cut fundamental thickness shear frequency to be affected by interfering modes at twice its frequency; Franx (1967) observed the same type of coupling due to a mode at three times the fundamental. Birch and Weston (1976) investigated both cases. Koga (1969) found the twenty-first harmonic of contour extension interfering nonlinearly with the thickness shear fundamental. Similar results were obtained by Fukuyo *et al.* (1967), who succeeded in measuring the temperature coefficients of a spectrum of interfering modes. In all cases the sensitivity of mode coupling to power levels is a characteristic of the nonlinearity. Hafner (1956) found that the anomalies encountered at the fifth and seventh harmonics were nonlinear in nature and depended on the electrode film as well as the quartz. This finding correlates with certain film stress results to be described in category d.

d. *Dynamic thermal and film stress effects.* Section IV discussed static frequency–temperature resonator characteristics; here transient thermal effects are reviewed. If an AT cut at thermal equilibrium experiences a small but abrupt temperature rise, the resonance frequency exhibits a sharp negative spike (order $10^{-7}/K$) followed by an asymptotic approach to its new equilibrium value (Warner, 1963). The BT cut behaves in opposite fashion. This dynamic thermal effect has been ascribed to thermal gradients, and explained on this basis by Holland (1974). He also disclosed (Holland, 1974, 1976) that the effect could be greatly reduced at the TS cut (thermal shock) of orientation $(YXwl)\varphi \simeq 22.8°/\theta \simeq 34.3°$. The prediction was quickly verified (Kusters, 1976; Kusters and Leach, 1977) with the experimental TTC cut (thermal transient compensated) having angles $\varphi = 21.93°$, $\theta = 33.93°$. For crystals of this orientation, the thermal transient effect for the c mode is reduced by a factor of more than one hundred, leading to promising applications for both fast warm-up oscillators and high precision frequency standards. It appears also possible that low cost temperature control could be provided for lower precision units by placing heating elements directly on the quartz surfaces.

The first actual disclosure of orientation angles for a doubly rotated cut having planar stress compensation was made by EerNisse (1975). This cut, designated the SC cut (stress compensated) is located at $(YXwl)\varphi \simeq 22.4°/\theta \simeq 34.3°$. The resonance frequencies of plates of this orientation are free from third-order elastic constant effects of mechanical stress bias, as would be caused by electrode films on the plate surfaces. It is no accident that the TS- and SC-cut orientations nearly coincide. Although the rationale in each case is different, the formulations are practically identical; they differ only in that electrode stresses are isotropic, whereas thermal stresses are not. Both problems are one-dimensional. Because "stress compensated" is a more general phrase, we choose to use it indifferently for the general family of cuts TS/TTC/SC in the neighborhood of which one or more planar stress effects are minimized. The acceleration problem mentioned in a is one-dimensional, and could also reasonably be expected to be improved for SC cuts. The influence of mechanical stress bias caused by electrode films may be seen from the experiments of Warner and may be related to Hafner's (1956) discovery of their role in causing activity dips. Warner (1963) investigated frequency transients caused by sudden temperature changes applied to resonators plated conventionally as well as to resonators plated for lateral field excitation. In the latter units the frequency spike was drastically reduced, probably because the most active central area of the resonator was devoid of plating, as is necessary for lateral field excitation. These results are discussed by Gerber and Sykes (1974). Another example of film stress bias is demonstrated by the experiments of Filler and Vig (1976), who used patches of electrobonded nickel applied diametrically at the periphery of circular AT-cut resonators to study the azimuthal dependence. Their results are similar to the azimuthal behavior found for the edge force effect, with frequency change replaced by apparent orientation angle shift.

The problem of aging, or long-term frequency drift, depends upon a number of causative factors, and electrode stress must be included among them. As electrode films relax slowly in time, the resonator frequency will similarly change, unless the cut is stress-compensated. Fortunately, for the SC cuts, the locus of zero stress sensitivity is nearly perpendicular to the locus of zero temperature coefficient, so these may be adjusted virtually independently (EerNisse, 1976). The static frequency–temperature behavior of these cuts has been presented in Fig. 10. It is adjusted by slight changes in θ; the stress coefficient is altered by changes in φ.

3. *Summary of Advantages and Disadvantages of Doubly Rotated Quartz Cuts*

We conclude the discussion of quartz resonators by summarizing the current status of pros and cons of doubly rotated resonators with orientations on the $\theta > 0°$ branch of the zero temperature coefficient locus. In all cases the comparison is with respect to an AT cut having the same frequency and electrode area. The doubly rotated cut in question usually lies in the neighborhood of the SC cut, although in some instances the data at present are too fragmentary to permit more than an indication of the orientation of optimal behavior.

Advantages

Lessened edge force sensitivity. The IT cut has about one-third the magnitude of the AT cut. This is of importance in shock and vibration environments.

Reduced acceleration sensitivity. One order of magnitude improvement is presently needed.

Improved amplitude–frequency behavior. This is of importance for high precision oscillators.

Lessened intermodulation effects. This is required for filters.

Fewer activity dips. Nonlinear elastic constants are compensated, reducing mode couplings.

Greatly improved transient frequency–temperature behavior. Two or three orders of magnitude improvement is found. This is important for fast warm-up oscillators.

Greatly improved planar stress behavior. Electrode film relaxation is a long-term aging factor that would be virtually eliminated. It is probable that current low aging crystals achieve their low aging behavior due to compensating aging processes. An SC cut fabricated using standard AT-cut cleaning procedures would probably show higher aging because the film component would be absent (J. R. Vig, private communication, 1976). Ultraclean procedures would nearly eliminate the contamination process and would lead the way to ultralow aging units.

Improved static frequency–temperature behavior. Inflection temperature is increased; frequency deviation is decreased. For high precision oven-controlled applications, operation at the inflection temperature or at the lower turning point is more stable than at the AT-cut upper turning point.

Thicker, less fragile plates; slightly improved quality factors.

Increased capacitance ratio. (See Section VI.) High precision applications call for reducing the influence of external circuitry to a minimum to maintain a good noise figure. Data on FC cuts indicate improvements in phase noise.

Disadvantages

More complex unwanted mode spectra. Additional contouring or beveling restraints are imposed; suitability for filter use is potentially impaired.

Proximity and strength of b mode. At the SC cut the undesired b mode is nearly of equal strength with the useful c mode, and about 10% above it in frequency. Additional circuitry is required to suppress the b mode in some applications; in others, proper contouring reduces its strength adequately.

Larger out-of-plane displacements. There are increased mounting losses with attendant Q degradation. Proper contouring can improve this situation.

Increased capacitance ratio and motional resistance. An increased ratio is an advantage in certain high precision applications, but is a drawback for temperature-compensated crystal oscillators (TCXOs) of the fast warm-up variety and for monolithic filters (Spencer, 1972).

Orienting and X raying doubly rotated cuts more difficult. The relatively strong $12\bar{3}1$ plane at $\varphi = 19.10°$, $\theta = +16.57°$ used for locating the IT cut can also be used for the SC cut.

Tighter φ, θ tolerances required. Most physical and electrical properties are more sensitive to angle misorientations when $\varphi > 0°$. For AT cuts, most quantities of interest have zero φ derivatives, so departures in φ are manifested only in second order.

The last two disadvantages are manufacturing considerations that will largely be obviated by new technology such as microprocessor control. Remaining entries are intrinsic and stem from the double rotation that lowers the plate symmetry from monoclinic to triclinic. Theoretical analyses are also rendered more difficult. The advantages enumerated arise largely from compensation of nonlinear effects. Additional considerations are given by Ballato and Iafrate (1976). At present there are insufficient data comparing AT and doubly rotated cuts with regard to ease of obtaining required surface finish, impurity migration rates, and X-ray/neutron resistance and recovery.

Additional work on doubly rotated orientations in quartz and other crystals is needed particularly in the areas of calculating the dispersion curves and developing approximate plate theories of various orders. These tasks have been carried out for piezoelectric crystals with monoclinic symmetry by Lee and Syngellakis (1975), and Syngellakis and Lee (1976). The study of static stress effects and acceleration sensitivity in triclinic plates is also of pressing interest, and measurement of the higher order piezoelectric and dielectric constants (Besson, 1974) is needed to fill in portions of the nonlinear treatments.

An additional nonlinear mechanism sufficiently large in quartz to cause discrepancies in high precision applications is the polarizing effect (Hruška and Kazda, 1968; Kusters, 1970; Hruška, 1971; Baumhauer and Tiersten, 1973; Tiersten, 1975a). In this effect (order 10^{-12} m/V) bias or stray electric fields produce changes in the effective elastic, piezoelectric, and dielectric material constants. The largest changes in quartz take place in the elastic constants, but the coefficient happens to vanish for the driven mode in rotated Y cuts. For doubly rotated cuts in general, the effect will be present. Ungrounded units used in probes, or in vibrators subject to oscillator circuitry biases, are susceptible but can be protected in a simple manner (J. A. Kusters, private communication, 1976).

B. ALUMINUM PHOSPHATE

Aluminum phosphate, $AlPO_4$, also known as berlinite, is a material in crystal class 32 very like quartz in structure and physical properties. It occurs naturally only as tiny crystals, but may be grown to large size by the hydrothermal method (Stanley, 1954). An early determination of the elastic and piezoelectric constants was reported by Mason (1950). More recent work by Chang and Barsch (1976) disclosed the presence of zero temperature coefficient thickness modes; zero coefficient SAW cuts have also been found (Carr and O'Connell, 1976, Jhunjhunwala *et al.*, 1976).

Using the recently determined material constants and temperature coefficients of Chang and Barsch (1976), Fig. 12 shows the calculated frequency constants $N_m = v_m/2$, the coupling factors $|k_m|$, and antiresonance temperature coefficients $T_{f_{mA}}^{(1)}$ for doubly rotated cuts $(YXwl)\varphi/\theta$, $\varphi = 0°(6°)30°$, and $|\theta| \leq 90°$. The similarity to Fig. 6 is readily apparent. Frequency constants for berlinite are somewhat lower; the coupling factors are roughly double that of quartz, and the temperature coefficients are very much the same. Degeneracy of the b and c modes takes place at $(YXl)\theta \simeq -24.5°$.

For plates $(YXw)\varphi$, as φ varies from 0° to 30°, $|k_a|$ rises almost linearly from 0 to 16%; $|k_b|$ arcs from 0 to 11% and back to 0%, while $|k_c|$ falls nearly linearly from 24 to 0%.

The loci of $T_{f_{mA}}^{(1)} = 0$ for $m = b, c$ are shown in Fig. 13. From Fig. 12 it is seen how shallow the $T_f^{(1)}$ surfaces are, particularly for $\varphi > 12°$. Small changes in the values of the measured constants could therefore affect the loci in Fig. 13 appreciably. For this reason, it is not appropriate to pursue more detailed calculations of angle gradients, etc., at this time. On the basis of current information, berlinite appears particularly attractive for mono-

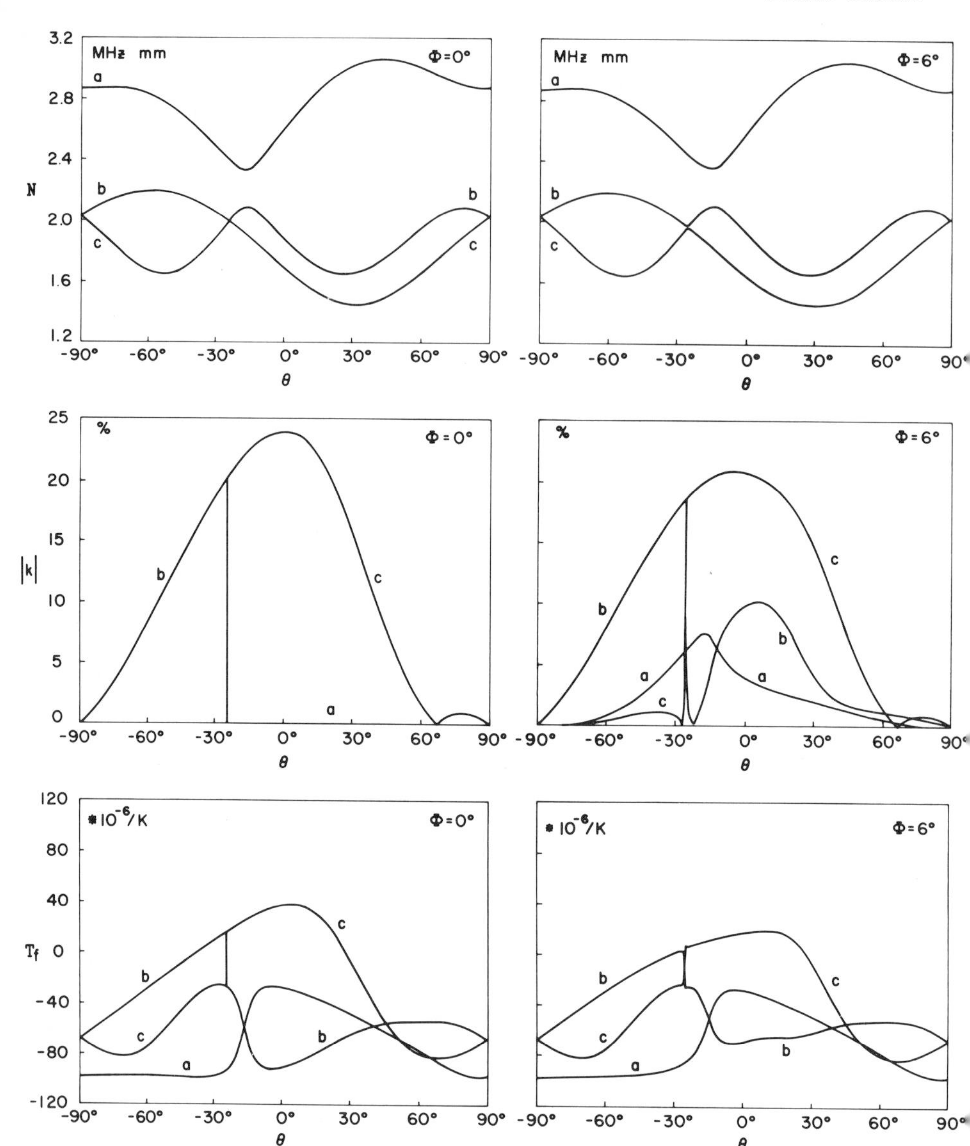

FIG. 12. Thickness mode properties of doubly rotated cuts of aluminum phosphate. Frequency constants N are in MHz mm, coupling factors are in percent, and temperature coefficients $T_{f_A}^{(1)}$ are in 10^{-6}/K.

lithic (Tiersten, 1969) and wider band conventional filters and for SAW applications. Additional measurements of the higher order temperature coefficients and the nonlinear elastic constants would be most desirable.

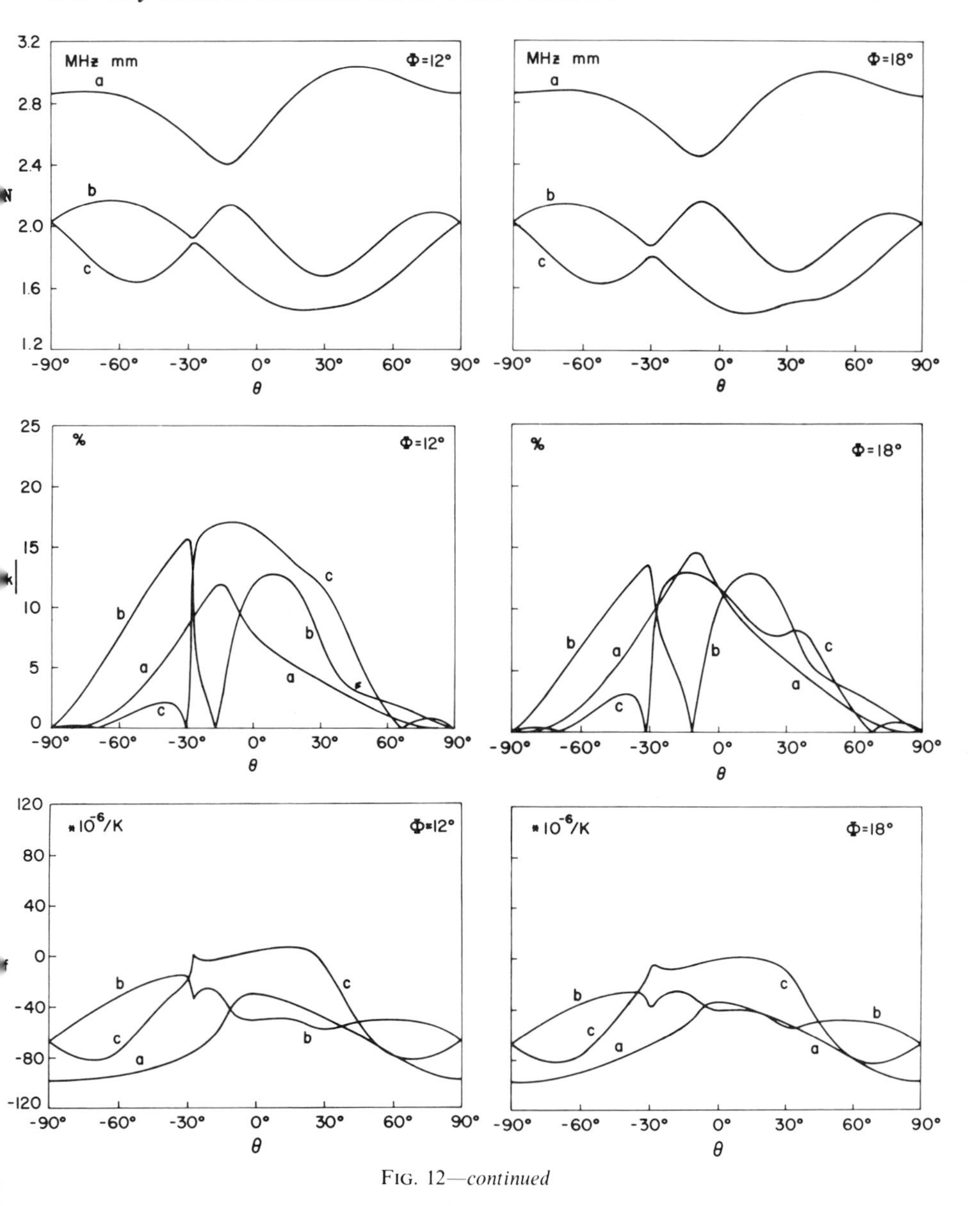

FIG. 12—*continued*

C. Lithium Tantalate

Lithium tantalate and lithium niobate (discussed in Section V,D) are refractory oxides in crystal class $3m$. A thorough investigation of their elastic, piezoelectric, and dielectric properties, including first and second

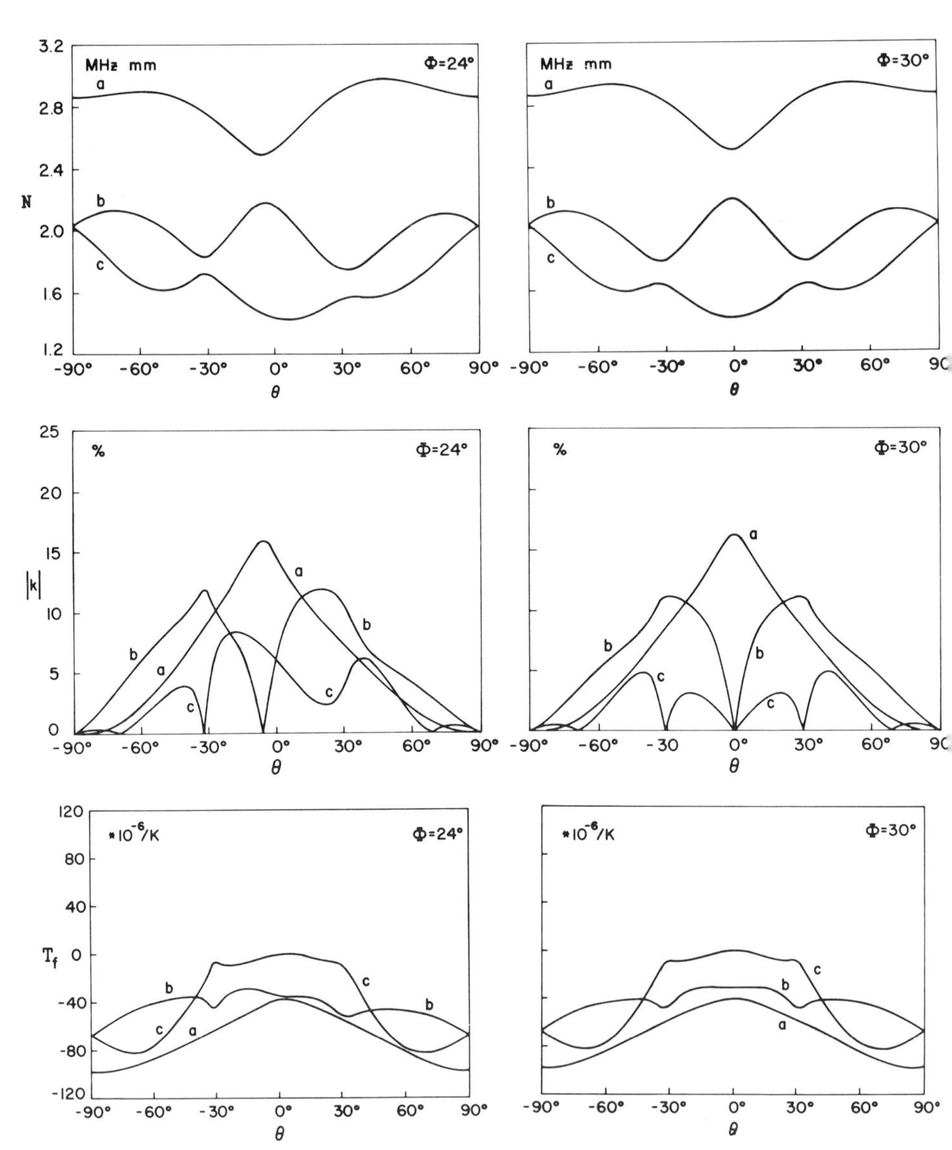

FIG. 12—*continued*

order temperature coefficients, was made by Smith and Welsh (1971). More recently. Détaint and Lançon (1976) investigated doubly rotated thickness modes of the tantalate and constructed altitude charts of coupling and temperature coefficients. In Fig. 14 are shown N_m, $|k_m|$ and $T^{(1)}_{fmA}$ for lithium

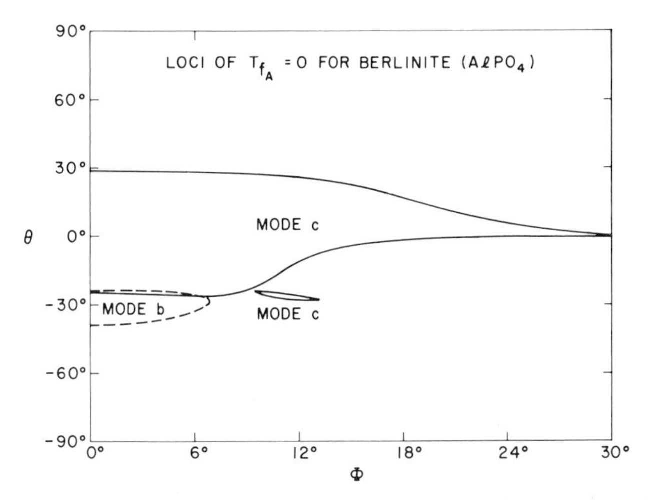

FIG. 13. Loci of zeros of first order temperature coefficient of antiresonance frequency in aluminum phosphate. Note the similarity to Fig. 5.

tantalate cuts $(YXwl)\varphi/\theta$, $\varphi = 0°(6°)30°$, and $|\theta| \leq 90°$. These have been computed using the data of Smith and Welsh (1971) with e_{33} taken from Warner *et al.* (1967) as suggested by Détaint and Lançon (1976). Points of accidental degeneracy of the b and c modes occur for cuts $(YXl)\theta \simeq -29.9°$, $+26.9°$, and $37.8°$.

Whereas quartz possesses a digonal axis of symmetry, $3m$ crystals have a mirror plane (Juretschke, 1974; Nye, 1957); taking this normal to the X_1 axis causes the elastic constants with indices 15, 16, 25, 26, 35, 36, 45, and 46 to vanish along with the piezoelectric constants with indices 11, 12, 13, 14, 25, 26, 35, and 36. Some of the differences between the 32 and $3m$ symmetries may be seen by a comparison of Fig. 14 with Fig. 6 or 12 for the case $\varphi = 30°$. For example, when $\theta = 0°$ for class $3m$ and when $|\theta| = 90°$ for class 32, $|k_a|$ is zero, while both $|k_b|$ and $|k_c|$ are finite; when $\theta = 0°$ for class 32 and when $|\theta| = 90°$ for class $3m$, $|k_a|$ is finite, while both $|k_b|$ and $|k_c|$ vanish. Also, for rotated Y cuts, the pure shear mode in class 32 is the only one piezoelectrically driven, while it is undriven in class $3m$, and the quasi-extensional and quasi-shear modes are driven instead. For lithium tantalate plates $(YXw)\varphi$, as φ varies from 0° to 30°, $|k_a|$ decreases from 21 to 0%; $|k_b|$ rises from 33 to 42.5%; and $|k_c|$ rises from 0% to a shallow maximum of 2.5% at $\varphi = 11°$ and levels off to 2%.

Because of the large piezoelectric coupling constants, the difference between the resonance and antiresonance fundamental frequencies is relatively large, as is the difference between their respective temperature coefficients. In Fig. 15 are given the loci of the zeros of $T_{f_A}^{(1)}$ and $T_{f_R}^{(1)}$ versus

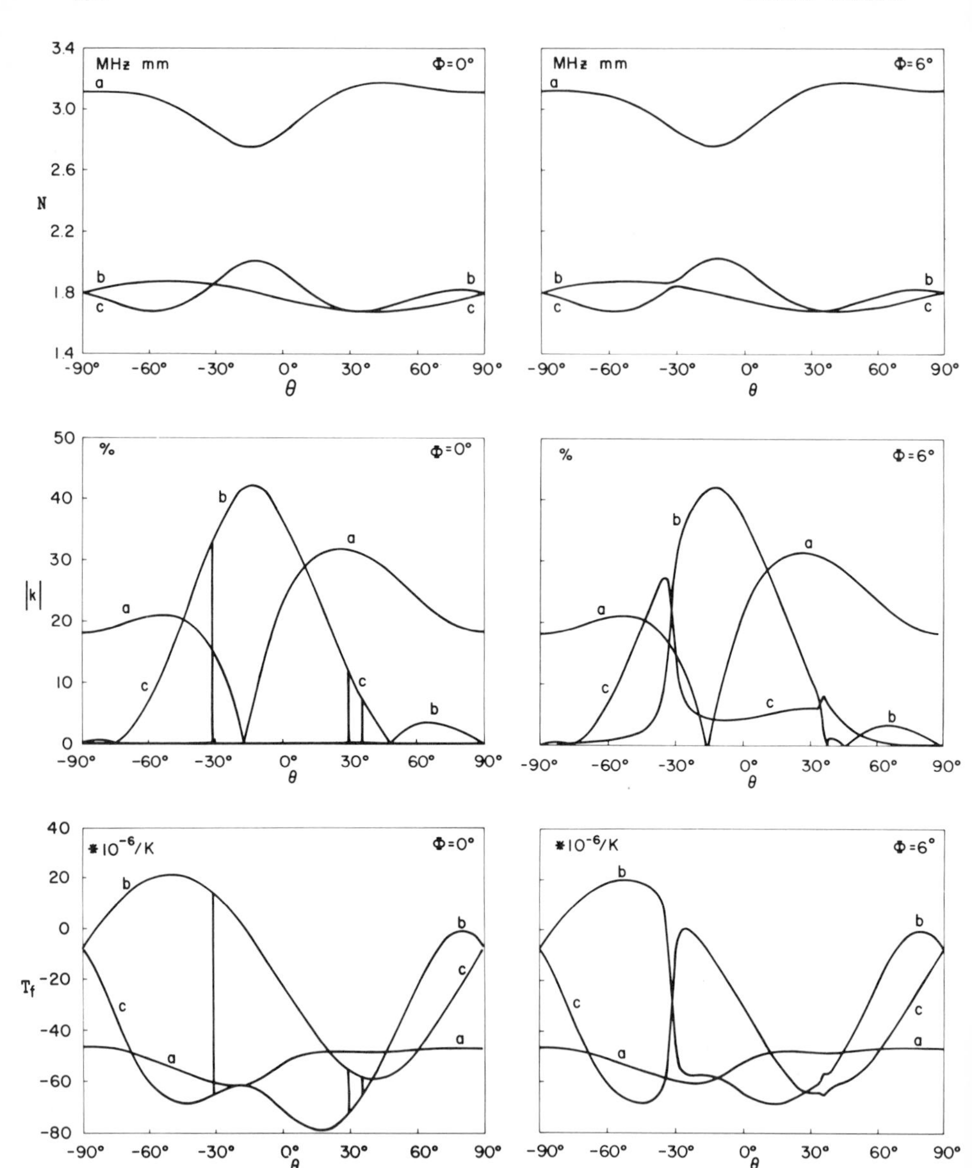

FIG. 14. Thickness mode properties of doubly rotated cuts of lithium tantalate. Frequency constants N are in MHz mm, coupling factors are in percent, and temperature coefficients $T_{f_A}^{(1)}$ are in 10^{-6}/K.

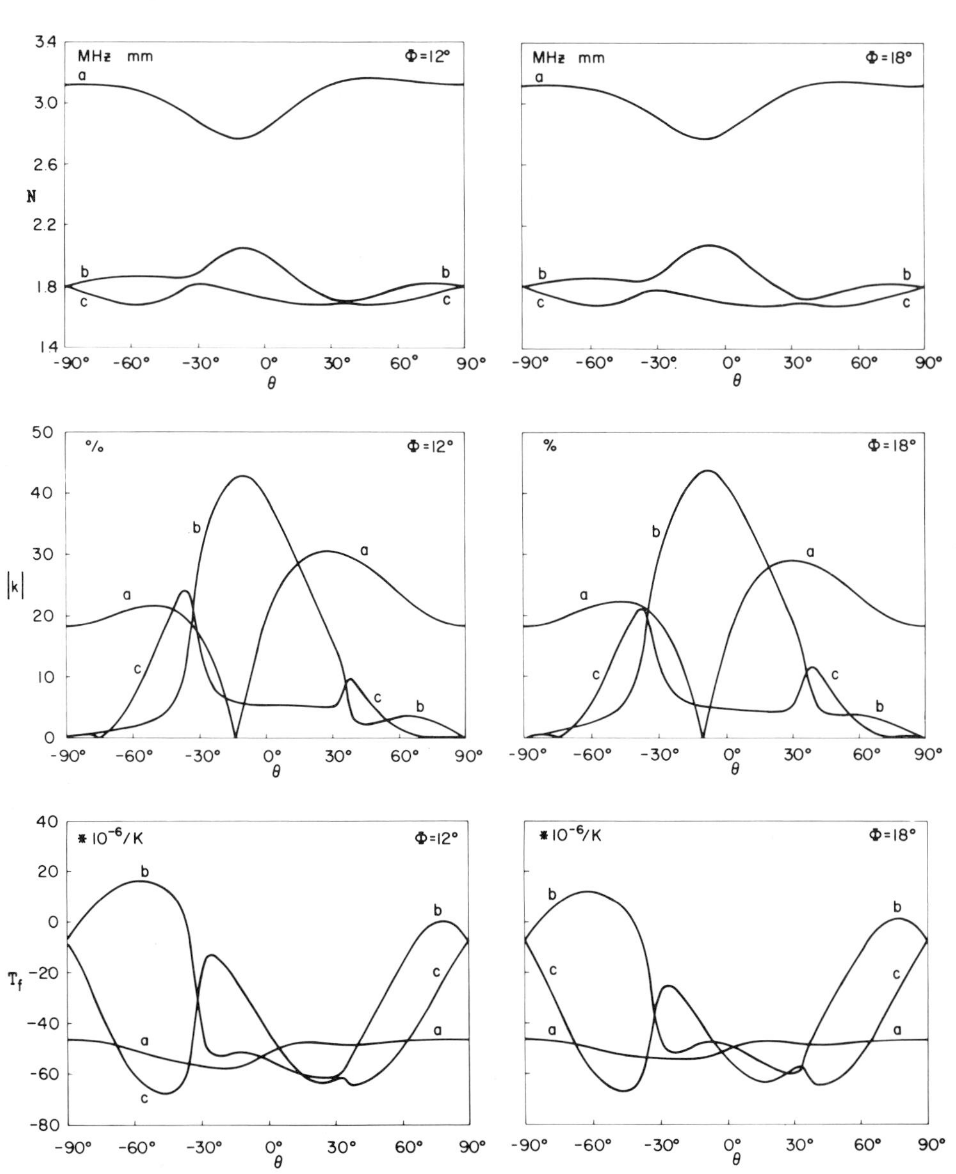

FIG. 14—*continued*

φ, θ for modes *b* and *c*, from Détaint and Lançon (1976). Mode *a* has no zero temperature coefficient. The curves are for $M = 1$; changing M, or using a series load capacitor C_L will shift the $T_{f_R}^{(1)}$ loci toward the invariant $T_{f_A}^{(1)}$ loci. The *b*-mode resonance frequency loop for $\varphi > 15^\circ$ has no corre-

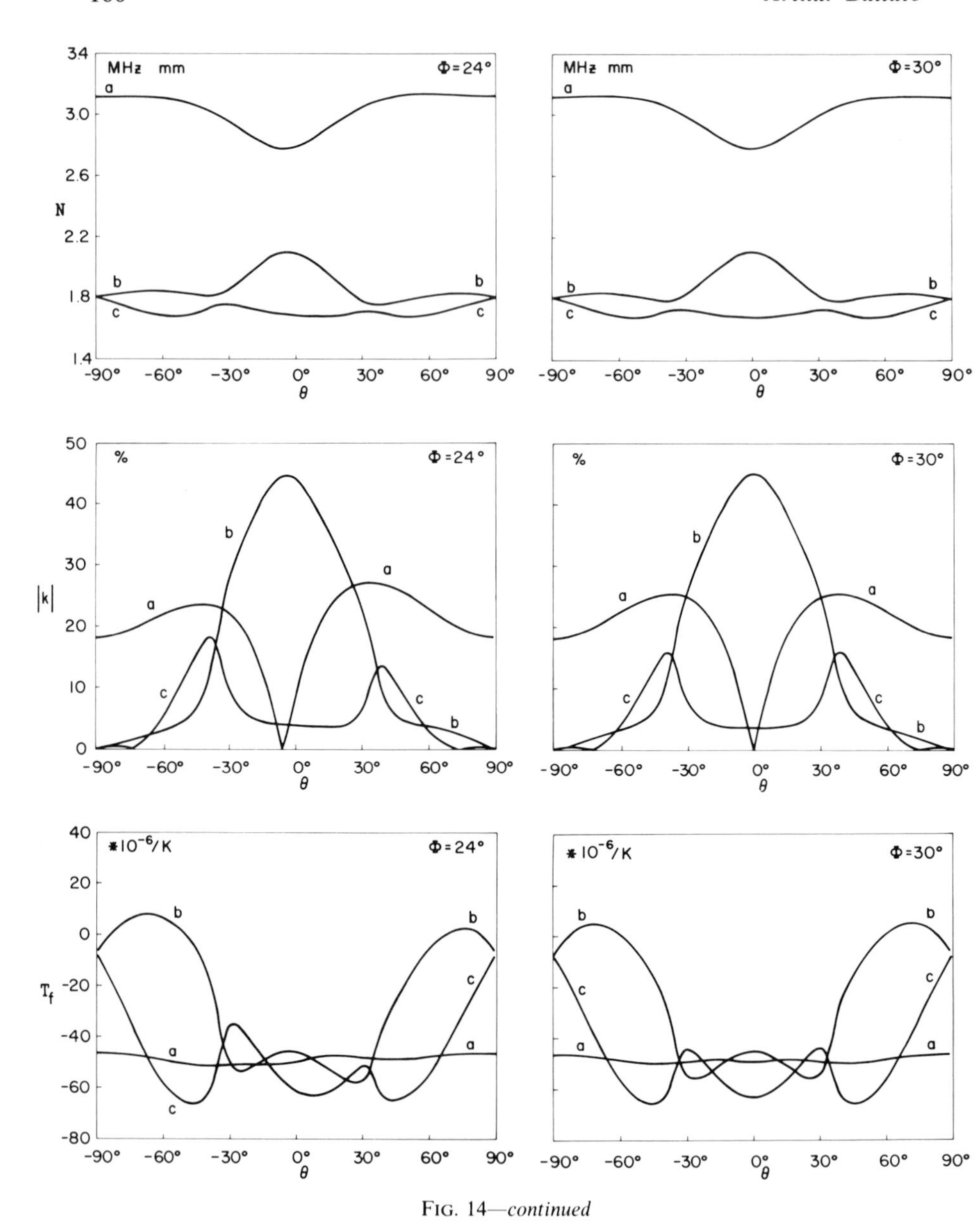

FIG. 14—*continued*

sponding $T_{f_A}^{(1)}$ curve; it occurs in the region of maximum $|k_b|$. An altitude chart of $|k_b|$ in percent, from Détaint and Lançon (1976), is shown in Fig. 16. When $\varphi = 0^\circ$, $|k_b|$ vanishes for $-90^\circ \leq \theta \leq -29.9^\circ$, and for $+26.9^\circ \leq \theta \leq +37.8^\circ$. Altitude charts of the fundamental resonance

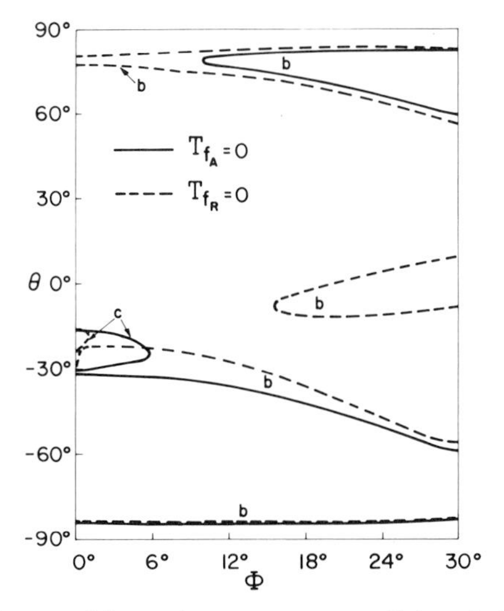

FIG. 15. Loci of zeros of first order temperature coefficients in lithium tantalate. Solid lines are for antiresonance frequencies; dashed are for resonance frequencies. All curves are for the fundamental harmonic. (After Détaint and Lançon, 1976).

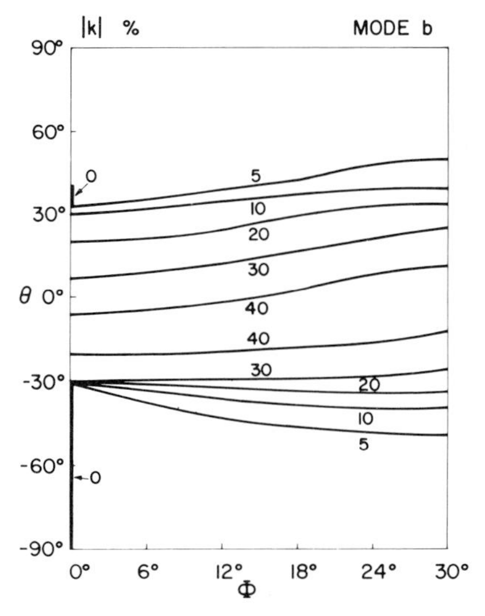

FIG. 16. Piezoelectric coupling coefficients for the b mode in lithium tantalate. (After Détaint and Lançon, 1976).

frequency constants, defined as in Eq. (41) with f_R substituted for $f_{mA}^{(M)}$, are given by Détaint and Lançon (1977).

The presence of a zero temperature coefficient locus in lithium tantalate for the fundamental and harmonics of the *b* and *c* modes, simultaneously with large piezoelectric coupling values, indicates excellent filter device potential. Three of the practical areas of work remaining for this material are: investigation of higher order temperature behavior $T_f^{(3)}$; measurement of the nonlinear elastic and piezoelectric constants with determination of possible "SC cuts"; computation of dispersion diagrams for doubly rotated orientations for both open- and short-circuited boundary conditions, and application of these spectra to the design of monolithic filters and other devices. Exact and approximate dispersion curves for these electrical conditions have already been determined for rotated Y cuts of $LiTaO_3$ (Lee and Syngellakis, 1975; Syngellakis and Lee, 1976).

D. Lithium Niobate

The data of Smith and Welsh (1971) have been used to compute the frequency constants N_m, piezoelectric coupling factors $|k_m|$, and first order temperature coefficients $T_{f_{mA}}^{(1)}$ for lithium niobate. Figure 17 presents the results for doubly rotated cuts $(YXwl)\varphi/\theta$, $\varphi = 0°(6°)30°$, $|\theta| \leq 90°$. Degeneracy of the *b* and *c* modes takes place for cuts $(YXl)\theta$, $\theta \simeq +24.1°$, $+48.9°$, and $+59.0°$. For plates $(YXw)\varphi$, as φ varies from 0° to 30°, $|k_a|$ decreases slowly from 31 to 0%; $|k_b|$ rises from 57 to 69%, and $|k_c|$ rises slowly from 0 to 7%. There are no orientations having $T_{f_{mA}}^{(1)}$ values greater than -40×10^{-6}/K; the main advantage of this material for piezoelectric applications is its suitability for transducer use owing to large $|k|$ values. Wideband filter applications are also possible.

VI. Electrical Characteristics of Plate Vibrators

The use of mechanical vibrators as undamped resonant elements for frequency control and as acoustic transducers has led to their representations as equivalent electrical circuits (Mason, 1948). This section gives expressions for the lumped equivalent circuit parameters of thickness mode plate vibrators with traction-free boundaries, obtained from the exact relation for input admittance. In terms of these parameters, values are given for doubly rotated quartz plates. Also provided are data on mode spectra for a variety of quartz cuts, showing the influences of multimode couplings.

A. Lumped Equivalent Circuit Parameters

Crystal vibrators are usually represented in the vicinity of a single resonance by an electrical circuit consisting of a shunt capacitance C_0 in

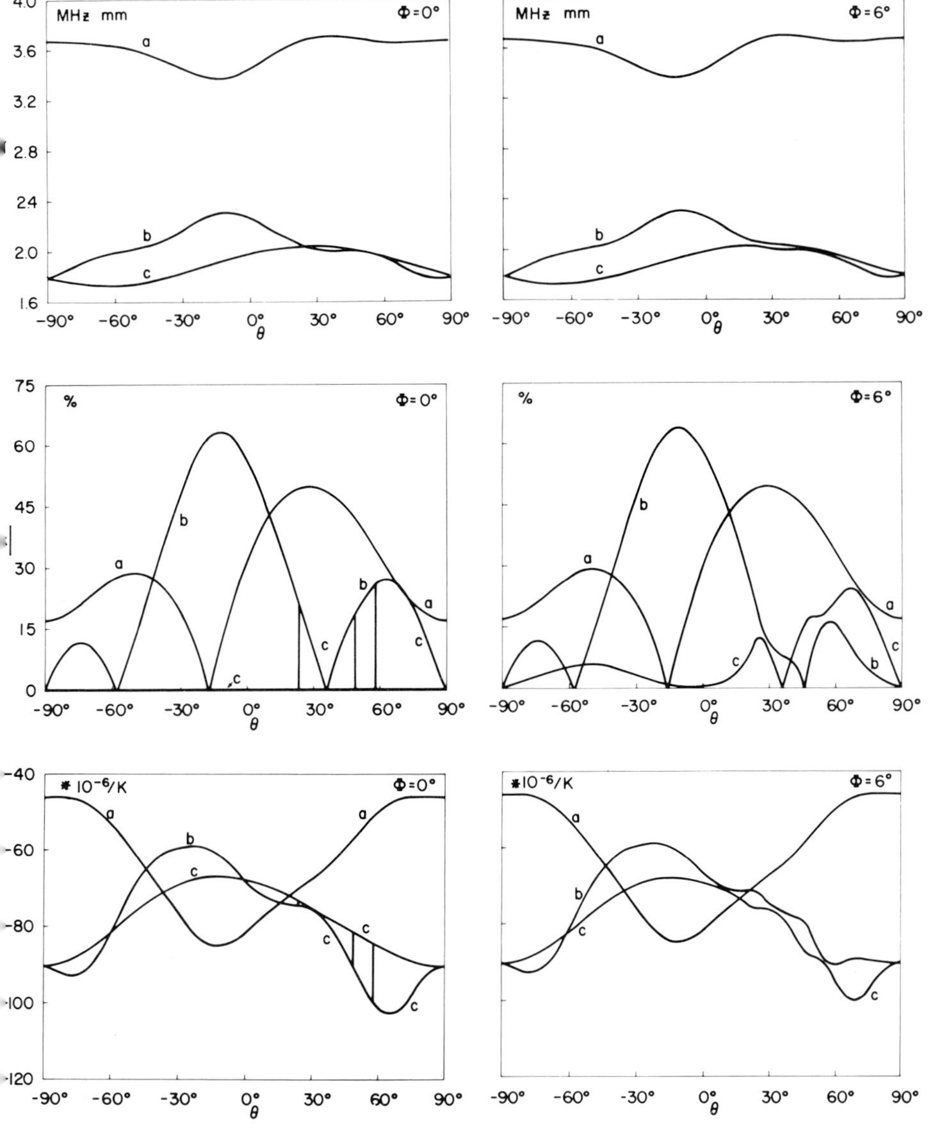

FIG. 17. Thickness mode properties of doubly rotated cuts of lithium niobate. Frequency constants N are in MHz mm, coupling factors are in percent, and temperature coefficients $T_{f_A}^{(1)}$ are in 10^{-6}/K.

parallel with a series C_1, R_1, L_1 chain (Mason, 1948). Reduction to this form starts from Eqs. (38) and (40). It is convenient to discuss the lossless case first, and to deal with a single mode. When Eq. (38), written for a single

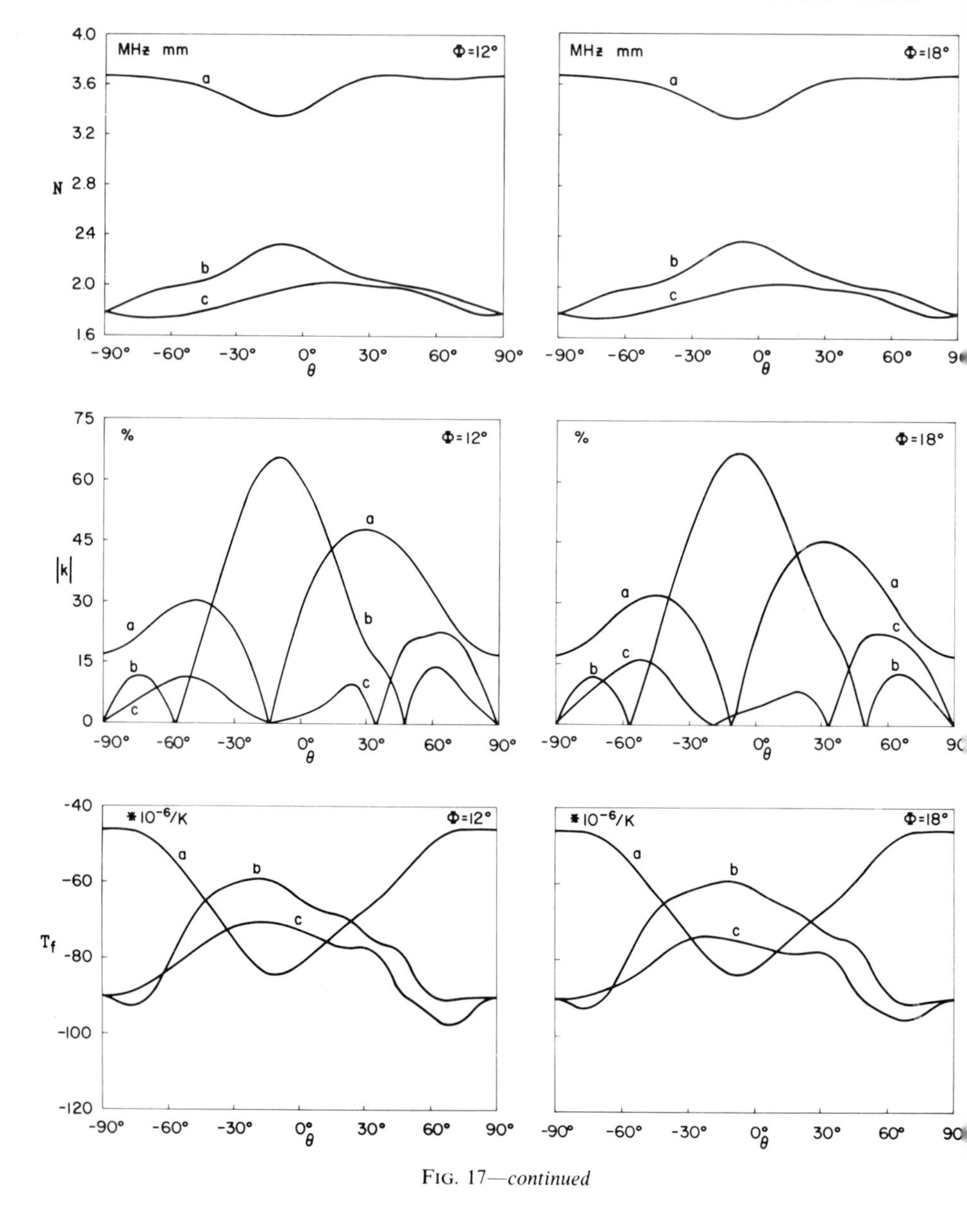

FIG. 17—*continued*

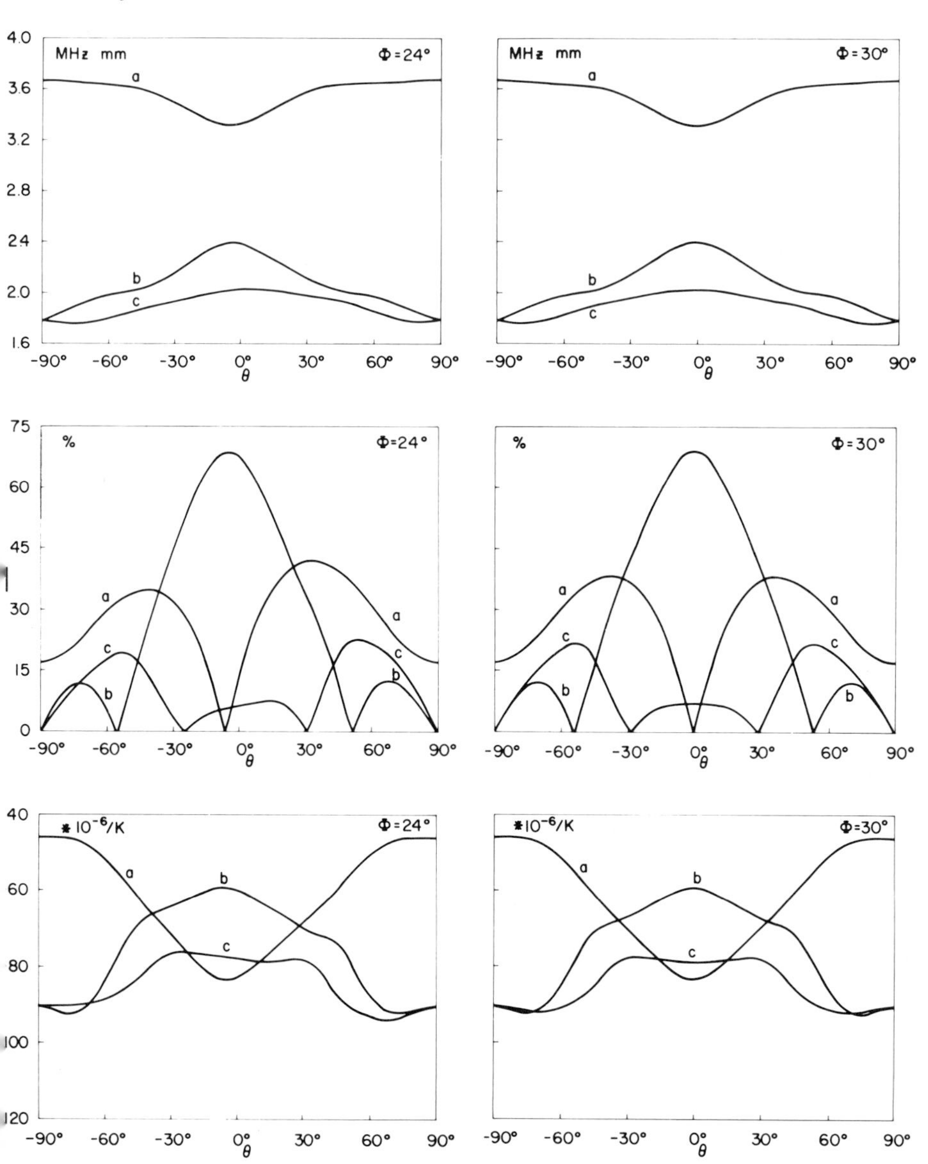

FIG. 17—*continued*

mode, has subtracted from it the immittances corresponding to a shunt capacitor C_0, followed by a series capacitor $-C_0$, the remaining admittance is

$$Y_{TL} = j\omega C_0 k^2 \tan X/X. \tag{91}$$

Expansion of the tangent function in partial fractions leads to a network realization of the individual terms as series LC circuits; Y_{TL} is then just the shunt connection of these series branches. Each series branch consists of an inductance of constant value

$$L' = h^2/2C_0 k^2 v^2 \tag{92}$$

and a capacitance that depends on the harmonic M

$$C'_M = 8C_0 k^2/\pi^2 M^2. \tag{93}$$

M is an odd integer, and v is obtained from Eq. (19). The complete and exact circuit for a single lossless mode is then a shunt capacitor C_0, followed by a series capacitor $-C_0$, followed in turn by an infinite number of parallel branches each consisting of a series $L'C'_M$ chain (Onoe and Jumonji, 1967). Loss is incorporated by adding to each $L'C'_M$ chain a resistance R'_M of value

$$R'_M = \tau_1/C'_M = \pi^2 M^2 \eta h/4e^2 A. \tag{94}$$

The piezoelectric constant e is given by Eq. (39), and the viscosity η by Eq. (21). Motional capacitance ratios r'_M are

$$r'_M = C_0/C'_M = \pi^2 M^2/8k^2. \tag{95}$$

In order to obtain the simple four-element circuit described earlier, valid in the vicinity of a single resonance, the influences of both the series negative capacitor and of the shunt harmonic branches $L'C'_M$ must be taken into account. We assume all shunt branches resonating above the branch of interest are represented as capacitors, and those below are represented as open circuits. Then the elements of the equivalent circuit become

$$\bar{C}_0 = C_0/F, \tag{96}$$

$$L = L'F^2, \tag{97}$$

$$C_M = C_0/r_M. \tag{98}$$

In Eqs. (96)–(98)

$$F = 1 - k^2 I, \tag{99}$$

$$I = 1 - 8\Sigma'/\pi^2, \tag{100}$$

$$\Sigma' = \sum_{K<M} K^{-2} \qquad (K, M \text{ odd}), \tag{101}$$

$$r_M = r'_M F - 1. \tag{102}$$

With loss included R_M becomes

$$R_M = R'_M F^2. \tag{103}$$

The four-element circuit conventionally used thus consists of R_M, L, and C_M connected in series, all shunted by $\bar{C}_0$; in the limit of small k these may be replaced by R'_M, L', C'_M, and C_0, respectively. Extension of the foregoing to more than one mode is straightforward, but usually not warranted. Equation (91) then consists of one term for each mode represented; the realization for each mode is the same as in the single mode case, with the infinite shunt branches in parallel. Reduction to a four-element network then follows. Temperature coefficients of the circuit elements are readily obtained from the material in Section IV and the defining Eqs. (92)–(103).

The piezoelectric coupling factors in quartz are not large, and the capacitance ratios r_m for each mode m can be obtained separately from Eq. (95). From Eqs. (43), (59), and (62) it follows that the fractional difference between antiresonance and resonance frequencies is approximately $1/2r_M$ for each mode; and from Eq. (71) it is seen that $G_0 \simeq 1/r_M$. Capacitance ratios are therefore important quantities in oscillator and filter design. For doubly rotated quartz cuts on the locus of Eq. (85), the ratios r_m for $m = a, b, c$ are given in Fig. 18 for the fundamental harmonic $M = 1$. At the SC cut the c-mode ratio is about three times that of the AT cut. Tables XII and XIII list

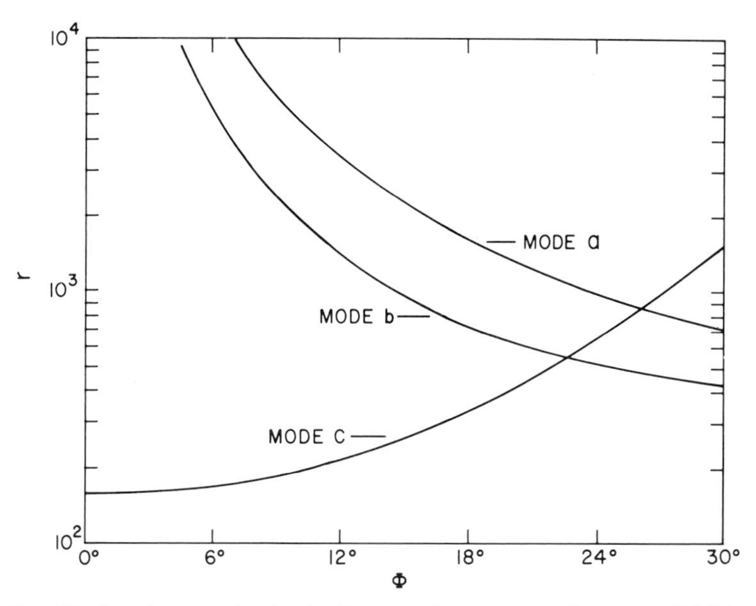

FIG. 18. Capacitance ratios for doubly rotated quartz cuts along the AT-SC locus of Fig. 5.

TABLE XII

CAPACITANCE RATIOS AND THEIR ANGLE GRADIENTS. QUARTZ CUTS $(YXwl)\varphi/\theta > 0^\circ$ [a]

Crystal cut	φ	θ	r_a	r_b	r_c	$\frac{\partial r_a}{\partial \theta}$	$\frac{\partial r_b}{\partial \theta}$	$\frac{\partial r_c}{\partial \theta}$	$\frac{\partial r_a}{\partial \varphi}$	$\frac{\partial r_b}{\partial \varphi}$	$\frac{\partial r_c}{\partial \varphi}$
	degrees		10^{+2}			10^{+2}/deg θ			10^{+2}/deg φ		
AT	0	35.25	—	—	1.59	—	—	0.11	—	—	0
5° V	5	34.94	191	74.3	1.66	11.2	14.7	0.10	−75.5	−28.8	0.04
10° V	10	34.64	48.1	19.3	1.94	2.80	3.73	0.09	−9.27	−3.36	0.09
13.9° V	13.90	34.40	25.4	10.7	2.41	1.49	1.97	0.06	−3.38	−1.18	0.16
FC	15	34.33	22.0	9.45	2.60	1.29	1.71	0.05	−2.68	−0.92	0.19
IT	19.10	34.08	14.1	6.59	3.68	0.83	1.12	−0.09	−1.26	−0.40	0.35
SC	21.93	33.93	11.1	5.55	4.96	0.65	0.90	−0.35	−0.82	−0.23	0.55
25° V	25	33.72	8.95	4.84	7.31	0.53	0.73	−1.07	−0.53	−0.13	0.92
30° V	30	33.42	6.90	4.19	14.9	0.42	0.55	−5.39	−0.27	−0.06	1.80
LC	11.17	9.39	11.9	2.11	1.46	0.50	−0.002	0.01	−2.08	−0.06	0.15

[a] $T^{(1)}_{fA} = 0$ for the c mode of these cuts.

values of r_m and their angle derivatives for selected quartz cuts. Note that the AT cut value of 159 given is for a laterally unbounded plate with a uniform distribution of motion over the plate surface area. This value is a lower bound; in practice, values in the range 180–210 are usually encountered. To account for the nonuniform amplitude distribution, Bechmann (1952) introduced a factor Ψ equal to the quotient of the square of the surface integral

TABLE XIII

CAPACITANCE RATIOS AND THEIR ANGLE GRADIENTS. QUARTZ CUTS $(YXwl)\varphi/\theta < 0^\circ$ [a]

Crystal cut	φ	θ	r_a	r_b	r_c	$\frac{\partial r_a}{\partial \theta}$	$\frac{\partial r_b}{\partial \theta}$	$\frac{\partial r_c}{\partial \theta}$	$\frac{\partial r_a}{\partial \varphi}$	$\frac{\partial r_b}{\partial \varphi}$	$\frac{\partial r_c}{\partial \varphi}$
	degrees		10^{+2}			10^{+2}/deg θ			10^{+2}/deg φ		
BT	0	−49.20	—	3.90	—	—	−0.27	—	—	0	—
	5	−46.56	92.2	3.47	302	−11.4	−0.22	−19.1	−33.2	0.08	−120
	10	−38.63	13.6	2.72	55.3	−1.22	−0.12	−1.09	−1.60	0.12	−10.6
	5	−24.98	11.8	1.64	6.31	−0.78	1.18	−18.0	−3.71	0.49	−5.91
	10	−32.23	7.99	2.12	55.4	−0.61	−0.07	1.20	−0.84	0.11	−8.70
	12.5	−33.33	7.04	2.50	38.0	−0.52	−0.09	1.68	−0.45	0.13	−4.48
RT	15	−34.50	6.98	3.01	27.0	−0.49	−0.12	1.39	−0.26	0.15	−2.68
	17.5	−35.78	6.90	3.54	20.6	−0.50	−0.15	0.89	−0.14	0.18	−1.80
	20	−36.79	7.19	4.21	16.0	−0.52	−0.20	0.57	−0.04	0.22	−1.28

[a] $T^{(1)}_{fA} = 0$ for the b mode of first three entries; for the c mode of remainder.

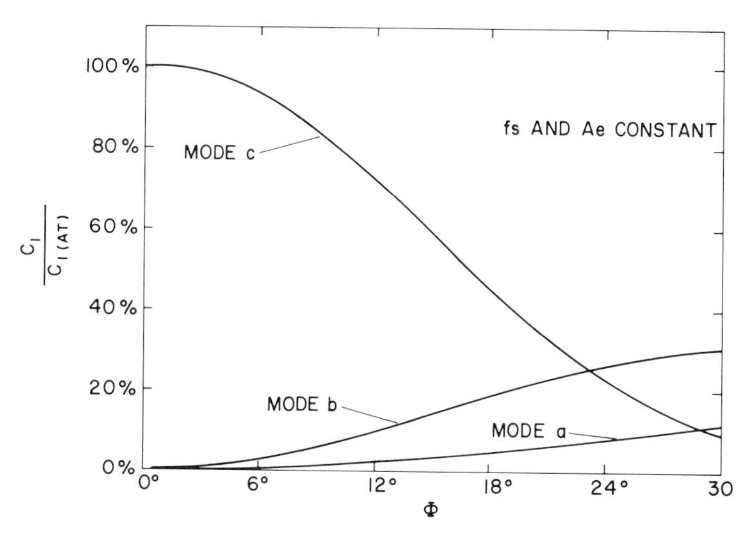

FIG. 19. Motional capacitances for doubly rotated quartz cuts along the AT-SC locus of Fig. 5, normalized to that of an AT cut having identical electrode area and frequency.

of the normalized amplitude function by the surface integral of its square. This factor, which is less than or equal to one, multiplies C_M and divides L and R_M to produce effective values. This approximate technique can be applied to doubly rotated cuts if the motional distribution is known for a given design; alternatively, one may derive more accurate expressions for any resonator geometry by using variational methods (Tiersten, 1969; Holland and EerNisse, 1969; Lee and Haines, 1974).

Motional capacitance C_M is proportional to electrode area and to fundamental frequency. Considering these to be fixed, we can determine the value of C_M normalized to that of the AT-cut c mode. For the locus of Eq. (85), the result is plotted in Fig. 19 from the harmonic-independent relation

$$\frac{C_M}{C_{M(\mathrm{AT})}} = \frac{C_1}{C_{1(\mathrm{AT})}} = \frac{(\varepsilon_\zeta^{\mathrm{S}}/r^{(m)}N_m)}{(\varepsilon_\zeta^{\mathrm{S}}/r^{(c)}N_c)_{\mathrm{AT}}}. \tag{104}$$

Since in crystals of trigonal symmetry $\varepsilon_\zeta^{\mathrm{S}}$ does not depend on angle φ and the zero temperature coefficient locus is nearly a constant in θ, the permittivities in Eq. (104) are virtually equal and the variations in Fig. 19 result from changes of coupling factor and modal velocity. The ratios shown will moreover hold well for finite plates since Ψ will be approximately constant for both the AT and doubly rotated plate, and will cancel in the quotient. Utilizing Eq. (94) plus the foregoing makes it possible to plot the ratio of the motional resistances along the locus; the results are given in Fig. 20. The results are similar to those in Fig. 19, but they depend, additionally, on the variations in $\tau_1^{(m)}$ with position on the locus. The same remark concerning finite plates holds here also. If C_0 is maintained constant, rather than electrode area and fundamental frequency, the resistance ratio curve for the *c*

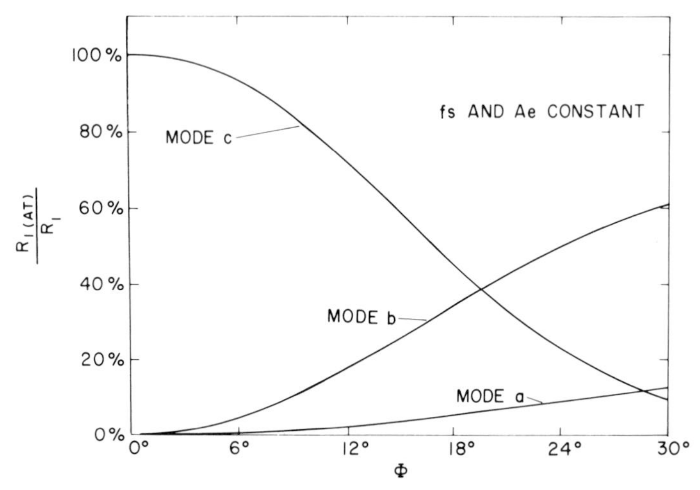

FIG. 20. Motional resistances for doubly rotated quartz cuts along the AT-SC locus of Fig. 5, reciprocally normalized to that of an AT cut having identical electrode area and frequency.

mode remains virtually unaltered; the *b*-mode curve is raised from about 60 to 75% at $\varphi = 30°$, while the *a*-mode curve rises to 25% at $\varphi = 30°$.

B. MODE SPECTROGRAPHS

Mode spectrographs are plots of resonator input susceptance versus frequency. Often measured in practice is a quantity proportional to the resonator plate susceptance when the shunt capacitance C_0 has been balanced out, e.g., by a bridge. We have plotted this quantity B_R normalized to the susceptance B_0 of the shunt capacitance in Fig. 21 for the FC cut of quartz at $\varphi = 15°$. The ordinate is found from

$$\left|\frac{B_R}{B_0}\right| = \left|\sum_m k_m^2 \frac{\tan X^{(m)}}{X^{(m)}}\right| \Bigg/ \left|1 - \sum_m k_m^2 \frac{\tan X^{(m)}}{X^{(m)}}\right|. \tag{105}$$

The abscissa is percentage frequency offset from the *c*-mode fundamental antiresonance frequency. Figure 21 is plotted for the lossless case, but both resonances are marked with a figure in percent to indicate the response level when losses are present, compared to the AT-cut *c*-mode fundamental reso-

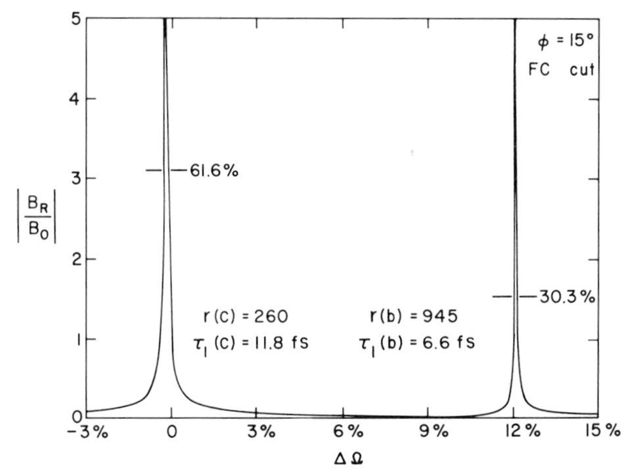

FIG. 21. Mode spectrograph for an FC-cut quartz resonator, showing proximity of *b* and *c* modes. Indicated on each resonance curve is the ratio of mode height, when loss is included, to that of an AT cut.

nance. The relative levels are given by

$$(r_1^{(c)}\tau_1^{(c)})_{\mathrm{AT}}/(r_M^{(m)}\tau_1^{(m)}). \tag{106}$$

Table XIV contains a list of modal parameters for additional cuts on the locus of Eq. (85); the entry $|B_R/B_0|$ denotes the maximum value. From this table one can visualize the shifting modal strengths and separations of the spectrographs with changing φ. In Fig. 22 is given a wider frequency plot for the SC cut. The ordinate is in decibels relative to the *b*-mode fundamental. Normalized frequency is the abscissa; the modes and their harmonics are labeled by the notation $m^{(M)}$. One sees the cluttering of the spectrum produced by the three series of harmonics. These are only the thickness modes; in practice, each mode will have associated with it anharmonic overtones due to lateral phase reversals across the plate, further crowding the spectrum. Contouring can reduce these effects very much. More important in many respects are the strength and proximity of the *b*- and *c*-mode fundamental resonances. Here contouring can reduce the *b*-mode strength somewhat, but the separation remains fixed.

When thin flat plates are required to have clean mode spectra, e.g., high frequency filter crystals and crystals for fix-tuned oscillators, the required amount of mode spectral purity is achieved by control of the electrode geometry. According to energy trapping considerations (Shockley *et al.*, 1967), optimal suppression of unwanted responses occurs when the electrode diameter to plate thickness ratio times $M \cdot \sqrt{\mu}$ equals K_M, a constant. For

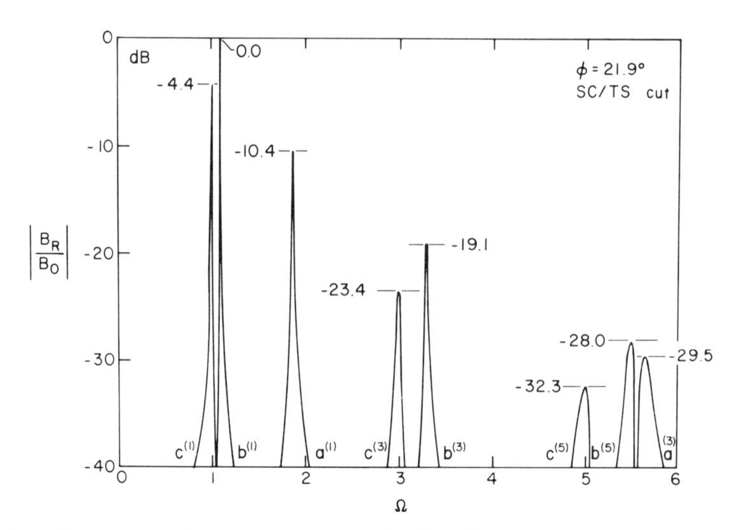

FIG. 22. Extended frequency spectrograph of an SC-cut resonator. Mode levels are normalized to the *b*-mode fundamental strength. Modes are labeled with superscript harmonic numbers. The ordinate scale is in decibels.

$M = 1$, K_M ranges from 1.8 to 2.4; $K_M = 1.8$, 1.6, and 1.4 for $M = 3$, 5, and 7, respectively, for AT-cut quartz. Strictly speaking, the electrode patch can be longer in one direction than in the other, and the best shape is nearly an ellipse (Mindlin, 1968). For SC-cuts, similar considerations appear to apply, with the possibility of even larger K_M. Spectra nearly devoid of

TABLE XIV

MODAL PARAMETERS OF SELECTED QUARTZ RESONATOR CUTS[a]

	Mode *b*			Mode *c*			
Cut	r	τ_1	$\lvert B_R/B_0\rvert$	r	τ_1	$\lvert B_R/B_0\rvert$	Mode separation
AT	∞	6.7	0	159	11.8	100	14.4
13.9° V	1071	6.6	26.6	241	11.8	66.4	12.4
FC	945	6.6	30.3	260	11.8	61.6	12.1
IT	659	6.4	47.3	368	11.8	39.9	10.6
SC	555	6.4	53.5	496	11.7	32.4	9.8
30° V	419	6.0	75.5	1493	11.6	10.9	6.9

[a] r for fundamental resonance ($M = 1$); τ_1 in femtoseconds; mode separation in percent; $|B_R/B_0|$ in percent, normalized to AT-cut, c-mode value.

anharmonic responses for all three modes up to $M = 5$ have been attained (T. J. Lukaszek, private communication, 1977).

VII. Analog Electric Network Models

In the preceding section lumped parameter networks were discussed. These are used to represent resonators over relatively narrow frequency ranges. When other than traction-free boundary conditions are to be employed, as in transducers, the lumped analysis can be extended to include the new circumstances. Operation over wider frequency bands, in the time domain, or with materials having large piezoelectric coupling, requires use of more exact networks. These contain transcendental functions representing wave transmission in the vibrator. Mason's book (1948) is the classical reference for such networks; Berlincourt *et al.* (1964) have provided a valuable survey of these equivalent circuits for many kinds of vibrators.

This section describes equivalent networks of the analog type. These have the distinction of being drawn in such a fashion as to match up, on a one-to-one basis, the circuit components with the physical features they represent. This description produces a schematic that is valid in a point-to-point manner rather than merely yielding a circuit that produces the correct port-to-port immittance matrix. Matching the physical features and their electrical counterparts leads to a building-block approach that heightens understanding and simplifies the representation and analysis of more complex structures; it also facilitates the use of computer aided design (CAD) techniques.

A. Transmission Line Equations for Waves in a Piezoelectric Crystal

Waves on a transmission line obey the Heaviside equations

$$V_{m,\zeta} = -j\omega I_m Z_m / v_m \tag{107}$$

and

$$I_{m,\zeta} = -j\omega V_m / Z_m v_m . \tag{108}$$

V_m and I_m are the modal voltage and current, respectively; Z_m is the modal impedance; and v_m is the wave velocity. Application of transmission line formalisms to acoustic wave problems has been made recently by Oliner (1969), Oliner *et al.* (1972a,b), and Auld (1973). We show here that acoustic plane waves propagating in an arbitrarily anisotropic, linear, piezoelectric medium satisfy Eqs. (107)–(108). In Section VII,B transmission lines are incorporated into analog networks for doubly rotated thickness mode plates.

Normal coordinate transformations of displacement and stress are defined by

$$u_m^\circ = \bar{\gamma}_{mj} u_j \tag{109}$$

and

$$T_{\zeta m}^\circ = \bar{\gamma}_{mj}(\bar{\Gamma}_{jk} u_{k,\zeta} + \Xi_j \mathscr{A}). \tag{110}$$

With the help of Eqs. (17), (39), and (109), Eq. (110) becomes

$$T_{\zeta m}^\circ = \bar{c}_m u_{m,\zeta}^\circ + e_m \mathscr{A}. \tag{111}$$

The first term on the right-hand side of Eq. (111) is due to the "wavy" stress component $\tilde{T}_{\zeta m}$, while the second term is a spatial constant, arising from the piezoelectric effect, that we shall denote as $\tilde{T}_{\zeta m}^\circ$. For the wave portion we have

$$\tilde{T}_{\zeta m}^\circ = \bar{c}_m u_{m,\zeta}^\circ. \tag{112}$$

Equation (1) is transformed into

$$\bar{\gamma}_{mj} T_{ij,i} = -\rho\omega^2 u_m^\circ. \tag{113}$$

The left-hand side of this equation is $T_{\zeta m,\zeta}^\circ$, but because $\bar{T}_{\zeta m,\zeta}^\circ$ vanishes within the crystal, we get finally

$$\tilde{T}_{\zeta m,\zeta}^\circ = -\rho\omega^2 u_m^\circ. \tag{114}$$

Equations (112) and (114) become isomorphic with Eqs. (107)–(108) by the definitions

$$V_m = A\tilde{T}_{\zeta m}^\circ \tag{115}$$

$$I_m = -j\omega u_m^\circ \tag{116}$$

$$Z_m = A\rho v_m \tag{117}$$

and identification of v_m with Eq. (19).

B. Equivalent Networks for Thickness Modes

Synthesis of Eq. (30), representing the electrical input admittance, begins as in Section VI,A with extraction of the static shunt capacitor C_0, followed by a negative C_0 in series. The negative capacitance results from the reactions between the driving electric field applied and the ζ-directed electric fields produced by the acoustic waves. With the capacitors extracted the remainder is [cf. Eq. (91)]

$$Y_{\text{TL}} = j\omega C_0 \sum_m k_m^2 \tan X^{(m)}/X^{(m)}. \tag{118}$$

The terms in Eq. (118) are now interpreted in the light of Section VII,A as leading to transmission lines. That is, each of the admissible plane waves for a given propagation direction ζ is represented by a transmission line supporting one of the wave stresses $\tilde{T}^{\circ}_{\zeta m}$ in the bulk of the crystal. The piezoelectric driving stresses $\tilde{T}^{\circ}_{\zeta m}$ come into play at the plate surfaces only. In order to have an analog network, the ideal transformers that are the circuit equivalents of the $\bar{T}^{\circ}_{\zeta m}$ must be removed to the surfaces. It has been shown that when this is done, the network that results is an exact analog of the physical workings of the plate vibrator (Ballato *et al.*, 1974).

For a single mode, the analog network is shown in Fig. 23 along with

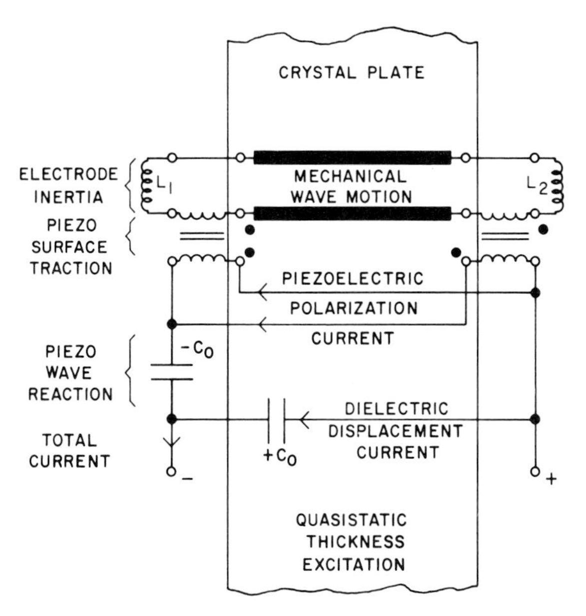

FIG. 23. Analog quasi-static exact equivalent network for a thickness mode crystal plate with one piezoelectrically excited mode. Inductors represent inertial tractions due to electrodes; transformers represent piezoelectric surface tractions driving the mode. Physical correspondences with network features are labeled.

the corresponding physical features. Lumped at the surfaces are identical piezoelectric transformers supplying δ functions of traction to drive the plate. The primaries are connected in parallel. Also located at the surfaces are the mechanical ports and the negative capacitor. In the traction-free case, the mechanical ports would be short-circuited; in Fig. 23 they have been provided with terminating inductors L_i that represent mass loadings caused by electrodes. The lumped electrode masses result in δ functions of inertial traction at the boundaries in series with the piezoelectric tractions. The inductances are of value

$$L_i = \bar{m}_i A, \tag{119}$$

where $\bar{m}_i$ is the mass per unit area on each surface. When the surface masses are equal, the schematic of Fig. 23 is mechanically symmetrical; the transmission line center is a current node, and may be bisected to yield the simplified network of Fig. 24. Figure 24 is no longer an analog circuit due to

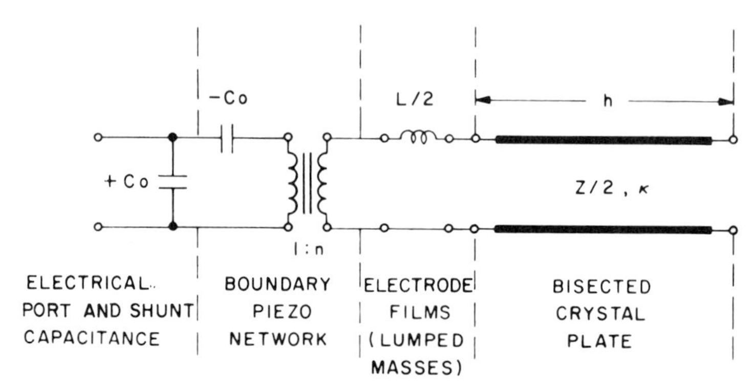

FIG. 24. Bisected version of Fig. 23 for the case of equal inductors.

the bisection, but its components are interpretable in physical terms. It has a transmission line of characteristic impedance $Z/2 = A\rho v_m/2$ and propagation constant $\kappa = \omega/v_m$; the inductor is $L/2 = \bar{m}A/2 = \rho h\mu A/2$, and the turns ratio is $n = e_m A/2h$. The input admittance of Eq. (55) is exactly realized, for one mode, by Fig. 24.

In the multimode case, where two or three modes are excited, the analog network is given by an augmented version of Fig. 23. Each mode has a transmission line of length $2h$, terminated at its mechanical ports by inductors if mass loaded or by short circuits if traction-free; in series with the mechanical ports are piezoelectric transformer secondaries. The primaries of all transformers are connected in parallel and lead to the $-C_0$ and shunt capacitor C_0. Each transformer has a turns ratio of

$$n_m = e_m A/2h = (C_0 A\bar{c}_m k_m^2/2h)^{1/2}. \tag{120}$$

The dot convention of Fig. 23 must be maintained for all transformer pairs; this is the manifestation of the polar nature of the piezoelectric effect. As long as the impedances terminating the mechanical ports are equal in pairs, the network may be bisected to yield a network like Fig. 24 having secondary circuits in parallel. For equal lumped masses on both surfaces, this circuit realizes the electrical port admittance Eq. (55) exactly.

Any of the analog networks described above may be used for time-domain analyses in addition to the usual frequency-domain uses. When a network is analog in nature, a solution of the circuit problem evaluated at any value of the thickness coordinate gives immediately the solution of the corresponding physics problem at that point.

C. INTERFACE NETWORK

Figure 23 and its generalization to three modes were presented in the preceding subsection simply as realizations of the electrical port input admittance of Eq. (38). In this instance the mechanical ports at the plate surfaces were shorted to satisfy traction-free boundary conditions. Termination of the mechanical ports by inductances realized the input admittance of Eq. (55), stemming from the boundary conditions of Eq. (53). The analog networks given are however capable of greater generality than the brief derivation might lead one to expect. Because there are three components of traction and of particle displacement at each surface as well as the electrical port, the complete representation of the overall electromechanical system takes the form of a seven-port network. This extended circuit equivalent is a necessary consequence of the multimode nature of doubly rotated crystal vibrators, and must, in general, be used to characterize their behavior unless the criteria of Section III,B are met.

At each surface the three mechanical ports in the equivalent network appear with the normal coordinate variables of Eqs. (109)–(110). Conversion to components referred to the crystallographic axes X_j requires that the three ports of each surface be connected to a network realizing a three-dimensional orthogonal transformation (Carlin and Giordano, 1964). This network, consisting solely of interconnected ideal transformers, is shown in Fig. 25. The secondary turns ratios are the components of the coordinate transformation; switching primary and secondary ports requires only that the transposed components be substituted in the ratios. If one network of Fig. 25 is attached to the mechanical ports on each side of the normal coordinate circuit, the result is a complete analog network that exactly realizes the seven-port impedance function of the vibrator. As drawn in Fig. 25, the primary ports would be attached to the normal mode ports, and the secondary ports would then represent the coordinates X_j. Then the β_{mj} appearing in the figure would be identified with the stiffened eigenvector components $\bar{\gamma}_{mj}$. For a number of reasons, such as the stacking of plates to be described in the next subsection, it is often more convenient to refer the secondary ports of Fig. 25 to a laboratory coordinate system coinciding with the plate thickness and lateral directions, rather than to the crystallographic axes. Since, by Eq. (7), the transformation to the plate axes from the crystal axes is determined by α_{ij}, and the transformation to normal coordinates is

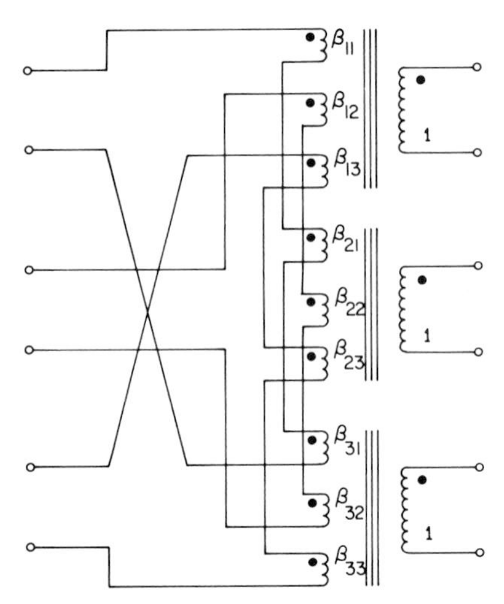

FIG. 25. Ideal transformer realization of a three-dimensional orthogonal transformation. This boundary, or interface, network converts between normal coordinates utilized within the plate and laboratory coordinates used at the surfaces for plate stacking. Turns ratios are the components of the transformation.

given by $\bar{\gamma}_{mj}$, one has

$$\beta_{mi} = \alpha_{ij}\bar{\gamma}_{mj}. \tag{121}$$

With these turns ratios the total equivalent circuit has its mechanical ports referred to the plate axes.

The network of Fig. 25 subjects immittances to similarity transformations. An impedance network, characterized by elements Z°_{mn}, that is attached to the primary ports will be perceived at the secondary ports as having components

$$Z_{ij} = \beta_{mi}\beta_{nj}Z^{\circ}_{mn}. \tag{122}$$

If the impedance is of diagonal form, no transformation takes place; the inductors of Eq. (119) are thus unchanged by the boundary network, which can be discarded in this instance. The network may be simplified in other cases as well because of inherent crystal symmetry and/or propagation direction. The role of these networks for stacks of plates will be discussed in the next subsection. Boundary and interface networks for waves at nonnormal incidence in isotropic media have been determined by Oliner *et al.* (1972b). Ballato (1974) has described networks for representing the inclusion of the electromagnetic modes when the quasi-static constraint is relaxed.

D. Stacking of Doubly Rotated Plates

Stacking of plates is used to form composite transducers, delay lines, and filters. These devices have almost invariably been of the single mode variety (Sittig, 1972). Use of doubly rotated crystal plates in a layered configuration opens the way for an extension of these devices to include the use of more than one mode. One example of such a use is the stacked-crystal filter (SCF) which operates on the multimode principle (Ballato *et al.*, 1974). Stacking of multimode plates was described by Onoe (1972) using Mason equivalent circuits.

Use of the complete analog circuit for the multimode plate, including the interface networks, permits the description of a stack of plates in a simple manner. The interface, or boundary, networks of Fig. 25 are referred for each plate in the stack to a common laboratory reference frame. This referral amounts to, at most, an additional rotation about the common thickness direction for each network; the rotation is manifested by a change in the turns ratios β_{mi} to account for the azimuthal angle ψ. Upon referring all networks to the common frame, stacking is accomplished by connecting corresponding mechanical ports together. That is, if one axis of the laboratory frame coincides with the mutual thickness coordinate of two adjacent plates, then the ports for this axis are joined, and so on. The entire stack then appears as a cascade connection of transmission line and interface networks. Simplifications arising from crystal symmetry or propagation direction that render certain of the β_{mi} zero are directly interpretable from the circuit schematic; so too are simplifications in the piezoelectric driving terms n_m. All of the multimode interactions are readily traced from the analog network picture. Analytically, each plate is representable by a set of seven-port impedance relations, with stacking carried out by a matrix procedure using extended A, B, C, D parameters which are themselves matrices.

References

Anonymous. (1949). *Proc. IRE* **37**, 1378–1395.

Auld, B. A. (1973). "Acoustic Fields and Waves in Solids," Vols. I and II. Wiley, New York.

Baldwin, C. F., and Bokovoy, S. A. (1936). U.S. Patent 2,254,866.

Ballato, A. (1960). *Proc. 14th Annu. Frequency Contr. Symp.* pp. 89–114.

Ballato, A. (1974). *Proc. Symp. Opt. & Acoust. Micro-Electron.*, pp. 599–615.

Ballato, A. (1976). *Proc. IEEE* **64**, 1449–1450.

Ballato, A., and Iafrate, G. J. (1976). *Proc. 30th Annu. Frequency Contr. Symp.* pp. 141–156.

Ballato, A., and Lukaszek, T. (1974). *IEEE Trans. Sonics Ultrason.* **21**, 269–274.

Ballato, A., and Lukaszek, T. (1975). *Proc. 29th Annu. Frequency Contr. Symp.* pp. 10–25.

Ballato, A., Bertoni, H. L., and Tamir, T. (1974). *IEEE Trans. Microwave Theory Tech.* **22**, 14–25.

Baumhauer, J. C., and Tiersten, H. F. (1973). *J. Acoust. Soc. Am.* **54**, 1017–1034.

Bechmann, R. (1951). *Arch. Elektr. Uebertr.* **5**, 89–90.

Bechmann, R. (1952). *Arch. Elektr. Uebertr.* **6**, 361–368.

Bechmann, R. (1956). *Proc. IRE* **44**, 1600–1607.

Bechmann, R. (1958). *Phys. Rev.* **110**, 1060–1061.

Bechmann, R. (1961). *Proc. IRE* **49**, 1454.

Bechmann, R., Ballato, A. D., and Lukaszek, T. J. (1962). *Proc. IRE* **50**, 1812–1822 and 2451.

Berlincourt, D. A., Curran, D. R., and Jaffe, H. (1964). *In* "Physical Acoustics: Principles and Methods" (W. P. Mason, ed.), Vol. 1, Part A, pp. 169–270. Academic Press, New York.

Besson, R. (1974). *Proc. 28th Annu. Frequency Contr. Symp.* pp. 8–13.

Birch, J., and Weston, D. A. (1976). *Proc. 30th Annu. Frequency Contr. Symp.* pp. 32–39.

Bokovoy, S. A., and Baldwin, C. F. (1935). U.K. Patent 457,342.

Bottom, V. E., and Ives, W. R. (1951). U.S. Patent 2,743,144.

Carlin, H. J., and Giordano, A. B. (1964). "Network Theory," pp. 190–207. Prentice-Hall, Englewood Cliffs, New Jersey.

Carr, P. H., and O'Connell, R. M. (1976). *Proc. 30th Annu. Frequency Contr. Symp.* pp. 129–131.

Chang, Z.-P., and Barsch, G. R. (1976). *IEEE Trans. Sonics Ultrason.* **23**, 127–135.

Christoffel, E. B. (1877). *Ann. Mat. Pura Appl.* [2] **8**, 193–243.

Détaint, J., and Lançon, R. (1976). *Proc. 30th Annu. Frequency Contr. Symp.* pp. 132–140.

Détaint, J., and Lançon, R. (1977). *Onde Electr.* **57** (in press).

EerNisse, E. P. (1975). *Proc. 29th Annu. Frequency Contr. Symp.* pp. 1–4.

EerNisse, E. P. (1976). *Proc. 30th Annu. Frequency Contr. Symp.* pp. 8–11.

Epstein, S. (1973). *Proc. 1973 Ultrason. Symp.* pp. 540–542.

Filler, R. L., and Vig, J. R. (1976). *Proc. 30th Annu. Frequency Contr. Symp.* pp. 264–268.

Franx, C. (1967). *Proc. 21st Annu. Frequency Contr. Symp.* pp. 436–454.

Fukuyo, H., Yoshie, H., and Nakazawa, M. (1967). *Bull. Tokyo Inst. Technol.* **82**, 53–64.

Gagnepain, J.-J. (1972). Doctoral Thesis, University of Besançon, France.

Gerber, E. A., and Miles, M. H. (1961). *Proc. IRE* **49**, 1650–1654.

Gerber, E. A., and Sykes, R. A. (1974). *Natl. Bur. Stand.* (*U.S.*), *Monogr.* **140**, 41–64.

Hafner, E. (1956). *Proc. 10th Annu. Frequency Contr. Symp.* pp. 182–189.

Hafner, E. (1974). *IEEE Trans. Sonics Ultrason.* **21**, 220–237.

Hammond, D. L., and Benjaminson, A. (1965). *Hewlett-Packard J.* **16**, 1–7.

Hammond, D. L., Adams, C., and Cutler, L. (1963). *Proc. 17th Annu. Frequency Contr. Symp.* pp. 215–232.

Holland, R. (1974). *Proc. 1974 Ultrason. Symp.* pp. 592–598.

Holland, R. (1976). *IEEE Trans. Sonics Ultrason.* **23**, 72–75.

Holland, R., and EerNisse, E. P. (1969). "Design of Resonant Piezoelectric Devices," MIT Press, Cambridge, Massachusetts.

Hruška, K. (1971). *IEEE Trans. Sonics Ultrason.* **18**, 1–7.

Hruška, K., and Kazda, V. (1968). *Czech. J. Phys.* **b18**, 500–503.

Jhunjhunwala, A., Vetelino, J. F., and Field, J. C. (1976). *Proc. 1976 Ultrason. Symp.* pp. 523–527.

Juretschke, H. J. (1974). "Crystal Physics." Benjamin, Reading, Massachusetts.

Koga, I. (1969). *Proc. 23rd Annu. Frequency Contr. Symp.* pp. 128–131.

Koga, I., Aruga, M., and Yoshinaka, Y. (1958). *Phys. Rev.* **109**, 1467–1473.

Kusters, J. A. (1970). *Proc. 24th Annu. Frequency Contr. Symp.* pp. 46–54.

Kusters, J. A. (1976). *IEEE Trans. Sonics Ultrason.* **23**, 273–276.

Kusters, J. A., and Leach, J. G. (1977). *Proc. IEEE* **65**, 282–284.

Lagasse, G., Ho, J., and Bloch, M. (1972). *Proc. 26th Annu. Frequency Contr. Symp.* pp. 148–151.

Lamb, J., and Richter, J. (1966). *Proc. R. Soc. London, Ser. A* **293**, 479–492.

Landolt-Börnstein (1966). "Numerical Data and Functional Relationships in Science and Technology," New Ser., Vol. III/1, pp. 40–123. Springer-Verlag, Berlin and New York.

Landolt-Börnstein (1969). "Numerical Data and Functional Relationships in Science and Technology," New Ser., Vol. III/2, pp. 40–101. Springer-Verlag, Berlin and New York.
Lawson, A. W. (1941). *Phys. Rev.* **59**, 838–839.
Lee, P. C. Y., and Haines, D. W. (1974). *In* "R. D. Mindlin and Applied Mechanics" (G. Herrmann, ed.), pp. 227–253. Pergamon, Oxford.
Lee, P. C. Y., and Syngellakis, S. (1975). *Proc. 29th Annu. Frequency Contr. Symp.* pp. 65–70.
Lee, P. C. Y., and Wu, K.-M. (1976). *Proc. 30th Annu. Frequency Contr. Symp.* pp. 1–7.
Lee, P. C. Y., Wang, Y. S., and Markenscoff, X. (1975). *J. Acoust. Soc. Am.* **57**, 95–105.
McSkimin, H. J., Andreatch, P., Jr., and Thurston, R. N. (1965). *J. Appl. Phys.* **36**, 1624–1632.
Mason, W. P. (1948). "Electromechanical Transducers and Wave Filters," 2nd ed., pp. 200–209 and 399–404. Van Nostrand-Reinhold, Princeton, New Jersey.
Mason, W. P. (1950). "Piezoelectric Crystals and Their Application to Ultrasonics," pp. 208–209. Van Nostrand-Reinhold, Princeton, New Jersey.
Mason, W. P. (1964). *In* "Physical Acoustics: Principles and Methods" (W. P. Mason, ed.), Vol. 1, Part A, pp. 335–416. Academic Press, New York.
Mindlin, R. D. (1968). *J. Acoust. Soc. Am.* **43**, 1329–1331.
Nye, J. F. (1957). "Physical Properties of Crystals." Oxford Univ. Press, London and New York.
Oliner, A. A. (1969). *IEEE Trans. Microwave Theory Tech.* **17**, 812–826.
Oliner, A. A., Bertoni, H. L., and Li, R. C. M. (1972a). *Proc. IEEE* **60**, 1503–1512.
Oliner, A. A., Li, R. C. M., and Bertoni, H. L. (1972b). *Proc. IEEE* **60**, 1513–1518.
Onoe, M. (1969). *Proc. IEEE* **57**, 702–703.
Onoe, M. (1972). *Trans. Inst. Elec. Eng. (Jpn.)* **55A**, 239–244.
Onoe, M., and Jumonji, H. (1967). *J. Acoust. Soc. Am.* **41**, 974–980.
Schulz, M. B., Matsinger, B. J., and Holland, M. G. (1970). *J. Appl. Phys.* **41**, 2755–2765.
Shockley, W., Curran, D. R., and Koneval, D. J. (1967). *J. Acoust. Soc. Am.* **41**, 981–993.
Sittig, E. K. (1972). *In* "Physical Acoustics: Principles and Methods" (W. P. Mason and R. N. Thurston, eds.), Vol. 9, pp. 221–275. Academic Press, New York.
Smith, R. T., and Welsh, F. S. (1971). *J. Appl. Phys.* **42**, 2219–2230.
Smythe, R. C. (1974). *Proc. 28th Annu. Frequency Contr. Symp.* pp. 5–7.
Spencer, W. J. (1972). *In* "Physical Acoustics: Principles and Methods" (W. P. Mason and R. N. Thurston, eds.), Vol. 9, pp. 167–220. Academic Press, New York.
Stanley, J. M. (1954). *Ind. Eng. Chem.* **32**, 1684–1689.
Syngellakis, S., and Lee, P. C. Y. (1976). *Proc. 30th Annu. Frequency Contr. Symp.* pp. 184–190.
Thurston, R. N. (1964). *In* "Physical Acoustics: Principles and Methods" (W. P. Mason, ed.), Vol. 1, Part A, pp. 1–110. Academic Press, New York.
Thurston, R. N., McSkimin, H. J., and Andreatch, P., Jr. (1966). *J. Appl. Phys.* **37**, 267–275.
Tiersten, H. F. (1963). *J. Acoust. Soc. Am.* **35**, 53–58.
Tiersten, H. F. (1969). "Linear Piezoelectric Plate Vibrations." Plenum, New York.
Tiersten, H. F. (1975a). *J. Acoust. Soc. Am.* **57**, 660–666.
Tiersten, H. F. (1975b). *J. Acoust. Soc. Am.* **57**, 667–681.
Tiersten, H. F. (1976). *J. Acoust. Soc. Am.* **59**, 866–878.
Valdois, M., Besson, J., and Gagnepain, J.-J. (1974). *Proc. 28th Annu. Frequency Contr. Symp.* pp. 19–32.
Warner, A. W. (1963). *Proc. 17th Annu. Frequency Contr. Symp.* pp. 248–266.
Warner, A. W., Onoe, M., and Coquin, G. A. (1967). *J. Acoust. Soc. Am.* **42**, 1223–1231.
Weinert, R. W., and Isaacs, T. J. (1975). *Proc. 29th Annu. Frequency Contr. Symp.* pp. 139–142.
Wood, A. F. B., and Seed, A. (1967). *Proc. 21st Annu. Frequency Contr. Symp.* pp. 420–435.
Yamada, T., and Niizeki, N. (1970). *Proc. IEEE* **58**, 941–942.
Zelenka, J., and Lee, P. C. Y. (1971). *IEEE Trans. Sonics Ultrason.* **18**, 79–80.

—6—

The Generalized Ray Theory and Transient Responses of Layered Elastic Solids

YIH-HSING PAO*

Department of Theoretical and Applied Mechanics
Cornell University, Ithaca, New York 14853

and

RALPH R. GAJEWSKI

Department of Civil Engineering, Engineering Mechanics and Materials
United States Air Force Academy, USAFA, Colorado 80840

* Research supported by the National Science Foundation through grants to the College of Engineering and Materials Science Center of Cornell University.

I. Introduction

Shortly after the publications of two monographs on waves in layered media (Ewing *et al.*, 1957; Brekhovskikh, 1960), the *theory of generalized ray* was developed for analyzing transient waves in a multilayered solid. In the theory the elastic waves that propagate along various ray paths because of multiple reflections and refractions are represented by a series of ray integrals, each of which can be evaluated exactly by applying the Cagniard method (Cagniard, 1939). Since the pulses that are represented by the ray integrals arrive at a point of observation in successive order, the theory furnishes an exact solution, up to the time of arrival of the next ray, for transient waves in layered media.

Although the theory was originally developed for geophysical applications (Spencer, 1960; Knopoff *et al.*, 1960; Pekeris *et al.*, 1965), it can readily be adopted for analyzing signals in *acoustic emission*. The acoustical signals radiated from the initiation or growth of defects and cracks in solid materials are not much different from the seismic waves caused by the movement of a fault or the underwater sound generated by an explosive. In all cases the waves radiated from the source are trapped inside a layer (a plate, a shell) or a multilayered medium. By following the ray paths of reflection and refraction, the theory of generalized ray enables one to analyze in detail the signals recorded by a receiver.

So far discussions of the theory of generalized ray are scattered in research periodicals, mainly in the field of geophysics. We attempt in this article to present systematically the essential elements of this mathematical theory and to collect in one place the formulas and results that are useful in applications. Part of this article is based on a previous report prepared by the authors and the late Professor Stephen A. Thau (Pao *et al.*, 1971).

In addition to its utilitarian value in acoustics and geophysics the theory of generalized ray is also a major development in the dynamic theory of elasticity. The analysis of transient waves in elastic solids began with the paper by Stokes (1849) on the "Dynamic theory of diffraction." Starting from the Stokes solution for a concentrated force in an unbounded solid, the method of generalized ray provides a unified treatment of transient waves in bounded media, including a half-space (Lamb, 1904; Lapwood, 1949; Pekeris, 1955; Chao, 1960), a plate (Knopoff, 1958a; Davids, 1959), two solids in welded contact (Cagniard, 1939; Brekhovskikh, 1960), one layer overlaying a half-space (Newlands, 1953; Pekeris *et al.*, 1965), and a medium composed of many parallel layers (Müller, 1969). Furthermore, the method has recently been extended to a medium bounded by spherical and cylindrical surfaces (Gilbert and Helmberger, 1972; Chen, 1977; Pao and Ceranoglu, 1977) and to inhomogeneous media (Chapman, 1974, 1976b). This is a remarkable accomplishment even though it took more than a century to achieve it.

In the next section we summarize the basic equations of elasticity and solutions derived by applying integral transforms. A sketch of the history of the theory of generalized ray together with a literature survey is also given. The essential elements of the theory are presented in two sections, III and IV. Section III discusses how an integral representation of the Laplace-transformed wave motion along a ray path, known as a *ray integral*, is constructed by assembling the *source function*, *reflection and refraction coefficients*, the *receiver function*, and the *phase function*. Section IV shows how the inverse Laplace transform of the ray integral is found in closed form by applying the *Cagniard method*.

In both Sections III and IV only plane strain problems are discussed. Although the methods of Cagniard and the generalized ray were originally developed for three-dimensional transient responses symmetric about an axis (axisymmetric response), the two-dimensional analysis shown in Sections III and IV is much simpler, and illustrates all essential points of the methods. Furthermore, the two-dimensional problems themselves are of interest in applied mechanics and soil mechanics. The analogy between plane strain problems and plane stress problems enables one to conduct experiments with a thin sheet instead of a bulky layered model (White, 1965).

The axisymmetric transient responses are discussed in Section V. The method of Cagniard is presented in its original form in this and previous sections. Since the publication of Cagniard's monograph in 1939, there have been numerous modifications or alterations of the method. However, many had missed the key point of the method, that is, the mapping of the integration variable in the phase function onto another variable in a complex plane, and the transforming of the new variable back to the original variable by the same mapping. It is through this ingenious mapping that one is able to analyze the transient waves in multilayered media and in a medium bounded by cylindrical and spherical surfaces. For this reason, considerable space in this paper (Sections IV and V) is devoted to this method.

Section VI presents the general solutions for transient waves generated by a point source, including a concentrated force applied obliquely to the surface of a layer, a double force, a center of rotation, and a single or double couple. The last-mentioned source is often used to represent the slipping of a fault which generates seismic waves. In acoustic emission the modeling of material defects or cracks in structures may require the superposition of several source functions.

The method of generalized ray is most effective in analyzing signals that arrive early at a receiver since the number of ray integrals increases phenomenally as the duration of observation lengthens. For source signals that are short in duration and high in frequency content, the ray integrals may be evaluated by approximate methods; these are discussed at the end of Section VI.

In general, the inverse transforms of the ray integrals for point sources are in the form of integrals along a complex contour. The numerical integration of these integrals is discussed in Section VII. In this last section we show in detail the results for two examples: a point source (center of dilatation) in a plate, and the same source in a layer overlaying a half-space.

II. Equations of Elasticity and Solutions

In this section we summarize the basic equations of elasticity and some general solutions for plane strain and axisymmetric problems. Only homogeneous and isotropic media are considered. An inhomogeneous medium with layered structure is subdivided into many layers, each layer is assumed homogeneous and isotropic.

For an unbounded medium, solutions for waves generated by various types of sources are well known; some of these are reviewed in this section. With one or several sources in a layered medium the waves generated by the sources are multiply reflected and refracted. Propagation of waves in a layered medium may be analyzed either by the method of normal mode or the method of generalized ray. The former is briefly summarized in this section; the latter is the theme subject of this chapter.

A. Equations of Elastic Waves

The displacement $\mathbf{u}$, or the velocity $\dot{\mathbf{u}} = \partial \mathbf{u}/\partial t$, of a particle in an elastic solid that occupies the position $\mathbf{x}(x, y, z)$ at time t satisfies the vector equation of motion,

$$(\lambda + \mu)\nabla\nabla \cdot \mathbf{u} + \mu \nabla^2 \mathbf{u} + \rho \mathbf{F} = \rho\, \partial^2 \mathbf{u}/\partial t^2. \tag{2.1}$$

In this equation ρ is the mass density, and λ and μ are Lame's constants of the material; $\mathbf{F}$ is the body force per unit mass. The stress tensor $\boldsymbol{\tau}$ in the solid is related to $\mathbf{u}$ by

$$\boldsymbol{\tau} = \lambda(\nabla \cdot \mathbf{u})\mathbf{I} + \mu(\nabla \mathbf{u} + \mathbf{u}\nabla), \tag{2.2}$$

where $\mathbf{I}$ is the isotropic tensor.

Equation (2.1) may be solved by introducing two displacement potentials ϕ and $\boldsymbol{\psi}$ and two body force potentials b and $\mathbf{B}$ such that

$$\mathbf{u} = \nabla\phi + \nabla \times \boldsymbol{\psi}, \tag{2.3}$$

$$\mathbf{F} = \nabla b + \nabla \times \mathbf{B}. \tag{2.4}$$

The original equation of motion is satisfied if

$$c^2\, \nabla^2\phi + b = \ddot{\phi}, \qquad c = [(\lambda + 2\mu)/\rho]^{1/2}, \tag{2.5}$$

$$C^2\, \nabla^2\boldsymbol{\psi} + \mathbf{B} = \ddot{\boldsymbol{\psi}}, \qquad C = (\mu/\rho)^{1/2}. \tag{2.6}$$

The c and C are respectively the speed of P waves (longitudinal, compressional) and S waves (transverse, shear) in the medium.

In cartesian coordinates (x, y, z) we attach a subscript x, y, or z to denote the corresponding component of a vector. Thus

$$u_x = \frac{\partial\phi}{\partial x} + \frac{\partial\psi_z}{\partial y} - \frac{\partial\psi_y}{\partial z}, \qquad \ldots, \tag{2.7}$$

$$F_x = \frac{\partial b}{\partial x} + \frac{\partial B_z}{\partial y} - \frac{\partial B_y}{\partial z}, \qquad \ldots, \tag{2.8}$$

and Eq. (2.6) reduces to three scalar wave equations

$$C^2\, \nabla^2\psi_x + B_x = \ddot{\psi}_x, \qquad \ldots, \tag{2.9}$$

where $\nabla^2 = \partial^2/\partial x^2 + \partial^2/\partial y^2 + \partial^2/\partial z^2$. In addition there is the auxiliary condition $\nabla \cdot \boldsymbol{\psi} = 0$. The stress components are

$$\tau_{zz} = \lambda\, \nabla^2\phi + 2\mu\left[\frac{\partial^2\phi}{\partial z^2} + \frac{\partial}{\partial z}\left(\frac{\partial\psi_y}{\partial x} - \frac{\partial\psi_x}{\partial z}\right)\right], \qquad \ldots,$$

$$\tau_{zx} = \mu\left[2\frac{\partial^2\phi}{\partial x\, \partial z} + \frac{\partial}{\partial z}\left(\frac{\partial\psi_z}{\partial y} - \frac{\partial\psi_y}{\partial z}\right) + \frac{\partial}{\partial x}\left(\frac{\partial\psi_y}{\partial x} - \frac{\partial\psi_x}{\partial y}\right)\right], \qquad \ldots. \tag{2.10}$$

For problems in cylindrical coordinates ((r, θ, z), $\boldsymbol{\psi}$ can be represented by two independent functions ψ and χ,

$$\boldsymbol{\psi} = \left(\frac{1}{r}\frac{\partial\psi}{\partial\theta}, -\frac{\partial\psi}{\partial r}, \chi\right), \tag{2.11}$$

where ψ and χ are solutions of scalar wave equations:

$$C^2\, \nabla^2\psi = \ddot{\psi}, \qquad C^2\, \nabla^2\chi = \ddot{\chi}, \tag{2.12}$$

and

$$\nabla^2 = \partial^2/\partial r^2 + r^{-1}\, \partial/\partial r + r^{-2}\, \partial^2/\partial\theta^2 + \partial^2/\partial z^2.$$

The components of **u** in terms of ϕ, ψ, and χ are

$$u_r = \frac{\partial \phi}{\partial r} + \frac{\partial^2 \psi}{\partial r\, \partial z} + \frac{1}{r}\frac{\partial \chi}{\partial \theta}, \qquad u_\theta = \frac{1}{r}\frac{\partial \phi}{\partial \theta} + \frac{1}{r}\frac{\partial^2 \psi}{\partial \theta\, \partial z} - \frac{\partial \chi}{\partial r},$$

$$u_z = \frac{\partial \phi}{\partial z} + \frac{\partial^2 \psi}{\partial z^2} - \frac{1}{C^2}\frac{\partial^2 \psi}{\partial t^2} \tag{2.13}$$

The stress components are

$$\tau_{zz} = \lambda\, \nabla^2 \phi + 2\mu \frac{\partial u_z}{\partial z} = \frac{\lambda}{c^2}\frac{\partial^2 \phi}{\partial t^2} + 2\mu \frac{\partial}{\partial z}\left[\frac{\partial \phi}{\partial z} + \frac{\partial^2 \psi}{\partial z^2} - \frac{1}{C^2}\frac{\partial^2 \psi}{\partial t^2}\right],$$

$$\tau_{zr} = \mu\left(\frac{\partial u_z}{\partial r} + \frac{\partial u_r}{\partial z}\right) = \mu\left[2\frac{\partial^2 \phi}{\partial r\, \partial z} + 2\frac{\partial^3 \psi}{\partial r\, \partial z^2} - \frac{1}{C^2}\frac{\partial^3 \psi}{\partial r\, \partial t^2} + \frac{1}{r}\frac{\partial^2 \chi}{\partial \theta\, \partial z}\right],$$

$$\tau_{z\theta} = \mu\left(\frac{\partial u_\theta}{\partial z} + \frac{1}{r}\frac{\partial u_z}{\partial \theta}\right) = \mu\left[\frac{2}{r}\frac{\partial^2 \phi}{\partial \theta\, \partial z} + \frac{2}{r}\frac{\partial^3 \psi}{\partial \theta\, \partial z^2} - \frac{1}{C^2 r}\frac{\partial^3 \psi}{\partial \theta\, \partial t^2} - \frac{\partial^2 \chi}{\partial r\, \partial z}\right]. \tag{2.14}$$

Expressions for $\tau_{\theta\theta}$, τ_{rr}, and $\tau_{r\theta}$ are omitted here.

For transient waves, the Laplace transform with respect to t is used. We denote the Laplace transform of a function by an overbar and the transform parameter s:

$$\bar{f}(\mathbf{x}, s) = \int_0^\infty f(\mathbf{x}, t) e^{-st}\, dt, \tag{2.15}$$

$$f(\mathbf{x}, t) = \frac{1}{2\pi i}\int_{\mathrm{Br}} \bar{f}(\mathbf{x}, s) e^{st}\, ds. \tag{2.16}$$

In the above Br means the Bromwich contour in the complex s plane, which is a line parallel to the imaginary axis with a constant positive real s.

B. General Solutions and Source Functions

General solutions for **u** are constructed from the solutions for scalar wave equations (2.5), (2.9), or (2.12). We assume that initially, both **u** and $\partial \mathbf{u}/\partial t$ vanish. Thus the initial conditions for the wave functions are

$$\phi = \partial\phi/\partial t = 0, \qquad \boldsymbol{\psi} = \partial\boldsymbol{\psi}/\partial t = 0 \qquad \text{at} \quad t = 0. \tag{2.17}$$

1. *Plane Strain Problems*

When $\mathbf{u} = [u_x, 0, u_z]$ and all field variables are independent of the y coordinate, the problem is *plane strain*. The displacement components are determined by two functions $\phi(x, z, t)$ and $\boldsymbol{\psi} = [0, \psi(x, z, t), 0]$.

Taking the Laplace transform of Eq. (2.5) and the second of Eq. (2.9), we obtain two equations for $\bar{\phi}(x, z, s)$ and $\bar{\psi}(x, z, s)$, respectively,

$$c^2 \nabla^2 \bar{\phi} - s^2 \bar{\phi} = -\bar{b}, \qquad C^2 \nabla^2 \bar{\psi} - s^2 \bar{\psi} = -\bar{B}, \tag{2.18}$$

where $\nabla^2 = \partial^2/\partial x^2 + \partial^2/\partial z^2$. These equations are reduced to ordinary differential equations by applying the Fourier transforms[1] (Sneddon, 1951):

$$\tilde{\phi}(\xi, z, s) = \int_{-\infty}^{\infty} \bar{\phi}(x, z, s) e^{-is\xi x}\, dx, \tag{2.19}$$

$$\bar{\phi}(x, z, s) = \frac{s}{2\pi} \int_{-\infty}^{\infty} \tilde{\phi}(\xi, z, s) e^{is\xi x}\, d\xi. \tag{2.20}$$

The solutions for $\tilde{\phi}$ and $\tilde{\psi}$ are

$$\tilde{\phi}(\xi, z, s) = A(\xi, s) e^{-s\eta z} + B(\xi, s) e^{s\eta z} + \tilde{\phi}_0,$$

$$\tilde{\psi}(\xi, z, s) = C(\xi, s) e^{-s\zeta z} + D(\xi, s) e^{s\zeta z} + \tilde{\psi}_0, \tag{2.21}$$

where $\tilde{\phi}_0$ and $\tilde{\psi}_0$ are particular solutions, not yet specified; A, B, C, D are unknown coefficients; and

$$\eta = (\xi^2 + c^{-2})^{1/2}, \qquad \zeta = (\xi^2 + C^{-2})^{1/2}. \tag{2.22}$$

Inverting the Fourier transforms, we obtain

$$\bar{\phi}(x, z, s) = \bar{\phi}_0 + \frac{s}{2\pi} \int_{-\infty}^{\infty} [A e^{-s\eta z} + B e^{s\eta z}] e^{is\xi x}\, d\xi,$$

$$\bar{\psi}(x, z, s) = \bar{\psi}_0 + \frac{s}{2\pi} \int_{-\infty}^{\infty} [C e^{-s\zeta z} + D e^{s\zeta z}] e^{is\xi x}\, d\xi. \tag{2.23}$$

Equations (2.23) are the general solutions of the Laplace transformed displacement potentials $\bar{\phi}$ and $\bar{\psi}$. The particular solutions $\bar{\phi}_0$ and $\bar{\psi}_0$ depend on the prescribed body force; the coefficients A, B, C, and D depend on boundary conditions. The following are three sample problems and solutions.

(i) *A line source of explosion in an infinite medium* (Fig. 1). We set in Eq. (2.18)

$$b = f(t)\delta(x)\delta(z - z_0), \qquad \mathbf{B} = [0, 0, 0]. \tag{2.24}$$

[1] These reduce to standard form of complex Fourier transform when $s\xi$ is replaced by another parameter, say k.

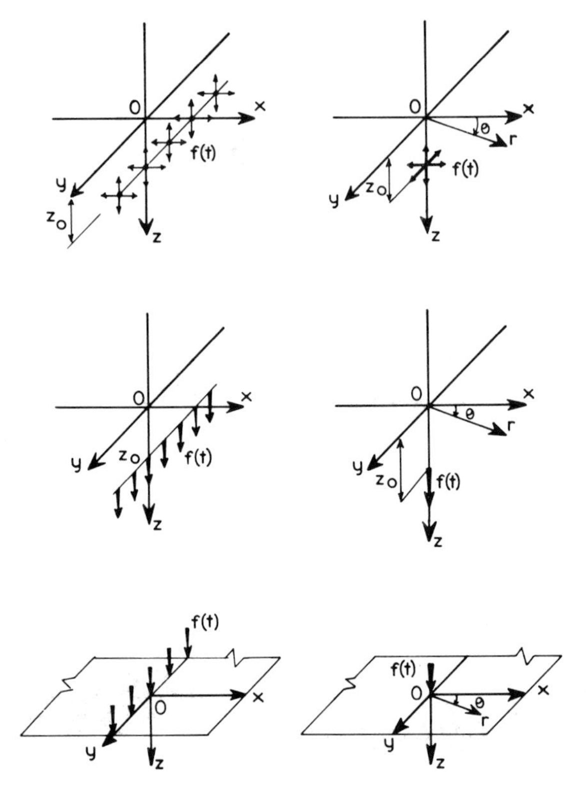

FIG. 1. Line and point sources.

In the above $\delta(x)$ is a delta function and $f(t)$ is the time function of the explosion, its Laplace transform being $\bar{f}(s)$.

(ii) *A line of vertical forces in an infinite medium.* Instead of prescribing body force, this problem can be solved by assuming (Achenbach, 1973, p. 295):

$$\text{for } z \geq z_0, \quad \tau_{zz} = -\tfrac{1}{2}\delta(x)f(t) \quad \text{and} \quad u_x = 0 \quad \text{at} \quad z = z_0;$$

$$\text{for } z \leq z_0, \quad \tau_{zz} = +\tfrac{1}{2}\delta(x)f(t) \quad \text{and} \quad u_x = 0 \quad \text{at} \quad z = z_0. \tag{2.25}$$

(iii) *A line of vertical forces at the free surface of a half-space.* The boundary conditions are ($z \geq 0$)

$$\tau_{zz} = -\delta(x)f(t) \quad \text{and} \quad \tau_{zx} = 0 \quad \text{at} \quad z = 0. \tag{2.26}$$

Solutions for these three problems can all be expressed as

$$\bar{\phi}(x, z, s) = s \int_{-\infty}^{\infty} \bar{F}(s) S_P(\xi) e^{s(i\xi x - \eta|z - z_0|)}\, d\xi$$

$$\bar{\psi}(x, z, s) = s \int_{-\infty}^{\infty} \bar{F}(s) S_V(\xi) e^{s(i\xi x - \zeta|z - z_0|)}\, d\xi \tag{2.27}$$

where S_P and S_V, which are the P and SV source functions respectively, are listed together with $\bar{F}(s)$ in Table I for each type of loading.

TABLE I

SOURCE FUNCTIONS FOR A LINE OR A POINT LOAD [EQS. (2.27) AND (2.35)][a]

Sources	$\bar{F}(s)$	$S_P(\xi)$	$S_V(\xi)$[b]
A point or a line of explosion	$\dfrac{\bar{f}(s)}{4\pi c^2 s}$	$\dfrac{1}{\eta}$	0
A point or a line of vertical force at interior	$\dfrac{\bar{f}(s)}{4\pi\rho s^2}$	$-\varepsilon$	$i\dfrac{\xi}{\zeta}$
A point or a line of vertical force at surface	$\dfrac{\bar{f}(s)}{2\pi\mu s^2}$	$\dfrac{\xi^2 + \zeta^2}{\Delta_r(\xi)}$	$-i\dfrac{2\xi\eta}{\Delta_r(\xi)}$

[a] $\varepsilon = \pm 1$ if the ray is in the $\pm z$ direction. $\eta = (\xi^2 + c^{-2})^{1/2}$ and $\zeta = (\xi^2 + C^{-2})^{1/2}$. $\Delta_r(\xi) = 4\eta\zeta\xi^2 - (\xi^2 + \zeta^2)^2$.

[b] The factor i should be omitted for a point source.

We have omitted here the case of *antiplane strain* $\mathbf{u} = [0, u_y(x, z, t), 0]$. It is usually solved by letting $\phi = 0$ and $\boldsymbol{\psi} = [0, 0, \psi_z(x, z, t)]$. In contrast to the vertical polarization of shear waves in plane strain (SV waves), the shear wave in this case is horizontally polarized (SH waves). The SH wave in cylindrical coordinates is discussed in Section VI,B.

2. *Plane Stress Problems*

Consider a thin plate bounded by two stress-free surfaces $y = \pm b$ (Fig. 1). If the excitation is such that the predominant motion is in the x-z plane and symmetric about the midsurface $y = 0$, the wave motion in the plate may be described also by $\phi(x, y, t)$ and $\psi(x, z, t)$ in Eq. (2.18). However, the P-wave speed should be replaced by the extensional plate wave speed with

$$c = [(\bar{\lambda} + 2\mu)/\rho]^{1/2}, \qquad \bar{\lambda} = (2\lambda\mu)/(\lambda + \mu). \tag{2.28}$$

The S-wave speed C remains unchanged.

The displacements $[u_x, 0, u_z]$ and stresses $[\tau_{xx}, \tau_{xz}, \tau_{zz}]$, as derived from Eqs. (2.7) and (2.10) respectively, are the averaged values of actual displacements and stresses across the thickness of the plate. Since in this case $\tau_{xy} = \tau_{yy} = \tau_{yz} = 0$, which is analogous to the plane stress solution in elastostatics, we call the problem *plane stress*. Because of the nature of approximation, the plane stress theory should be applied when the frequencies (in hertz) of waves in a plate are less than $C/4b$ (Pao and Kaul, 1974).

With these understandings the solutions given by Eqs. (2.19)–(2.27) are also applicable to extensional waves in thin plates (Achenbach, 1973, Section 6.12.3).

3. *Axisymmetric Problems*

For wave motion symmetric about the z axis, the displacements are independent of the θ coordinate. Excluding the rotational motion $\mathbf{u} = [0, u_\theta, 0]$, the displacement field $\mathbf{u} = [u_r, 0, u_z]$ is determined from two potentials $\phi(r, z, t)$ and $\psi(r, z, t)$ and $\chi = 0$ in Eq. (2.13), where $\bar{\phi}$ and $\bar{\psi}$ satisfy Eq. (2.18) with $\nabla^2 = \partial^2/\partial r^2 + (1/r)\,\partial/\partial r + \partial^2/\partial z^2$.

We then apply the Hankel transform to these equations, which are represented by (Sneddon, 1951, p. 48)

$$\hat{\phi}(\xi, z, s) = \int_0^\infty \bar{\phi}(r, z, s) J_0(s\xi r) r\, dr,$$

$$\bar{\phi}(r, z, s) = s^2 \int_0^\infty \hat{\phi}(\xi, z, s) J_0(s\xi r)\xi\, d\xi. \tag{2.29}$$

For convenience, $s\xi$ is taken as the parameter of transform. The solution for the twice transformed potentials are

$$\hat{\phi}(\xi, z, s) = A(\xi, s)e^{-s\eta z} + B(\xi, s)e^{s\eta z} + \hat{\phi}_0,$$

$$\hat{\psi}(\xi, z, s) = \frac{C(\xi, s)}{-s\xi} e^{-s\zeta z} + \frac{D(\xi, s)}{-s\xi} e^{s\zeta z} + \hat{\psi}_0, \tag{2.30}$$

where η and ζ are defined in Eq. (2.22), and A, B, C, D are unknown coefficients. By applying the inverse Hankel transform, we obtain[2]

$$\bar{\phi}(r, z, s) = \bar{\phi}_0 + s^2 \int_0^\infty [Ae^{-s\eta z} + Be^{s\eta z}] J_0(s\xi r)\xi\, d\xi,$$

$$\bar{\psi}(r, z, s) = \bar{\psi}_0 + s^2 \int_0^\infty \frac{1}{-s\xi} [Ce^{-s\zeta z} + De^{s\zeta z}] J_0(s\xi r)\xi\, d\xi. \tag{2.31}$$

[2] In the original papers by Spencer (1960) and Müller (1968) the potential function $A = \partial\psi/\partial r$ and the Hankel transform of A (Fourier–Bessel transform) with respect to $J_1(s\xi r)$

The particular solutions $\bar{\phi}_0$ and $\bar{\psi}_0$ again are determined from the prescribed body forces.

As in the previous subsection, we consider three types of point sources:

(i) *A point source of explosion in an infinite medium* (Fig. 1):

$$b = f(t)\delta(z - z_0)\delta(r)/2\pi r, \qquad \mathbf{B} = [0, 0, 0]. \tag{2.32}$$

(ii) *A concentrated force in an infinite medium:*

$$\begin{aligned} &\text{for } z \geq z_0; \quad \tau_{zz} = -\tfrac{1}{2}f(t)\delta(r)/2\pi r \quad \text{and} \quad u_r = 0 \quad \text{at} \quad z = z_0; \\ &\text{for } z \leq z_0; \quad \tau_{zz} = +\tfrac{1}{2}f(t)\delta(r)/2\pi r \quad \text{and} \quad u_r = 0 \quad \text{at} \quad z = z_0. \end{aligned} \tag{2.33}$$

(iii) *A vertical force at the free surface of a half-space:*

$$\tau_{zz} = -f(t)\delta(r)/2\pi r \quad \text{and} \quad \tau_{zr} = 0 \quad \text{at} \quad z = 0. \tag{2.34}$$

The Laplace-transformed solutions for these problems are

$$\begin{aligned} \bar{\phi}(r, z, s) &= s^2 \int_0^\infty \bar{F}(s)S_{\mathrm{P}}(\xi)e^{-s\eta|z-z_0|}J_0(s\xi r)\xi \, d\xi, \\ \bar{\psi}(r, z, s) &= s^2 \int_0^\infty \frac{\bar{F}(s)}{-s\xi} S_{\mathrm{V}}(\xi)e^{-s\zeta|z-z_0|}J_0(s\xi r)\xi \, d\xi. \end{aligned} \tag{2.35}$$

With the insertion of the factor $-1/s\xi$, the source functions S_{P} and S_{V} in the preceding equations are the same as those in plane strain problems (Table I).

The three problems mentioned here are classical problems in elastodynamics and are discussed in many texts (e.g., Achenbach, 1973). The inverse Laplace transforms of the solutions in Eq. (2.27) or Eq. (2.35) can be carried out by Cagniard's method which will be discussed later.

4. *Asymmetric Problems*

If the load applied to the layered medium is neither symmetric about the x-y plane (plane problems) nor symmetric about the z axis (axisymmetric problem), the solutions are rather complicated. We postpone the discussion of solutions for such a problem until Section VI, where additional source functions are also discussed.

---

are used. With potentials ϕ and A we have $u_r = \partial\phi/\partial r - \partial A/\partial z$ and $u_z = \partial\phi/\partial z + \partial A/\partial r + A/r$. Since $\partial J_0(s\xi r)/\partial r = (-s\xi)J_1(s\xi r)$, the second of Eqs. (2.31) can be neatly written as

$$\bar{A}(r, z, s) = \bar{A}_0 - s^2 \int_0^\infty [Ce^{-s\zeta z} + De^{+s\zeta z}]J_1(s\xi r)\xi \, d\xi.$$

The insertion of $-s\xi$ in Eqs. (2.30) and (2.31) proves very convenient in defining the source functions (in this section) and the reflection coefficients (Section V). The potential function A is not convenient for problems without symmetry.

C. The Theory of Normal Modes

The complexity of solutions increases as the number of boundary conditions increases. Consider, for example, the plane strain problem of a layer overlaying a semi-infinite space (Newlands, 1952). At the free surface $z = 0$ (see Fig. 12 in Section V), the boundary conditions may be either traction free ($\tau_{zz} = \tau_{zx} = 0$) or the same as that given in Eq. (2.26). At the interface $z = h$ there are two conditions of continuity for displacements ($u_{x1} = u_{x2}$, $u_{z1} = u_{z2}$) and two for stresses ($\tau_{zz1} = \tau_{zz2}$, $\tau_{zx1} = \tau_{zx2}$). The appended subscript distinguishes layer 1 from the semi-infinite space 2. In addition, there may be sources inside the layer or in the semi-infinite space. A total of six boundary conditions must be satisfied for this problem.

The solutions are obtained by constructing four wave functions $\bar{\phi}_1$, $\bar{\psi}_1$, $\bar{\phi}_2$, and $\bar{\psi}_2$, as given by Eq. (2.23). The six constants A_1, B_1, C_1, D_1, and A_2, C_2 ($B_2 = D_2 = 0$) can then be determined from the six boundary conditions (Ewing *et al.*, 1957, p. 189). Omitting all details here, we express these solutions symbolically by

$$A_1 = \Delta_A/\Delta, \qquad B_1 = \Delta_B/\Delta, \qquad C_1 = \Delta_C/\Delta, \qquad D_1 = \Delta_D/\Delta,$$
$$A_2 = \delta_A/\Delta, \qquad C_2 = \delta_C/\Delta. \tag{2.36}$$

Where $\Delta_A, \ldots, \Delta_D$, δ_A, δ_C, and Δ are 6×6 determinants. After expanding the determinant, the common denominator Δ may be expressed as (Pekeris *et al.*, 1965, Eqs. 33 and 110)

$$\begin{aligned}\Delta(s\xi, s) = e^{s(\eta_1 + \zeta_1)h}[&R^{PP}(R_{PP}e^{-2s\eta_1 h} - R_{SS}e^{2s\zeta_1 h}) \\ &+ (R_{PP}R_{SS} - R_{SP}R_{PS})e^{-2s(\eta_1 + \zeta_1)h} \\ &+ 2R^{SP}R_{PS}e^{-s(\eta_1 + \zeta_1)h} - 1].\end{aligned} \tag{2.37}$$

In the preceding equation, $\eta_1 = (\xi^2 + c_1^{-2})^{1/2}$ and $\zeta_1 = (\xi^2 + C_1^{-2})^{1/2}$. The R^{PP}, R^{SP} are reflection coefficients of plane waves (P to P mode and S to P mode) at the free surface [see Eq. (3.11)], whereas $R_{PP}, \ldots, R_{SS}$ are those at the lower surface [see Eq. (3.9)].

In Eq. (2.37), if $s\xi$ is replaced by k, and s by $i\omega$, the equation

$$\Delta(k, \omega) = 0 \tag{2.38}$$

is the frequency equation for the Rayleigh wave in the top layer. The roots of this transcendental equation determine the dispersion relation between the wave number k and the frequency ω, or phase velocity ω/k (Ewing *et al.*, 1957, p. 195). Each pair of numbers (ω, k) gives rise to a *normal mode* of the Rayleigh wave.

For a medium with many layers, the coefficients $\Delta_A, \ldots,$ and Δ can be determined by the matrix method (Thomson, 1950; Haskell, 1953). Further discussions of this method are found in the papers by Gilbert and Backus (1966), and Richards (1971).

Substituting Eqs. (2.36) and (2.37) into Eq. (2.23) and then taking the inverse Laplace transform, we obtain

$$\begin{aligned}\phi(x, z, t) &= \phi_0 + \frac{1}{2\pi i}\int_{\mathrm{Br}} ds\, e^{st} \int_{-\infty}^{\infty} \frac{1}{2\pi\Delta} \\ &\quad \times [\Delta_A e^{-\alpha_1 z} + \Delta_B e^{\alpha_1 z}] e^{ikx}\, dk, \\ \psi(x, z, t) &= \psi_0 + \frac{1}{2\pi i}\int_{\mathrm{Br}} ds\, e^{st} \int_{-\infty}^{\infty} \frac{1}{2\pi\Delta} \\ &\quad \times [\Delta_C e^{-\beta_1 z} + \Delta_D e^{\beta_1 z}] e^{ikx}\, dk, \end{aligned} \tag{2.39}$$

where

$$\alpha_1 = s\eta_1 = (k^2 + s^2/c_1^2)^{1/2}, \qquad \beta_1 = s\zeta_1 = (k^2 + s^2/C_1^2)^{1/2},$$

and $s\xi$ has been changed to k.

The final solutions [Eq. (2.39)] are in the form of double integrals. Changing the order of integration, we first evaluate the s integral by the calculus of residues:

$$\phi(x, z, t) = \phi_0 + \frac{1}{2\pi}\int_{-\infty}^{\infty} e^{ikx} \sum_{n=1}^{\infty} \left[\left(\frac{\partial\Delta}{\partial s}\right)^{-1} (\Delta_A e^{-\alpha_1 z} + \Delta_B e^{\alpha_1 z}) e^{st} \right]_{s = \pm i\omega_n} dk. \tag{2.40}$$

The $\omega_n(k)$ are the frequencies of normal modes, and $s = \pm i\omega_n$ are simple poles in the complex s plane. A similar expression may be obtained for $\psi(x, z, t)$. The remaining integration in k may be carried out numerically (fast Fourier transform) or asymptotically by the method of stationary phase.

The normal modes analysis as outlined above has been applied to media with many layers (Harkrider 1964, 1970; Fuchs, 1966; Dunkin and Corbin, 1970). The numerical work involved in this analysis is long and difficult; and the method is more effective for long-time transient responses at remote observation points. It is for this reason that an alternative is desirable.

Newlands (1952) applied the "Bromwich expansion method" and "Sommerfeld contour distorsion" to evaluate these double integrals. The former involves expanding Δ^{-1} in Eq. (2.39) into an infinite series. Each term in the series forms a ray integral which is then evaluated by applying the Sommerfeld method. It was Bromwich (1916) who showed in a one-dimensional problem of wave propagation that expansion in negative powers of exponentials expresses the motion in a series of pulses.

D. The Development of the Theory of Generalized Ray

In the theory of geometrical optics a light ray is defined to be the orthogonal trajectories to the wave fronts. The intensity of light along a ray path may be determined from the asymptotic solutions of wave equations at high frequencies (Born and Wolf, 1975). This is occasionally referred as the ray theory of optics.

Without the application of asymptotic analysis the general solution for waves in a bounded medium can also be sorted out into many parts by the Bromwich expansion or the like, each part identifiable as waves traveling along a ray path. In the literature this is also known as ray theory, but it is quite different from the ray theory of geometrical optics. Such a ray theory was developed early by Debye (1909) for investigating the diffraction of light inside a spherical particle, and later by Van der Pol and Bremmer (1937) to calculate the diffraction of radio waves by the earth. Their analyses dealt only with monochromatic waves (steady state response).

Applications of the normal mode theory and the ray theory to underwater sound were made by Pekeris in the early 1940s. His analysis of sound pulses in two- or three-layered liquid media was reported later in a memoir (Ewing *et al.*, 1948) in which the integrals based on the normal mode theory were evaluated by the method of stationary phase. The ray integrals were not evaluated until much later (Pekeris and Longman, 1958; Pekeris *et al.*, 1959).

These ray integrals for transient waves were evaluated by applying the so-called "Cagniard's method," a method to which Pekeris (1940) has also made original contributions. A historical account of Cagniard's method is given in the preface to the English translation of Cagniard's (1939) monograph prepared by Flinn and Dix in 1962.

For waves in elastic solids, applications of the ray theory and the Cagniard method were made to determine the transient response of a plate (Mencher, 1953; Knopoff, 1958a; Davids and Lawhead, 1965) and the transient Love waves in a layer (Knopoff, 1958b; Pekeris *et al.*, 1963). The more difficult problem of transient Rayleigh's waves in a layer was solved by Knopoff *et al.* (1960) and Pekeris *et al.* (1965).

In all papers cited the ray integrals are derived from the Bromwich expansion of the exact solutions for waves in a layered medium. As shown in Eq. (2.37), such an expansion requires the evaluation of a 6×6 determinant for a two-layered solid (a layer overlaying a half-space). The size of determinant increases to 10×10 for waves in a three-layered solid with axial symmetry and to 15×15 if without symmetry. Thus the Bromwich expansion, which is a cornerstone of the ray theory, becomes a stumbling block in the application of the ray theory to media with many layers.

The publication of three papers by Spencer (1960, 1965a,b) finally removed this stumbling block. He introduced, without recourse to the Bromwich expansion, the concept of *generalized ray paths* and showed how the ray integrals can be constructed directly from the known source functions and reflection and transmission coefficients for plane waves along each path. Since the ray integrals can be evaluated by the Cagniard method, and the solution is exact up to the time of arrival of the next ray, the theory is called by Spencer the "exact elastic ray theory." However, in later publications, the title of "generalized ray theory" prevailed.

The theory was applied by Berry and West (1966) to analyze waves in stratified media. A comprehensive presentation of the theory along with numerical examples was made by Müller (1968, 1969). In his papers a Fourier transform of the time variable and a modified Cagniard's method developed by Bortfeld (1962) were adopted. More recent literature on the theory is cited in the paper by Chapman (1976b).

III. The Theory of Generalized Ray

As shown in the previous section, the general solutions of the Laplace transformed potentials for a homogeneous layer are given by Eqs. (2.23) and (2.31). We shall append an index i ($i = 1, 2, 3, \ldots$) for physical quantities in each layer.

Consider a medium composed of three layers, the thickness of the lowest layer being very large. A line source is placed in layer 2 (Fig. 2). The problem is then plane strain and all field variables are independent of the y coordinate.

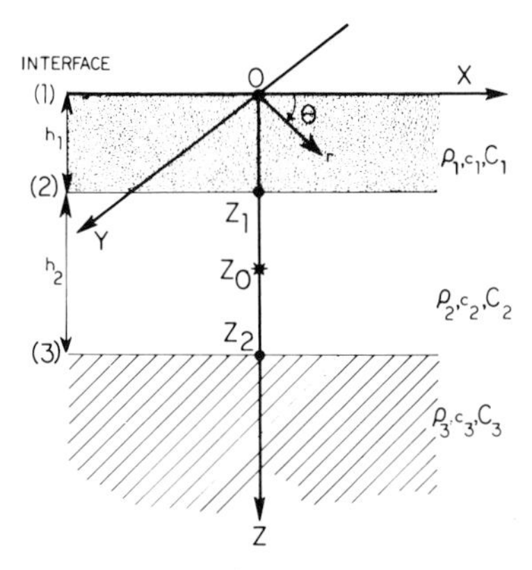

FIG. 2. Geometry of a three-layer medium.

The coefficients A_i, B_i, C_i, D_i ($i = 1, 2, 3$) are determined from the source functions in Eqs. (2.27) and the boundary conditions at the interfaces:

$$(\tau_{zx})_1 = (\tau_{zz})_1 = 0 \quad \text{at} \quad z = 0,$$

$$\begin{matrix} (\tau_{zx})_i = (\tau_{zx})_{i+1}, & (\tau_{zz})_i = (\tau_{zz})_{i+1} \\ (u_x)_i = (u_x)_{i+1}, & (u_z)_i = (u_z)_{i+1} \end{matrix} \quad (i = 1, 2) \quad \text{at} \quad z = z_i. \tag{3.1}$$

There are a total of ten unknown coefficients and ten boundary conditions for a three-layered medium ($B_3 = D_3 = 0$).

Instead of determining these coefficients from a system of ten simultaneous equations and then applying Bromwich expansion to sort out individual rays, we shall construct the ray integrals by the method of generalized ray (Knopoff *et al.*, 1960; Spencer, 1960). The inverse transform of these ray integrals is discussed in the next section.

A. THE UNDERLYING PRINCIPLE

1. *Reflection and Transmission of Cylindrical Waves*

Consider a line source in layer 2. Throughout this section we consider only the P waves emitted by the source, which, according to Eq. (2.27), are represented by $\bar{\psi}_0 = 0$ and

$$\bar{\phi}_0(x, z, s) = s\bar{F}(s) \int_{-\infty}^{\infty} S_{P2} e^{s(i\xi x - \eta_2 |z - z_0|)} \, d\xi, \tag{3.2}$$

where

$$\eta_i = (\xi^2 + c_i^{-2})^{1/2}, \qquad \zeta_i = (\xi^2 + C_i^{-2})^{1/2} \qquad (i = 1, 2, 3),$$

and $\bar{F}(s)$ and S_P are given in Table I.

The solution represents a cylindrical wave with axis parallel to the y coordinate, which expands with wave speed c_2 from the source at $z = z_0$. Physically, this cylindrical wave may be regarded as a superposition of plane waves passing through the source simultaneously in all directions. This is seen by noting that the Laplace transform of a plane wave $f[t - (mx + nz)/c]$ is $\bar{f}(s) \exp[-(s/c)(mx + nz)]$, m and n being the direction cosines of the wave normal and $m^2 + n^2 = 1$. A change of variable from m to $-i\xi c$ gives rise to $\bar{f}(s) \exp[s(i\xi x - \eta z)]$. When each plane wave is assigned an amplitude $S_{P2}(\xi)$ and plane waves in all directions are summed through an integration with respect to ξ, the resultant is a cylindrical wave.

When a plane wave component impinges on the interface of two solids

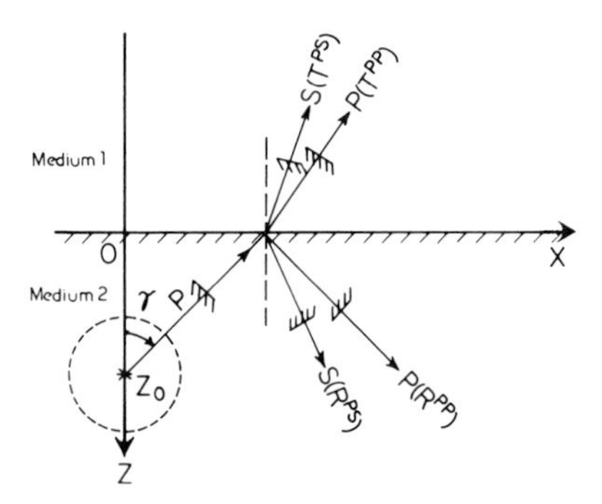

FIG. 3. Reflection and transmission of a plane wave component at the interface.

(Fig. 3), two plane waves, P and S, are reflected into the source medium (medium 2) and two are transmitted into the adjacent medium (medium 1). Let the ratios of the displacement potentials for the reflected waves to those of the incident waves be R^{PP}, R^{PS} (P wave reflects as P or S wave) and R^{SP}, R^{SS} (S wave reflects as P or S wave). Similarly the ratios for the transmitted waves are designated by T^{PP}, T^{PS}, T^{SP}, and T^{SS}. These reflection and transmission coefficients, which are usually given as functions of angle of incidence and angle of emergence (Ewing *et al.*, 1957), can be expressed in terms of the wave slownesses ξ, η_i, and ζ_i [see Eqs. (3.9)]. Hence for an incident cylindrical wave $\bar{\phi}_0$, the compressional and shear parts of the reflected wave in medium 2 are obtained by superposing all reflected plane P waves or S waves respectively:

$$\bar{\phi}_2 = s\bar{F}(s) \int_{-\infty}^{\infty} S_{P2}(R^{PP}) e^{s(i\xi x - \eta_2 z - \eta_2 z_0)}\, d\xi,$$

$$\bar{\psi}_2 = s\bar{F}(s) \int_{-\infty}^{\infty} S_{P2}(R^{PS}) e^{s(i\xi x - \zeta_2 z - \eta_2 z_0)}\, d\xi. \tag{3.3}$$

The total wave in the source medium is $\bar{\phi}_0 + \bar{\phi}_2$ and $\bar{\psi}_2$. Similarly, the two transmitted waves in medium 1 are

$$\bar{\phi}_1 = s\bar{F}(s) \int_{-\infty}^{\infty} S_{P2}(T^{PP}) e^{s(i\xi x + \eta_1 z - \eta_2 z_0)}\, d\xi,$$

$$\bar{\psi}_1 = s\bar{F}(s) \int_{-\infty}^{\infty} S_{P2}(T^{PS}) e^{s(i\xi x + \zeta_1 z - \eta_2 z_0)}\, d\xi. \tag{3.4}$$

In Eqs. (3.2)–(3.4), each integral that is a superposition of all plane waves is called a "ray." Thus $\bar{\phi}_0$ is the "incident ray" or "source ray," $\bar{\phi}_2$ (P wave to P wave) and $\bar{\psi}_2$ (P wave to S wave) are the "reflected rays," and $\bar{\phi}_1$

and $\bar{\psi}_1$ the "transmitted rays." Spencer (1960) showed that the solutions so constructed indeed satisfy the boundary conditions at the interface. The coefficients R^{PP} and R^{PS} together with the factor $s\bar{F}(s)S_{P2}\exp(-s\eta_2 z_0)$ replace the coefficients A_2 and C_2 respectively in Eq. (2.23). Similarly, T^{PP} and T^{PS} replace B_1 and D_1, respectively. With reference to Fig. 3, coefficients B_2 and D_2 in Eq. (2.23) vanish because the z axis is unbounded in medium 2, and A_1, C_1 are zero because medium 1 is unbounded in the $-z$ direction.

Between the source $(0, z_0)$ and receiver (x, z), there is a least-travel-time path for each ray. Since the wave propagates with finite speed, no signal will be detected at the receiver before the least time of arrival. The least-time paths for the source ray ($\bar{\phi}_0$) and the two reflected rays ($\bar{\phi}_2$ and $\bar{\psi}_2$) are shown in Fig. 4. Reflected waves along different ray paths (angles different

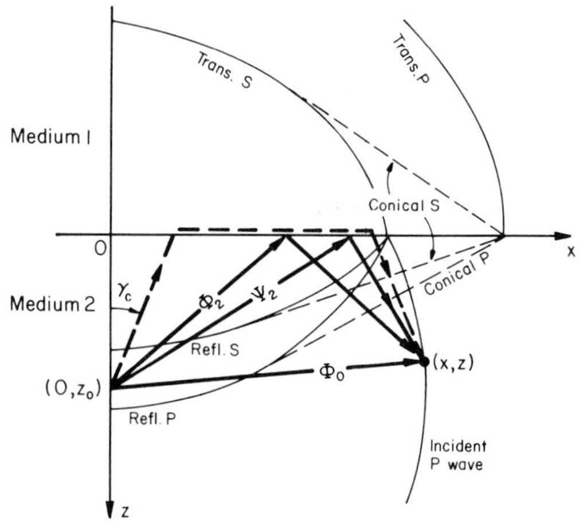

FIG. 4. Incident, reflected, and refracted waves and rays ($c_1 > c_2 > C_1 > C_2$).

from those shown) will arrive at later times. However, we know that for plane waves, if the angle of incidence exceeds a critical angle γ_c, and when the wave speeds in the adjacent medium are faster than those in the source medium, the transmitted waves are refracted and move along the interface with the faster wave speed. Conical waves moving along this "refracted ray path" may arrive earlier than those following the direct reflected ray path. A refracted P ray (P wave along the interface) is also shown in Fig. 4 in dashed lines where $\gamma_c = \sin^{-1}(c_2/c_1)$. The existence of such a "refracted ray" depends on the velocity ratios of the two media, the depths of the source and receiver, and the horizontal distance between them. Calculations of the least travel time for the direct ray and refracted ray will be discussed in Section IV.

2. *Receiver Functions*

Once the potentials $\bar{\phi}$ and $\bar{\psi}$ are constructed for the waves along each ray, the displacements and stresses are calculated according to Eqs. (2.7) and (2.10), respectively. For instance, displacements due to the reflected P-P and P-S rays are

$$\begin{aligned}\bar{u}_{x2} &= s\bar{F}(s)\int_{-\infty}^{\infty} sS_{P2}R^{PP}e^{-s(\eta_2 z_0+\eta_2 z)+is\xi x}(i\xi)\,d\xi \\ &\quad + s\bar{F}(s)\int_{-\infty}^{\infty} sS_{P2}R^{PS}e^{-s(\eta_2 z_0+\zeta_2 z)+is\xi x}(\zeta_2)\,d\xi, \\ \bar{u}_{z2} &= s\bar{F}(s)\int_{-\infty}^{\infty} sS_{P2}R^{PP}e^{-s(\eta_2 z_0+\eta_2 z)+is\xi x}(-\eta_2)\,d\xi \\ &\quad + s\bar{F}(s)\int_{-\infty}^{\infty} sS_{P2}R^{PS}e^{-s(\eta_2 z_0+\zeta_2 z)+is\xi x}(i\xi)\,d\xi. \qquad (3.5)\end{aligned}$$

Similarly, those in medium 1 due to the transmitted P-P and P-S rays are

$$\begin{aligned}\bar{u}_{x1} &= s\bar{F}(s)\int_{-\infty}^{\infty} sS_{P2}T^{PP}e^{-s(\eta_2 z_0-\eta_1 z)+is\xi x}(i\xi)\,d\xi \\ &\quad + s\bar{F}(s)\int_{-\infty}^{\infty} sS_{P2}T^{PS}e^{-s(\eta_2 z_0-\zeta_1 z)+is\xi x}(-\zeta_1)\,d\xi, \\ \bar{u}_{z1} &= s\bar{F}(s)\int_{-\infty}^{\infty} sS_{P2}T^{PP}e^{-s(\eta_2 z_0-\eta_1 z)+is\xi x}(\eta_1)\,d\xi \\ &\quad + s\bar{F}(s)\int_{-\infty}^{\infty} sS_{P2}T^{PS}e^{-s(\eta_2 z_0-\zeta_1 z)+is\xi x}(i\xi)\,d\xi. \qquad (3.6)\end{aligned}$$

Aside from the parameter s, the extra factors $i\xi$, ζ_2, $-\eta_2$, $-\zeta_1$, and η_1 (all within parentheses) will be designated as the *receiver functions* for displacements. Thus we may construct ray integrals for any type of responses (displacements, velocities, stresses, etc.) by using the same reflection and transmission coefficients for the displacement potentials, together with a proper source function and a receiver function.

Since only a P source is considered here, each displacement is composed of two rays, P-P and P-S. If the source at z_0 emits S waves as well as P waves, there would be four ray integrals (P-P, P-S, S-P, S-S). The receiver functions are, however, the same for both P and S sources, except when the receiver is at the free surface or the interface.

If an observational point is at the free surface of a half-space, all transmitted rays vanish. Furthermore, when $z = 0$ the three rays shown in Fig. 4, the source P ray and the reflected P-P and P-S rays, coalesce into one P ray. By letting $z \to 0$ in Eq. (3.5) and adding the contribution from $\bar{\phi}_0$ we obtain

$$\bar{u}_x(x, 0, s) = s^2 \bar{F}(s) \int_{-\infty}^{\infty} S_P[i\zeta + i\xi R^{PP} + \zeta R^{PS}] e^{s(i\xi x - \eta z_0)}\, d\xi,$$

$$\bar{u}_z(x, 0, s) = s^2 \bar{F}(s) \int_{-\infty}^{\infty} S_P[\eta - \eta R^{PP} + i\xi R^{PS}] e^{s(i\xi x - \eta z_0)}\, d\xi. \tag{3.7}$$

Similarly, with a line S source at $z = z_0$, the responses at $z = 0$ due to the coalesced S ray are

$$\bar{u}_x(x, 0, s) = s^2 \bar{F}(s) \int_{-\infty}^{\infty} S_V[-\zeta + i\xi R^{SP} + \zeta R^{SS}] e^{s(i\xi x - \zeta z_0)}\, d\xi,$$

$$\bar{u}_z(x, 0, s) = s^2 \bar{F}(s) \int_{-\infty}^{\infty} S_V[i\xi - \eta R^{SP} + i\xi R^{SS}] e^{s(i\xi x - \zeta z_0)}\, d\xi. \tag{3.8}$$

In the above the subscripts 2 for S, η, and ζ have been dropped.

The quantities inside the brackets of Eqs. (3.7) and (3.8) will be called the *surface receiver functions* for the coalescent ray. For quick reference, we summarize the receiver functions for displacements and stresses in Table II.

TABLE II

RECEIVER FUNCTIONS FOR PLANE STRAIN [D IN EQ. (3.18)]

Displacement or stress	q	P Wave	S Wave
		Receiver at an interior point	
$\bar{u}_z$	1	$-\varepsilon\eta$	$i\xi$
$\bar{u}_x$	1	$i\xi$	$\varepsilon\zeta$
$\bar{\tau}_{zz}$	2	$\mu(\zeta^2 + \xi^2)$	$-2\mu\varepsilon(i\xi)\zeta$
$\bar{\tau}_{xx}$	2	$\mu(\zeta^2 - \xi^2 - 2\eta^2)$	$2\mu\varepsilon(i\xi)\zeta$
$\bar{\tau}_{zx}$	2	$-2\mu\varepsilon(i\xi)\eta$	$-\mu(\zeta^2 + \xi^2)$
		Receiver at the free surface $z = 0$	
$\bar{u}_z$	1	$\eta - \eta R^{PP} + i\xi R^{PS}$	$i\xi - \eta R^{SP} + i\xi R^{SS}$
$\bar{u}_x$	1	$i\xi + i\xi R^{PP} + \zeta R^{PS}$	$-\zeta + i\xi R^{SP} + \zeta R^{SS}$

B. REFLECTION AND TRANSMISSION COEFFICIENTS

The reflection and transmission coefficients in the preceding equations are the same as those for plane P and SV waves (Ewing *et al.*, 1957, p. 87)[3]:

$$R^{PP} = [(l_1 - l_3)(m_2 + m_4) - (m_1 - m_3)(l_2 + l_4)]\Delta^{-1},$$

$$R^{PS} = i\varepsilon 2(l_2 m_4 - l_4 m_2)\Delta^{-1},$$

$$T^{PP} = 2(m_2 + m_4)\Delta^{-1}, \qquad T^{PS} = i\varepsilon 2(l_2 + l_4)\Delta^{-1},$$

[3] The factor i should be omitted for axial symmetric waves.

$$R^{SP} = \varepsilon 2(m_3 l_1 - m_1 l_3)(i\Delta)^{-1},$$
$$R^{SS} = [(l_2 - l_4)(m_1 + m_3) - (m_2 - m_4)(l_1 + l_3)]\Delta^{-1},$$
$$T^{SP} = \varepsilon 2(m_1 + m_3)(i\Delta)^{-1}, \qquad T^{SS} = 2(l_1 + l_3)\Delta^{-1}, \tag{3.9}$$

where

$$\Delta = (l_1 + l_3)(m_2 + m_4) - (l_2 + l_4)(m_1 + m_3).$$

The elements $l_1, l_2, \ldots, m_3, m_4$ may be expressed in terms of shear modulus μ_i, and ξ, η_j, ζ_j as defined in Eq. (3.2) where the subscript j (=1, 2) is appended to indicate the source (2) and adjacent (1) medium:

$$\begin{aligned}
l_1 &= C_2^2[\bar{\mu}(\xi^2 + \zeta_1^2) - 2\xi^2],\\
l_2 &= C_2^2(\xi/\zeta_2)[\bar{\mu}(\xi^2 + \zeta_1^2) - (\xi^2 + \zeta_2^2)],\\
l_3 &= C_2^2(\eta_1/\eta_2)[\xi^2 + \zeta_2^2 - 2\bar{\mu}\xi^2],\\
l_4 &= C_2^2(2\xi\eta_1)(1 - \bar{\mu}),\\
m_1 &= -C_2^2(2\xi\zeta_1)(\bar{\mu} - 1),\\
m_2 &= C_2^2(\zeta_1/\zeta_2)[\xi^2 + \zeta_2^2 - 2\bar{\mu}\xi^2],\\
m_3 &= -C_2^2(\xi/\eta_2)[\xi^2 + \zeta_2^2 - \bar{\mu}(\xi^2 + \zeta_1^2)],\\
m_4 &= C_2^2[\bar{\mu}(\xi^2 + \zeta_1^2) - 2\xi^2],
\end{aligned} \tag{3.10}$$

where

$$\bar{\mu} = \mu_1/\mu_2 \qquad \text{and} \qquad C_2^2 = \mu_2/\rho_2.$$

If medium 1 is a vacuum, the transmission coefficients are zero and the reflection coefficients at a free surface are (subscripts for η and ζ are omitted)[4]

$$R^{PP} = [4\eta\zeta\xi^2 + (\xi^2 + \zeta^2)^2]\Delta_r^{-1}, \qquad R^{SS} = R^{PP},$$
$$R^{PS} = \varepsilon(i4\xi\eta)(\xi^2 + \zeta^2)\Delta_r^{-1}, \qquad R^{SP} = \varepsilon 4\xi\zeta(\xi^2 + \zeta^2)(i\Delta_r)^{-1}, \tag{3.11}$$

where

$$\Delta_r = 4\eta\zeta\xi^2 - (\xi^2 + \zeta^2)^2.$$

In Eqs. (3.9) and (3.11) the mode-converting reflection or transmission coefficients (R^{PS}, R^{SP}, T^{PS}, T^{SP}) contain the direction factor ε which is $+1$ or -1 according as the incident wave propagates in the $+z$ or $-z$ direction. The signs for mode-preserving coefficients (R^{PP}, R^{SS}, T^{PP}, T^{SS}) remain unchanged.

[4] The factor i should be omitted for axial symmetric waves.

Note that all elements l_i, m_i are either real, even functions of ξ, or real, odd ones of ξ when ξ is a real variable. Thus, for real ξ, all mode-preserving coefficients are real, even functions of ξ, and all mode-converting coefficients are imaginary, odd functions of ξ.

C. Receiver Functions and Generalized Rays

In a multilayered solid the wave radiated by the source undergoes a series of reflections and transmissions before it reaches a receiver. A generalized ray path can be constructed to connect the source and receiver, which specifies the vertical distance traversed by each mode of waves (P or S wave) in each layer, the total horizontal distance, and the direction of propagation. With each reflection or transmission the amplitudes of the plane wave components are altered by the corresponding reflection or transmission coefficients. Thus for a multiply reflected and transmitted cylindrical wave, we may determine the resultant response by the same principle applied in deriving Eqs. (3.5) and (3.6).

Two examples of generalized ray paths are shown in Fig. 5. The Laplace transforms of the ray integrals for each example are given below ($d = z - h_1$).

Example A A line source in layer 2, receiver in layer 2 (P wave):

$$\bar{u}_x = s^2\bar{F}(s)\int_{-\infty}^{\infty} S_{P2}R^{PP}_{(2)}R^{(3)}_{PS}R^{SS}_{(2)}R^{(3)}_{SP}(i\xi)e^{sg(x,\,z;\,\xi)}\,d\xi$$

$$g(x, z; \xi) = i\xi x - [\eta_2(z_0 - h_1 + h_2) + 2\zeta_2 h_2 + \eta_2(h_2 - d)]$$

$\bar{u}_z$: Replace the receiver function $(i\xi)$ in the above integral by $+\eta_2$. (3.12)

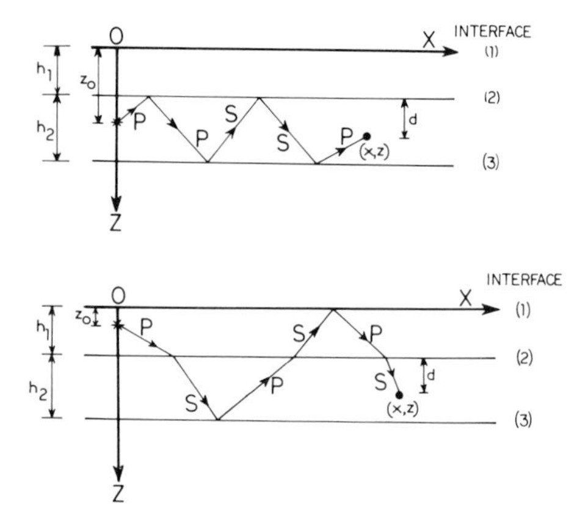

Fig. 5. Paths of generalized rays: (a) a line source in layer 2, P wave at receiver; (b) a line source in layer 1, S wave at receiver.

Since the first segment is a P mode and the rest can be either a P or a S mode, there are a total of $1 \times 2^4 = 16$ ray integrals for all combinations of modes along this path, eight P-wave arrivals and eight S-wave arrivals.

Example B A line source in layer 1, receiver in layer 2 (S-wave):

$$\bar{u}_x = s^2 \bar{F}(s) \int_{-\infty}^{\infty} S_{P1}\, T_{PS}^{(2)} R_{SP}^{(3)} T_{(2)}^{PS} R_{(1)}^{SP} T_{PS}^{(2)} (\zeta_2) e^{sg(x,\, z;\, \xi)}\, d\xi$$

$$g(x, z; \xi) = i\xi x - [\eta_1(h_1 - z_0) + \zeta_2 h_2 + \eta_2 h_2 + \zeta_1 h_1 + \eta_1 h_1 + \zeta_2 d]$$

$\bar{u}_z$: Replace the receiver function (ζ_2) in the above integral by $i\xi$. (3.13)

The total number of ray integrals along this path is $1 \times 2^5 = 32$.

In the above equations the numerals in parentheses attached to R or T indicate the interface at which the reflection or transmission takes place; the PP, PS, ... letters indicate the change of modes, superscripts meaning that the incident ray is propagating upward ($\varepsilon = -1$) and subscripts meaning downward ($\varepsilon = +1$).

In general the Laplace transform of the displacements corresponding to each of the generalized rays is

$$(\bar{u}_x, \bar{u}_z) = s^2 \bar{F}(s) \int_{-\infty}^{\infty} S(\xi) \pi(\xi) D(\xi) e^{sg(x,\, z;\, \xi)}\, d\xi, \tag{3.14}$$

where

$$g(x, z; \xi) = i\xi x - \sum_j (\eta_j z_{pj} + \zeta_j z_{sj}). \tag{3.15}$$

In the phase function $g(x, z; \xi)$, z_{pj} and z_{sj} are the total vertical distances (projections on the z axis) traversed by the P-wave segments or the S-wave segments respectively in the jth layer. In Eq. (3.14) $\bar{F}(s)$ is the transform of the time function of a line source, $S(\xi)$ is the source function, $D(\xi)$ the receiver function, and $\pi(\xi)$ is a continuous product of the reflection and transmission coefficients at the interfaces that are struck by the generalized ray.

The source function $S(\xi)$ may be either a P source or an SV source as given in Table I. The receiver functions $D(\xi)$, which depend solely on the mode (P or S wave) and direction (up or down) of the last segment of the generalized ray are listed in Table II. For $\bar{u}_x$, $D(\xi)$ is imaginary and odd if the last segment is a P mode, and real and even for an S mode. The converse is true for $\bar{u}_z$.

The properties for $\pi(\xi)$ are determined by the modes of the first and the last segments of the ray, and the properties of the reflection and transmission coefficients. For real ξ, all mode-preserving coefficients (R^{PP}, R^{SS}, T^{PP}, T^{SS}) are real, even functions of ξ and all mode-conversion coefficients (R^{PS}, R^{SP}, T^{PS}, T^{SP}) are imaginary, odd functions of ξ. Since the first segment initiated at a P source is always a P mode, $\pi(\xi)$ is then a real, even function in ξ if the last segment is a P mode (mode preserving); and imaginary, odd function if the last one is a S mode (mode converting).

Based on previous discussions, we conclude that for a P source, the product $S(\xi)\pi(\xi)D(\xi)$ is always an imaginary and odd function of ξ for $\bar{u}_x(-i\xi E_x)$, and a real and even one for $\bar{u}_z(E_z)$. This enables us to rewrite Eq. (3.14) as

$$\bar{u}_x(x, z, s) = 2s^2\bar{F}(s) \operatorname{Im} \int_0^\infty \xi E_x(\xi) e^{sg(x, z; \xi)} \, d\xi, \tag{3.16}$$

$$\bar{u}_z(x, z, s) = 2s^2\bar{F}(s) \operatorname{Re} \int_0^\infty E_z(\xi) e^{sg(x, z; \xi)} \, d\xi, \tag{3.17}$$

where Im and Re stand for "imaginary part" and "real part," respectively, and $E_x(\xi)$ and $E_z(\xi)$ stand for even functions of ξ, real when ξ is real.

To summarize, the procedure to derive the integral representation of a generalized ray is as follows:

(1) Determine the ray path by connecting the source and receiver with a straight (source ray) or broken line (reflected and transmitted ray). For a multireflected ray, each segment of the broken line must traverse the entire thickness of a layer except at both ends.

(2) Specify the mode (P or S mode) of the traveling waves along each segment of the path in each layer.

(3) Assign the reflection coefficient R and transmission coefficient T at each interface touched or traversed by the ray.

(4) Assemble the integrand as a product of the source function S (Table I), reflection and transmission coefficients π [Eqs. (3.9)–(3.11)], receiver function D (Table II), and a phase function $\exp[is\xi x - sh(\xi, z)]$, where, as in Eq. (3.15), $h(\xi, z) = \sum (\eta_j z_{pj} + \zeta_j z_{sj})$.

The final form of a ray integral for either displacements or stresses is

$$(\bar{\tau}_{ij}, \bar{u}_i) = s\bar{F}(s) \int_{-\infty}^{\infty} s^q S(\xi)\pi(\xi)D(\xi) e^{-sh(\xi, z) + is\xi x} \, d\xi, \tag{3.18}$$

where the power q is also given in Table II. The limits of integration are then changed from $(-\infty, \infty)$ to $(0, \infty)$ as done in Eqs. (3.16) and (3.17).

IV. The Cagniard Method and Transient Waves due to a Line Load

To find the inverse Laplace transform of $\bar{u}_x$ and $\bar{u}_z$ represented by Eqs. (3.16) and (3.17), respectively, we consider first the integral

$$\bar{u}(s) = \int_0^{\infty} \xi E(\xi) e^{sg(x, z; \xi)} \, d\xi. \tag{4.1}$$

Note that the function $g(\xi)$ has the dimension of time. When it is replaced by the variable, say, $-t$, and if the inverse relation $\xi = g^{-1}(t)$ can be uniquely determined, the integral is then the Laplace transform of the function $\xi(t)E[\xi(t)](d\xi/dt)$. The inverse Laplace transform can be obtained by inspection. This is, in essence, the Cagniard method.

There have been several versions of Cagniard's method in the literature. Our presentation follows the original work by Cagniard (1939, Chapter 5) and the paper by Garvin (1956).

We begin here by writing the phase function [Eq. (3.15)] as

$$g(x, z; \xi) = i\xi x - \sum_{j=1}^{4} z_j(\xi^2 + a_j^2)^{1/2} \tag{4.2}$$

with $a_4 = 1/c_2$, $a_3 = 1/C_2$, $a_2 = 1/c_1$, $a_1 = 1/C_1$; z_j is the total vertical projection of all segments associated with the slowness a_j. We may assume without losing generality that $c_2 > C_2 > c_1 > C_1$. If the source is in layer 1 and the generalized ray does not traverse through layer 2, then the summation is only up to 2. On the other hand, if a ray passes to a receiver in layer 3, the velocities of the waves in the third layer c_3 $(1/a_6)$ and C_3 $(1/a_5)$ should be included in $g(\xi)$ and the summation would be extended to $j = 6$. We assume $a_6 < a_5 < a_4 < a_3 < a_2 < a_1$.

The factor $E(\xi)$ in Eq. (4.1) is in general a function of ξ^2 and $(\xi^2 + a_j^2)^{1/2}$ for $j = 1, 2, \ldots, 6$ because $R_{\mathrm{PP}}^{(3)}$ and $T_{\mathrm{PS}}^{(3)}$, etc., depend on a_5 and a_6. $E(\xi)$ is real when ξ is real.

A. The Cagniard Method

A transformation of ξ to t by

$$-t = g(x, z; \xi) = i\xi x - \sum_{j} z_j(\xi^2 + a_j^2)^{1/2} \tag{4.3}$$

maps the complex variable ξ onto a complex t plane (Figs. 6 and 7). The functions $E(\xi)$ and $g(\xi)$ are made single-valued with the branch cuts shown.

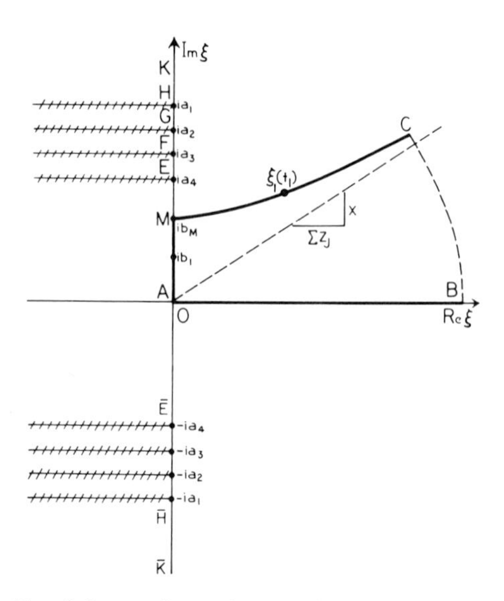

FIG. 6. Integration path AB and complex ξ plane.

In Fig. 6 the branch points at $\pm ia_5$ and $\pm ia_6$ and the corresponding cuts are omitted. Their significance will be discussed in Section IV,B,2.

The function $E(\xi)$ also has simple poles on the imaginary ξ axis, corresponding to the wave slownesses of the Stoneley waves at interfaces (2) and (3) and of the Rayleigh wave at the free surface (1) (see Fig. 2). Since Stoneley

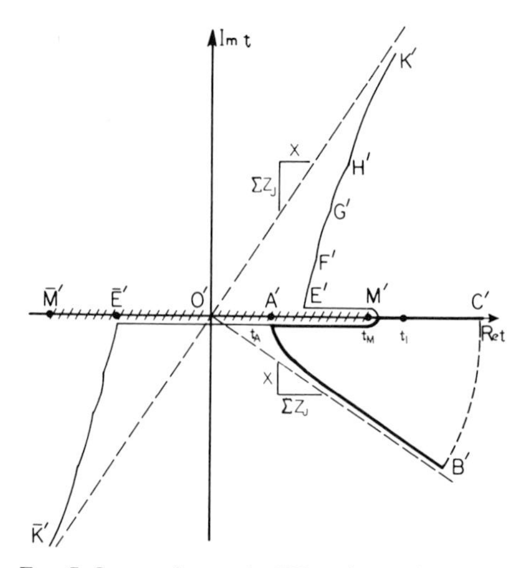

FIG. 7. Integration path $A'B'$ and complex t plane.

wave velocities are less than the P and S wave speeds in the adjacent media and the Rayleigh wave velocity is less than both wave speeds in the top layer, all these poles will be located above (and below) the branch point $+ia_4$ $(-ia_4)$ in the upper (and lower) ξ plane. These poles are omitted in the figures.

Equation (4.3) maps the real ξ axis, which is the original path of integration (line AB in Fig. 6), into a curve in the complex t plane (curve $A'B'$ in Fig. 7). The curve originates at A' which corresponds to $t = t_A$ with

$$t_A = -g(x, z; 0) = \sum_j (z_j a_j).$$

It has the asymptote $O'B'$ shown in dashed lines.

The mapping of the imaginary ξ axis is more complicated and requires a study of the inverse transformation of Eq. (4.3), that is, $\xi = g^{-1}(t)$. If there are only one or two radicals in the equation, ξ can be solved explicitly in terms of t. For the case of more than two radicals, the inverse transformation can be evaluated numerically. The inverse function $g^{-1}(t)$ may have singularities when $dt/d\xi = 0$. At these singularities the transformation in Eq. (4.3) has stationary values.

Along the imaginary axis, let $\xi = iw$, w being real. Equation (4.3) may then be written as

$$t = xw + \sum_j z_j(a_j^2 - w^2)^{1/2}.$$

Thus t is complex when $|w| > a_4$ (points above E, or below $\bar{E}$ in Fig. 6), and real when $|w| < a_4$ (points between E and $\bar{E}$). As $w \to \pm\infty$, t approaches an asymptote $K'\bar{K}'$ (shown in Fig. 7) because $t \to \pm w(x + i \sum z_j)$.

The stationary points of the function $t = -g(\xi)$ are obtained from the equation

$$\frac{dt}{d\xi} = -ix + \sum_j \frac{z_j \xi}{(\xi^2 + a_j^2)^{1/2}} = 0. \tag{4.4}$$

It can be shown that this equation has only one root, $\xi = ib_M$, where b_M is a positive, real number, less than a_4 for all positive values of x. This point is labeled M on the ξ plane (Fig. 6) and it is mapped into the point M' on the real t axis. At this stationary point, the inverse function $g^{-1}(t)$ is singular of half-order. Thus a branch cut is introduced from M' along the real t axis to the point $\bar{M}'$ on Fig. 7 to render $t(\xi)$ single-valued. The second point $\bar{M}'$ is the mapping of $\xi = -ib_M$ which would be a root of Eq. (4.4) if different branches of the radicals were chosen.

Hence, along the positive imaginary ξ axis, the AME segment is mapped into a curve $A'M'E'$ on the t plane, with $A'M'$ below the branch line and $M'E'$ above the branch line (Fig. 7). At M' the value for t is

$t_M = -g(x, z, ib_M)$, which is the maximum of $t(\xi)$ between the points A and E. The remainder of the axis maps into the complex path $E'K'$ with discontinuous slopes at E', etc. The points E', F', G', and H' are the mapping of the corresponding branch points. The path has the asymptote $O'K'$ and all poles on the imaginary ξ axis are also mapped onto this path (poles are not shown in the figures). The negative imaginary ξ axis is mapped to the $A'\bar{E}'\bar{K}'$ curve on the t plane.

Since there are no singular points in the region enclosed by the curve $A'B'$ (B' at infinity) and the real t axis, and the contribution along $B'C'$ can be shown to vanish, an integration along the curve $A'B'$ can be replaced by one along $A'M'C'$, staying below the branch cut from A' to M'. As can be seen from Eq. (4.3), the $M'C'$ portion of the real t axis maps into MC in the complex ξ plane. Again there are no singular points within $AMCB$ (CB at infinity). Thus the integration along the original path AB in the ξ plane can be replaced by either one of the following three paths: (i) $A'B'$ in the t plane, (ii) $A'M'C'$ ($A'M'$ below the cut) in the t plane, (iii) AMC in the ξ plane.

If the path $A'M'C'$ in the t plane is followed, the integration in Eq. (4.1) is transformed to the real variable

$$\bar{u}(s) = \int_{t_A}^{\infty} \xi(t)E[\xi(t)] \frac{d\xi}{dt} e^{-st}\, dt. \tag{4.5}$$

The inverse Laplace transform of $\bar{u}(s)$ is then, by inspection,

$$u(t) = \xi(t)E[\xi(t)] \frac{d\xi}{dt} H(t - t_A), \tag{4.6}$$

where $H(t)$ is the Heaviside step function and $\xi(t)$ is found from the inverse function $\xi = g^{-1}(t)$.

Since $u(t)$ in Eq. (4.6) is the inverse Laplace transform of $\bar{u}(s)$ in Eq. (4.1), the inverse transform of $\bar{u}_x$ and $\bar{u}_z$ in Eqs. (3.16) and (3.17) are then found by applying the convolution theorem. Table I shows that $s^2F(s) = As^n f(s)$ where A is a constant and $n = 0$ or 1. Hence

$$u_x(x, z, t) = 2A \int_0^t f^{(n)}(t - \tau)\, \mathrm{Im}\left[\xi E_x(\xi) \frac{d\xi}{d\tau} H(\tau - \tau_A)\right] d\tau,$$

$$u_z(x, z, t) = 2A \int_0^t f^{(n)}(t - \tau)\, \mathrm{Re}\left[E_z(\xi) \frac{d\xi}{d\tau} H(\tau - \tau_A)\right] d\tau, \tag{4.7}$$

where the condition $f(0^+) = 0$ has been assumed and $f^{(n)}(\theta) = d^n f(\theta)/d\theta^n$. Equation (4.7) yields the exact solution for waves along a generalized ray.

Note that in Eq. (4.7) ξ is an implicit function of τ, defined by Eq. (4.3). Furthermore, $d\xi/dt$ is singular at $t = t_M$ [Eq. (4.4)]. For numerical evaluation of these integrals, it is more convenient to carry out the integration first with respect to the ξ variable. First, an integration of Eq. (4.6) yields

$$\int_0^t u(\tau)\,d\tau = \int_0^t \xi(\tau)E[\xi(\tau)]\frac{d\xi}{d\tau}H(\tau - t_A)\,d\tau$$

$$= H(t - t_A)\int_{t_A}^t \xi(\tau)E[\xi(\tau)]\frac{d\xi}{d\tau}\,d\tau.$$

The last integration can be reverted to one with respect to the original variable ξ if the transformation Eq. (4.3) is reapplied. Let $\xi = \xi_1$ when the time t takes a value t_1, say. Then, because $\xi = 0$ when $t = t_A$, we obtain

$$\int_0^{t_1} u(\tau)\,d\tau = H(t_1 - t_A)\int_0^{\xi_1} \xi E(\xi)\,d\xi. \tag{4.8}$$

The path of integration for ξ is along AMC in Fig. 6.

The upper limit $\xi_1(x, z; t_1)$ is either pure imaginary ($\xi_1 = ib_1$) or complex, depending on whether $t_1 < t_M$ or $t_1 > t_M$, respectively. It is seen from Eq. (4.3) that when $t_A < t_1 < t_M$, ξ will take on two possible values: a smaller one between A and M or a larger one between M and E. However, the smaller one must be chosen here because the mapping of AM corresponds to the integration path $A'M'$ below the branch cut on the t plane (Fig. 7). Such a point is labeled ib_1 in Fig. 6. For $t_1 > t_M$, the corresponding point in Fig. 6 is labeled ξ_1.

Next, we carry out the integration by parts of Eq. (4.7) and make use of Eq. (4.8), assuming the source function is a smooth function,

$$u_x(x, z, t) = 2AH(t - t_A)\int_{t_A}^t f^{(n+1)}(t - \tau)\left[\operatorname{Im}\int_0^{\xi_1(\tau)} \xi E_x(\xi)\,d\xi\right]d\tau,$$

$$u_z(x, z, t) = 2AH(t - t_A)\int_{t_A}^t f^{(n+1)}(t - \tau)\left[\operatorname{Re}\int_0^{\xi_1(\tau)} E_z(\xi)\,d\xi\right]d\tau. \tag{4.9}$$

In the above A is a constant and $f(t)$ is the time function of the source as listed in Table I; $n = 1$ for the explosive source and $n = 0$ for the vertical force, which are determined from $s^2\bar{F}(s) = As^n\bar{f}(s)$. The function $f(t)$ and its first derivative are continuous for $t \geq 0$ and $f(0) = f'(0) = 0$.

Equations (4.7) and (4.9) complete the analysis of the line load response for an arbitrary ray in a multilayered solid. The total response is the sum of the responses corresponding to all generalized rays. Since there is a least-time path and an initiation time for each ray, before which it makes no contribution, it follows that only a finite number of rays are required to describe the total response over a finite time interval. The least time of arrival for each ray is discussed in Sections IV,B and VII,A. The total response for waves traveling along all possible generalized ray paths is discussed in Section VII.

B. Direct Rays and Refracted Rays

In Eq. (4.9) the ray integrals are either in the form of an imaginary part of a complex integral (u_x),

$$u_{\mathrm{I}} = \mathrm{Im} \int_{AMC} \xi E(\xi)\, d\xi, \tag{4.10}$$

or a real part of a similar integral (u_z)

$$u_{\mathrm{R}} = \mathrm{Re} \int_{AMC} E(\xi)\, d\xi, \tag{4.11}$$

The integrand $E(\xi)$ is a function of ξ^2 and $(\xi^2 + a_j^2)^{1/2}$ ($j = 1, 2, \ldots, 6$), being real when ξ and all $(\xi^2 + a_j^2)^{1/2}$ are real. The integration path AMC is defined by the transformation (4.3) and it is shown in Fig. 6. The path of integration terminates at ξ_1 which is calculated for given values of x and z_j at the time of observation.

The integration along AM on the imaginary axis is further complicated by the possibility of having additional branch points $\xi = ia_{6,5}$ between points A and M as mentioned previously. Their effect will now be discussed.

1. *Direct Rays*

Along the path AM, let $\xi = iw$ where w is a real variable. Then Eq. (4.3) is changed to

$$t - wx - \sum_{j=1} z_j(a_j^2 - w^2)^{1/2} = 0. \tag{4.12}$$

The left-hand side of the above equation is just the phase of a "generalized plane wave," x and $\sum z_j$ being the horizontal and vertical coordinates, respectively, and w the slowness of the phase along the horizontal direction. The complete equation represents the advancement of the plane wave front.

The time required for the cylindrical wave defined by each ray to reach the receiver is determined from that value of w that makes $t(w)$ maximal. This is because different values of w indicate different tangent planes to the actual curved wave front, and the extended plane at the intersecting point of the wave front and the radial line joining the source and receiver will be the last to reach the receiver. Thus the value of w with which a ray reaches the receiver directly (including multireflections and transmissions) is the unique root of

$$\frac{dt}{dw} = x - \sum_j \frac{z_j w}{(a_j^2 - w^2)^{1/2}} = 0. \tag{4.13}$$

The root was defined earlier as $\xi = iw = ib_M$ with $0 < b_M < a_4$ (point M in Fig. 6). The corresponding value for t is t_M,

$$t_M = xb_M + \sum_j z_j(a_j^2 - b_M^2)^{1/2}. \tag{4.14}$$

Therefore, at $t = t_M$, the wave front of the regularly reflected and transmitted ray arrives at the receiver. Rays with regular reflection and transmission are called *direct rays* (Fig. 8a).

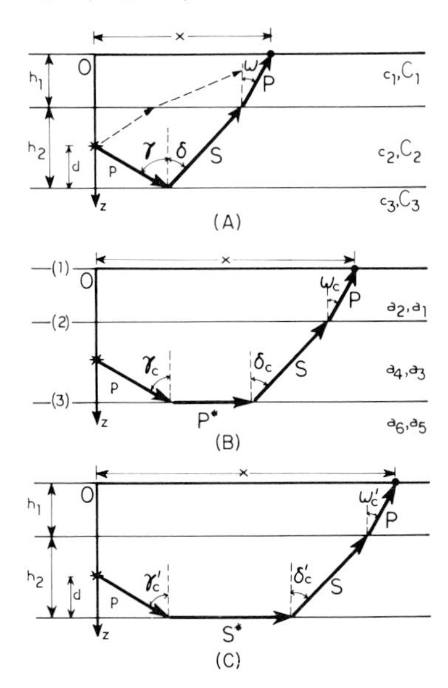

FIG. 8. (A) Direct ray; (B) refracted P ray along the fast bottom; (C) refracted S ray along the fast bottom.

2. *Refracted Rays*

It was mentioned in Section III that a plane wave may be refracted at an interface when the angle of incidence exceeds a critical angle γ_c, and a cylindrical wave represented by a ray integral may propagate along such a *refracted ray path* to arrive first at a receiver (Fig. 4). This refracted wave is called a "head wave" or a "conical wave" (Cagniard, 1939, Chapter 7). In our case, a refracted wave may occur when a downward incident P or S wave impinges at the interface (2) or (3) because wave speeds are assumed to be faster in lower layers (Fig. 2).

For refracted waves to exist, the slowness w of the phase as given in (4.12) must equal one of the slownesses of the fast bottom. Consider the case of a three-segment generalized ray emitted by a P source in layer 2.

Figure 8A depicts the direct pSP ray where a lowercase letter indicates a downward segment. From Eq. (4.12), the phase of this ray is

$$t = xw + \sum_j z_j(a_j^2 - w^2)^{1/2}$$
$$= xw + d(a_4^2 - w^2)^{1/2} + h_2(a_3^2 - w^2)^{1/2} + h_1(a_2^2 - w^2)^{1/2}. \quad (4.15)$$

When $w = a_5$ or a_6, a part of this ray is refracted at the interface (3), and waves along the refracted ray arrive at time t_5 or t_6 with (Fig. 8B,C):

$$t_f = xa_f + \sum_j z_j(a_j^2 - a_f^2)^{1/2} \qquad (f = 5, 6). \quad (4.16)$$

Since $a_6 < a_5 < a_4$ ($c_3 > C_3 > c_2$), the necessary and sufficient condition for the arrival of the refracted ray prior to a direct ray ($t_{5,6} < t_M$ or $a_{5,6} < b_M$) is

$$(dt/dw)_{w=a_{5,6}} > 0 \quad (4.17)$$

because t_M is the maximum value of $t(w)$ for $0 < w < a_4$. The above condition then yields, in view of Eq. (4.13),

$$x > x_f = a_f \sum_j \frac{z_j}{(a_j^2 - a_f^2)^{1/2}} \qquad (f = 5, 6). \quad (4.18)$$

Since $a_6 < a_5$, it follows that $x_6 < x_5$. The distances x_5 and x_6 are the two critical ranges, less than which the angles of incidence for a transmitted ray will be less than the critical angle so that no refracted waves are generated.

The evaluation of the complex integrals in (4.10) and (4.11) clearly depends on the location of the receiver (x, z), the time of observation as compared with the critical ranges x_5 and x_6, and the critical times t_M, t_5, and t_6, respectively. For both equations, the integration may be divided into two parts, the first from A to M along the imaginary ξ axis ($=iw$) and the second from M to $\xi_1(t)$ along MC in the complex plane (Figs. 6 and 9).

We first evaluate x_f ($f = 5, 6$) according to Eq. (4.18), knowing only the vertical projections z_j of a generalized ray, and then compare the horizontal distance x with $x_{5,6}$.

Case A $x < x_6$ The waves arrive through the direct pSP ray as shown in Fig. 8A, and there is no refraction. The angles of incidence, reflection, and refraction satisfy Snell's law:

$$\frac{\sin\gamma}{c_2} = \frac{\sin\delta}{C_2}, \qquad \frac{\sin\delta}{C_2} = \frac{\sin\omega}{c_1}. \quad (4.19)$$

The w in Eq. (4.15) is the slowness of the plane waves projected along the x axis with the relation $w = 1/c$, and

$$\sin\gamma = c_2/c, \qquad \sin\delta = C_2/c, \qquad \sin\omega = c_1/c. \quad (4.20)$$

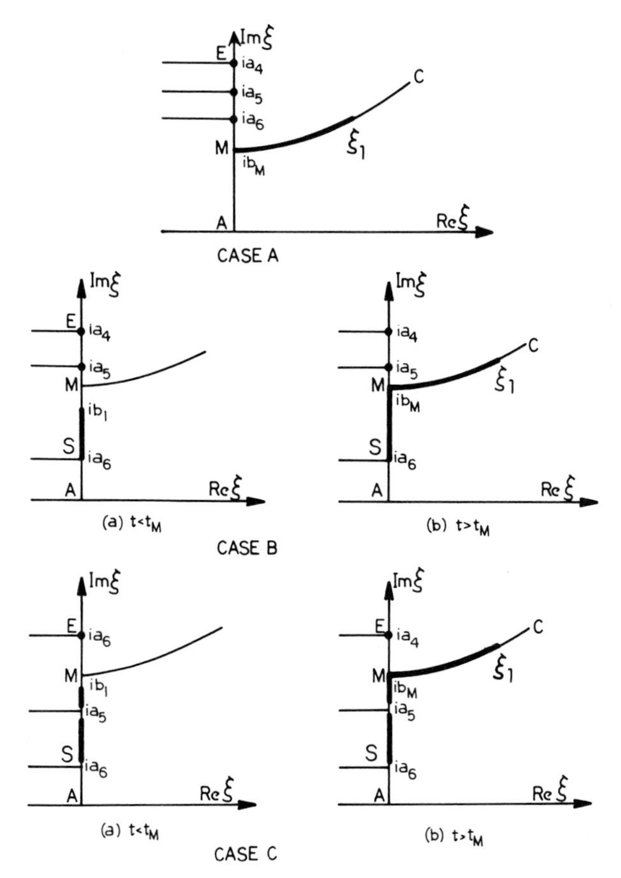

FIG. 9. Paths of integration for direct and refracted rays.

Note that Snell's law can directly be verified from the above equations. The value of w or c depends on the incident angle γ, and vice versa. When $w = b_M$ [Eq. (4.13)], the travel time along this path is minimal, and the arrival time of this ray is t_M.

Since in this case, $x < x_6$ $(<x_5)$, the branch points $ia_{5,\,6}$ lie above the stationary point M as shown in Fig. 9. In Eqs. (4.10) and (4.11), the values of u_I and u_R along AM vanish,

$$\operatorname{Im} \int_{AM} \xi E(\xi)\, d\xi = \operatorname{Im} \int_0^{b_M} (-w) E(iw)\, dw = 0,$$

$$\operatorname{Re} \int_{AM} E(\xi)\, d\xi = \operatorname{Re} \int_0^{b_M} E(iw) i\, dw = 0. \tag{4.21}$$

The reason is that $E(iw)$ is a real function for $0 < w < b_M$. Thus only the second part of integration along $M\xi_1$ contributes to the total response when $t > t_M$.

Case B $x_5 > x > x_6$ When the observational distance is longer than the critical range x_6, part of the ray is refracted along the interface (3) as P wave (Fig. 8B). This will be denoted as the pP*SP ray. In this case, $w = a_6 = 1/c_3$, and the angles γ, δ, and ω are fixed by Eq. (4.20) with

$$\sin \gamma_c = c_2/c_3, \qquad \sin \delta_c = C_2/c_3, \qquad \sin \omega_c = c_1/c_3. \tag{4.22}$$

These are the critical angles of refraction. From the above equations, it is seen that the critical angles exist only when $c_3 > c_2 > C_2$.

The arrival time along this path is, from Eq. (4.16),

$$t_6 = xa_6 + d(a_4^2 - a_6^2)^{1/2} + h_2(a_3^2 - a_6^2)^{1/2} + h_1(a_2^2 - a_6^2)^{1/2}. \tag{4.23}$$

We can verify this formula from the geometry in Fig. 8B by noting that the total time required to travel the path is

$$t = \frac{a_4 x_p}{\sin \gamma_c} + a_6 x^* + \frac{a_3 x_S}{\sin \delta_c} + \frac{a_2 x_P}{\sin \omega_c},$$

where $x_p + x^* + x_S + x_P = x$, and each term is the x projection of the respective segment.

As shown in Fig. 9B, the branch point ia_6 in this case lies below the point M but ia_5 still lies above it. The integration along the imaginary axis from A to ia_6 vanishes because $E(iw)$ is real for $0 < w < a_6$. Thus the response at the receiver begins at $t = t_6$. It is followed by the arrival of the direct ray at $t = t_M$. The terminal point ξ_1 of the path of integration depends on two possible values for t: (a) If $t < t_M$, then $\xi_1 = ib_1$ ($b < b_M$), a pure imaginary number. Hence the second part of integration should be dropped from consideration. (b) If $t > t_M$, ξ_1 is a complex number on the path MC and both parts of integration contribute to the total response.

Case C $x > x_5$ Since $x_5 > x_6$, there is, in addition to the pP*SP ray, the refracted S ray (pS*SP) as shown in Fig. 8C. The critical angles are

$$\sin \gamma_c' = c_2/C_3, \qquad \sin \delta_c' = C_2/C_3, \qquad \sin \omega_c' = c_1/C_3. \tag{4.24}$$

The arrival time is t_5 which is the same as that in Eq. (4.23) when a_6 is replaced by a_5.

As shown in Fig. 9 for this case, both branch points ia_6, ia_5 lie below the stationary point M. The value of the integral still vanishes from A to ia_6. The response begins with the arrival of the pP*SP at $t = t_6$ and undergoes an abrupt change upon the arrival of the pS*SP ray at $t = t_5$. These two refracted rays are followed by the direct ray at $t = t_M$. The terminal points of integration are the same as those in case B.

3. *Buried and Surface Forces and Head Waves*

Another ray path which is analogous to the refracted ray along a fast bottom is associated with the head wave generated by a force on an otherwise traction free surface (Fig. 10). Solutions for this problem are discussed in Eqs. (2.26) and (2.27). The phase functions for a P ray and S ray are, respectively,

$$-t = i\xi x - z(\xi^2 + a_2^2)^{1/2}, \qquad a_2 = 1/c_1,$$

$$-t = i\xi x - z(\xi^2 + a_1^2)^{1/2}, \qquad a_1 = 1/C_1. \tag{4.25}$$

The arrival times are given by $\xi = iw = ib_M$ [Eq. (4.13)], where $b_M = a_2 x/(x^2 + z^2)^{1/2}$ for the P ray, and $b_M = a_1 x/(x^2 + z^2)^{1/2}$ for the S ray.

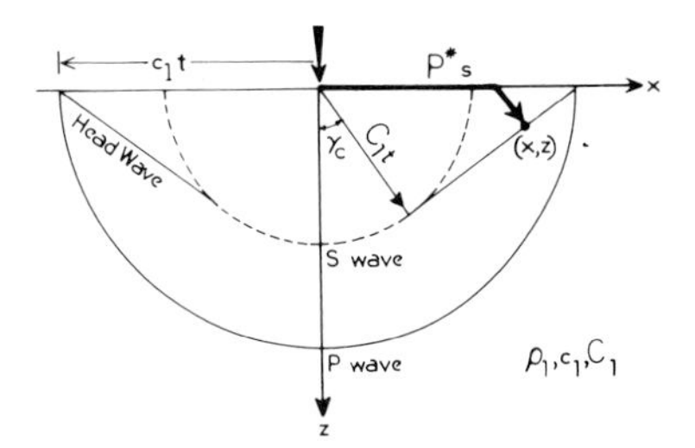

FIG. 10. P, S, and head waves generated by a force at the surface of a half-space ($\sin \gamma_c = C_1/c_1$).

There is however the possibility that the x projection of the phase velocity of the S ray equals the P wave speed of the medium. Setting $\xi = iw$ and letting $w = a_2$ in the second of Eq. (4.24), we obtain

$$t_2 = xa_2 + z(a_1^2 - a_2^2)^{1/2}. \tag{4.26}$$

This is the arrival time of the S ray that is "refracted" along the surface, and it is indicated by P*s in Fig. 10. Such a ray exists only when the observation point (x, z) is outside the sector of the critical angle $\gamma_c = \sin^{-1}(C_1/c_1)$.

The path of integration for this case is similar to that of case B in Fig. 9. There are only two branch points ia_1 and ia_2, which replace ia_5 and ia_6 respectively in the figure. The integration path starts from ia_2 (arrival time t_2), moves along the imaginary axis to ib_M for the S ray, and then turns to the complex plane along path MC.

If the concentrated force is in the interior, it generates a spherical P and S wave front. In addition to the Pp, Ps, Sp, Ss reflected waves, there exists the head wave front (conical S wave) associated with the refracted P wave along the free surface. The SP*s ray and all wave fronts are depicted in Fig. 11. The integrand of the SP*s ray is the same as that of Ss ray, and the path of the former starts at the branch point of the P-wave slowness.

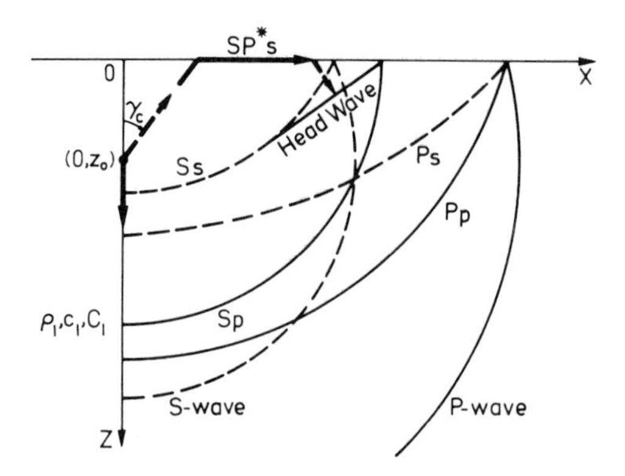

FIG. 11. Incident, reflected, and head waves generated by a buried force in a half-space ($\sin \gamma_c = C_1/c_1$).

Note that when the vertical force is moved to the surface ($z_0 \to 0$ in Fig. 11), the reflected Pp wave front coalesces with the incident P wave front, and the Ss wave coincides with the incident S wave. The reflected Ps wave degenerates to the head wave P*s shown in Fig. 10. Thus the surface force may be treated as a degenerate case of a buried force (see Section VI,B,1).

V. Axisymmetric Waves due to a Point Load

For wave motion symmetric about the z axis (Fig. 2), all field variables are independent of θ. As discussed in Section II,B,2, the displacements $[u_r, 0, u_z]$ may be calculated from two potentials $\phi(r, z, t)$ and $\psi(r, z, t)$. The general solutions of $\bar{\phi}(r, z, s)$ and $\bar{\psi}(r, z, s)$ are represented by Eq. (2.31).

For a three-layered medium, the ten unknown coefficients $A_1, \ldots, D_1$; $A_2, \ldots, D_2$; A_3; and C_3 are determined from the ten boundary conditions in Eq. (3.1) when τ_{zx} and u_x are replaced by τ_{zr} and u_r, respectively. These conditions may be replaced by requiring the continuity of the following Laplace transformed potentials at the interfaces:

$$\bar{\phi} + \frac{\partial \bar{\psi}}{\partial z}, \qquad \frac{\partial \bar{\phi}}{\partial z} + \frac{\partial^2 \bar{\psi}}{\partial z^2} - \frac{s^2}{C^2}\bar{\psi},$$

$$2\mu\left[\left(\frac{s^2}{2c^2} - \frac{s^2}{C^2}\right)\bar{\phi} + \frac{\partial^2 \bar{\phi}}{\partial z^2} + \frac{\partial}{\partial z}\left(\frac{\partial^2 \bar{\psi}}{\partial z^2} - \frac{s^2}{C^2}\bar{\psi}\right)\right], \; 2\mu\left[\frac{\partial \bar{\phi}}{\partial z} + \frac{\partial^2 \bar{\psi}}{\partial z^2} - \frac{s^2}{2C^2}\bar{\psi}\right]. \tag{5.1}$$

These conditions are particularly convenient because they involve only derivatives with respect to z.

As shown in Section II,B the Laplace–Hankel transformed potentials $\hat{\phi}$ and $\hat{\psi}$ of the axisymmetric solutions Eq. (2.30) have the same dependence in z as the Laplace–Fourier transformed potentials $\tilde{\phi}$ and $\tilde{\psi}$ of the plane strain solutions Eq. (2.21). Thus, the generalized rays for axisymmetric waves can be determined analogously.

With reference to Fig. 3 a point source at $z = z_0$ emits a P wave with spherical wave front. It is reflected and refracted at the interface $z = 0$. The waves in the two adjacent media are[5]:

$$\bar{\phi}_i(r, z, s) = s^2 \int_0^\infty \hat{\phi}_i(\xi, z, s) J_0(s\xi r)\xi \, d\xi$$

$$\bar{\psi}_i(r, z, s) = s^2 \int_0^\infty \frac{-1}{s\xi} \hat{\psi}_i(\xi, z, s) J_0(s\xi r)\xi \, d\xi \qquad (i = 1, 2), \quad (5.2)$$

where

$$\hat{\phi}_2 = \bar{F}(s) S_P(\eta_2)[e^{-s\eta_2|z-z_0|} + R^{PP} e^{s(-\eta_2 z - \eta_2 z_0)}],$$
$$\hat{\psi}_2 = \bar{F}(s) S_P(\eta_2) R^{PS} e^{s(-\zeta_2 z - \eta_2 z_0)}, \quad (5.3)$$

and

$$\hat{\phi}_1 = \bar{F}(s) S_P(\eta_2) T^{PP} e^{s(\eta_1 z - \eta_2 z_0)},$$
$$\hat{\psi}_1 = \bar{F}(s) S_P(\eta_2) T^{PS} e^{s(\zeta_1 z - \eta_2 z_0)}. \quad (5.4)$$

The first term of $\hat{\phi}_2$ is the source ray [Eq. (2.35) and Table I]. The remaining terms of Eqs. (5.3) and (5.4) correspond to the Fourier transforms of Eqs. (3.3) and (3.4), respectively. The factor $-s\xi$ is inserted in Eq. (5.2) so that the reflection and transmission coefficients $R^{PP}, \ldots, T^{PS}$ are the same as those in the plane strain case [Eqs. (3.9)–(3.11)] provided the i's $(\sqrt{-1})$ in these equations are omitted.

It should now become obvious how to derive the generalized rays like Eqs. (3.14) for axially symmetric waves in layered media. Examples are given in Section V,A. However because of the Bessel functions in the integrand, the inverse Laplace transform of $\bar{\phi}$ and $\bar{\psi}$ cannot be carried out as in Section IV without some modifications. This is discussed in Section V,B.

A. Axisymmetric Waves in a Single Layer

The problems of transient waves in a plate or in a layer overlaying a half-space are of great interest in applications. Since the waves are confined to a single layer, and those leaked into the half-space are not reflected into

[5] See footnote 2, p. 192.

the layer, the generalized rays in this layer can be developed systematically by a matrix formulation.

1. *Matrix Formulation of Generalized Ray Groups*

Consider again a point source in the layer (Fig. 12) at a depth $z = z_0$.

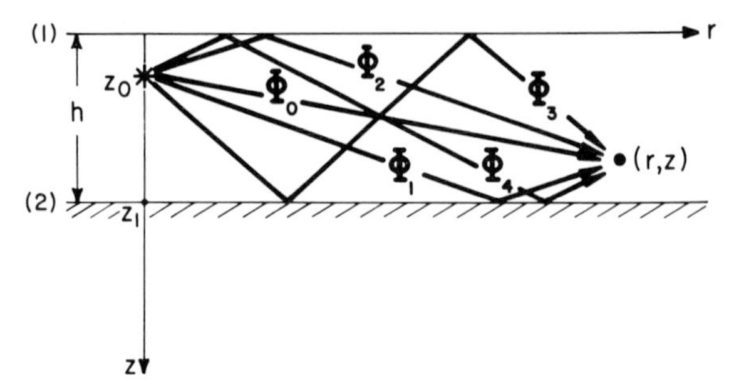

FIG. 12. Ray groups in a single layer.

The Laplace–Hankel transformed waves emitted by the source may be expressed as

$$\mathbf{\Phi}_0(\xi, z, s) = \mathbf{H}(|z - z_0|)\mathbf{S}(\xi)\bar{F}(s), \tag{5.5}$$

where

$$\mathbf{\Phi}_0(\xi, z, s) = \begin{bmatrix} \hat{\phi}_0(\xi, z, s) \\ \hat{\psi}_0(\xi, z, s) \end{bmatrix}, \qquad \mathbf{S}(\xi) = \begin{bmatrix} S_P(\xi) \\ S_V(\xi) \end{bmatrix},$$

$$\mathbf{H}(z) = \begin{bmatrix} e^{-s\eta z} & 0 \\ 0 & e^{-s\zeta z} \end{bmatrix}. \tag{5.6}$$

$\mathbf{\Phi}$ and $\mathbf{S}$ are two column matrices. The former designates the ray groups and the latter represents the source function. The subscript 0 of $\mathbf{\Phi}_0$ signifies that it is the zeroth ray group, or the source ray group. The diagonal square matrix $\mathbf{H}(z)$ is a part of the phase functions of waves.

In component form Eq. (5.5) gives the same as the Hankel transforms of Eq. (2.35). For a point source of explosion and for a concentrated force, $\mathbf{S}$ is respectively (Table I)

$$\mathbf{S}(\xi) = \begin{bmatrix} 1/\eta \\ 0 \end{bmatrix}, \qquad \mathbf{S}(\xi) = \begin{bmatrix} -\varepsilon \\ \xi/\zeta \end{bmatrix}.$$

Furthermore, if we let

$$\mathbf{\Phi}_0(\xi, z, s) = \begin{bmatrix} \tilde{\phi}_0(\xi, z, s) \\ \tilde{\psi}_0(\xi, z, s) \end{bmatrix}, \tag{5.7}$$

Eq. (5.5) is the same as the Fourier transforms of Eq. (2.27). Thus the results in this subsection, Eqs. (5.5)–(5.15), are also applicable to plane strain problems with line sources (Norwood, 1975; Schmuely, 1975).

The source ray is now treated as an incident wave that is reflected at the bottom surface ($z = h$) with reflection coefficients $\mathbf{R}^{(2)}$, or at the upper surface $z = 0$ with reflection coefficients $\mathbf{R}^{(1)}$. We call these once-reflected waves the first ($\mathbf{\Phi}_1$) and second ray groups ($\mathbf{\Phi}_2$), respectively:

$$\begin{aligned}\mathbf{\Phi}_1(\xi, z, s) &= \mathbf{H}(h - z)\mathbf{R}^{(2)}\mathbf{\Phi}_0(\xi, h, s) \\ &= \mathbf{H}(h - z)\mathbf{R}^{(2)}\mathbf{H}(h - z_0)\mathbf{S}(\xi)\bar{F}(s),\end{aligned} \tag{5.8}$$

$$\begin{aligned}\mathbf{\Phi}_2(\xi, z, s) &= \mathbf{H}(z)\mathbf{R}^{(1)}\mathbf{\Phi}_0(\xi, 0, s) \\ &= \mathbf{H}(z)\mathbf{R}^{(1)}\mathbf{H}(z_0)\mathbf{S}(\xi)\bar{F}(s).\end{aligned} \tag{5.9}$$

The matrices for the reflection coefficients at interfaces (1) and (2) respectively are

$$\mathbf{R}^{(1)} = \begin{bmatrix} R^{\text{PP}}_{(1)} & R^{\text{SP}}_{(1)} \\ R^{\text{PS}}_{(1)} & R^{\text{SS}}_{(1)} \end{bmatrix}, \qquad \mathbf{R}^{(2)} = \begin{bmatrix} R^{(2)}_{\text{PP}} & R^{(2)}_{\text{SP}} \\ R^{(2)}_{\text{PS}} & R^{(2)}_{\text{SS}} \end{bmatrix}. \tag{5.10}$$

The components $R^{\text{PP}}, \ldots, R_{\text{SS}}$ are given in Section III,B with proper modifications for axisymmetric waves.

Upon the second reflection $\mathbf{\Phi}_1$ is reflected at $z = 0$ as $\mathbf{\Phi}_3$, and $\mathbf{\Phi}_2$ is reflected at $z = h$ as $\mathbf{\Phi}_4$. The results are

$$\begin{aligned}\mathbf{\Phi}_3(\xi, z, s) &= \mathbf{H}(z)\mathbf{R}^{(1)}\mathbf{\Phi}_1(\xi, 0, s) \\ &= \mathbf{H}(z)\mathbf{R}^{(1)}\mathbf{H}(h)\mathbf{R}^{(2)}\mathbf{H}(h)\mathbf{H}(-z_0)\mathbf{S}(\xi)\bar{F}(s),\end{aligned} \tag{5.11}$$

where use has been made of the relation $\mathbf{H}(h - z_0) = \mathbf{H}(h)\mathbf{H}(-z_0)$; and

$$\begin{aligned}\mathbf{\Phi}_4(\xi, z, s) &= \mathbf{H}(h - z)\mathbf{R}^{(2)}\mathbf{\Phi}_2(\xi, h, s) \\ &= \mathbf{H}(h - z)\mathbf{R}^{(2)}\mathbf{H}(h)\mathbf{R}^{(1)}\mathbf{H}(z_0)\mathbf{S}(\xi)\bar{F}(s).\end{aligned} \tag{5.12}$$

Note that $\mathbf{\Phi}_0, \ldots, \mathbf{\Phi}_4$ are called ray groups since each one of them is composed of several rays. The source ray group $\mathbf{\Phi}_0$ is composed of two rays (P and S rays); $\mathbf{\Phi}_1$ and $\mathbf{\Phi}_2$ are composed of four rays (PP, SP, SS, and PS rays). For $\mathbf{\Phi}_1$, the first segment of the ray is downward from the source, whereas for $\mathbf{\Phi}_2$, it is upward from the source. Analogously, $\mathbf{\Phi}_3$ and $\mathbf{\Phi}_4$ are composed of eight rays (PPP, SPP, ..., SSS rays).

The higher order ray groups all contain the factor

$$\mathbf{M} = \mathbf{R}^{(1)}\mathbf{H}(h)\mathbf{R}^{(2)}\mathbf{H}(h) \tag{5.13}$$

which represents the portion of the ray that traverses twice through the thickness of the layer and is reflected once at the lower surface ($\mathbf{R}^{(2)}$ coefficient) and once at the upper surface ($\mathbf{R}^{(1)}$ coefficient). The complete result is

$$\begin{aligned}\mathbf{\Phi}(\xi, z, s) &= \sum_{m=0}^{\infty} \mathbf{\Phi}_m(\xi, z, s)\\ &= \mathbf{\Phi}_0 + \bar{F}(s)[\mathbf{H}(h - z)\mathbf{R}^{(2)}\mathbf{H}(h) + \mathbf{H}(z)\mathbf{M}]\\ &\qquad \times \sum_{n=0}^{\infty} \mathbf{M}^n\mathbf{H}(-z_0)\mathbf{S}(\xi)\\ &\quad + \bar{F}(s)[\mathbf{H}(z) + \mathbf{H}(h - z)\mathbf{R}^{(2)}\mathbf{H}(h)]\\ &\qquad \times \sum_{n=0}^{\infty} \mathbf{M}^n\mathbf{R}^{(1)}\mathbf{H}(z_0)\mathbf{S}(\xi).\end{aligned} \tag{5.14}$$

The first term is the source ray group [Eq. (5.5)]. The second term is the sum of all odd-numbered ray groups that are originated from the source with a downward first segment; and the third term is the sum of all even numbered ray groups with the first segment pointing upward.

The above formulation of generalized rays in matrix form is advantageous for at least two reasons:

(1) It systematically enumerates all generalized rays, including all rays due to mode conversions, and automatically combines rays into groups. The inverse transforms of $\mathbf{\Phi}$ will be carried out for each group. This combination of rays into groups is necessary in order to prevent divergence of numerical integration of each individual ray (Spencer, 1965a).

(2) The $\mathbf{M}$ matrix is a square matrix of second degree, and it satisfies its own characteristic equation according to the Cayley–Hamilton theorem. Thus

$$\mathbf{M}^2 = -a\mathbf{M} - b\mathbf{I}, \tag{5.15}$$

where $\mathbf{I}$ is the identity matrix, and a and b are coefficients of the characteristic equation of the $\mathbf{M}$ matrix. Furthermore, by repeatedly applying the above equation, $\mathbf{M}^m$ can be expressed as a linear combination of $\mathbf{M}$ and $\mathbf{I}$ (Norwood, 1975).

The inverse Hankel transform should be carried out according to Eq. (5.2). With $\mathbf{\Phi}_m(\xi, z, s) = [\hat{\phi}_m, \hat{\psi}_m]$, we have

$$\begin{aligned}\bar{\phi}_m(r, z, s) &= s^2 \int_0^{\infty} \hat{\phi}_m(\xi, z, s)J_0(s\xi r)\xi\, d\xi,\\ \bar{\psi}_m(r, z, s) &= s^2 \int_0^{\infty} \frac{-1}{s\xi}\hat{\psi}_m(\xi, z, s)J_0(s\xi r)\xi\, d\xi.\end{aligned} \tag{5.16}$$

2. *Potentials, Displacements, and Stresses*

From the matrix products, we find that $\hat{\phi}_m$ or $\hat{\psi}_m$ of Eq. (5.14) is composed of two parts, one due to the S_P source, one due to S_S. Each part is of the form

$$(\hat{\phi}_m, \hat{\psi}_m) = \bar{F}(s)S\pi e^{-sh(\xi, z)}, \tag{5.17}$$

where the phase function, as that in Eq. (3.18), is given by

$$h(\xi, z) = \zeta z_1 + \eta z_2 = \sum_{j=1} z_j(\xi^2 + a_j^2)^{1/2}. \tag{5.18}$$

z_2 is the total projection on the z axis of all P segments of a ray with slowness a_2 $(= 1/c)$, and z_1 is the total z projection of all S segments with slowness a_1. The last segment of $\hat{\phi}_m$ must be a P mode and that of $\hat{\psi}_m$ an S mode. In Eq. (5.17), the S can be either a P source, or an S source, and π represents the continuous products of reflection coefficients.

Substituting Eq. (5.17) into (5.16), and then into Eqs. (2.13) and (2.14), we obtain the Laplace transformed displacements and stresses, respectively. The final expression for each quantity is a sum of two integrals, one derived from $\bar{\phi}_m$, one from $\bar{\psi}_m$; and each integral is of the form

$$(\bar{u}_i, \bar{\tau}_{ij})_m = s^{2+q}\bar{F}(s)\int_0^\infty S(\xi)\pi(\xi)D(\xi)e^{-sh(\xi, z)}J_n(s\xi r)\xi\, d\xi. \tag{5.19}$$

The power q, order n, and the receiver functions $D(\xi)$ are given in Table III.

TABLE III

q, n, AND $D(\xi)$ FOR DISPLACEMENTS AND STRESSES [EQ. (5.19)]

	q	n	D (P wave)	D (SV wave)
u_z	1	0	$-\varepsilon\eta$	$-\xi$
u_r	1	1	$-\xi$	$-\varepsilon\zeta$
τ_{zz}	2	0	$\mu(\xi^2 + \zeta^2)$	$2\mu\varepsilon\xi\zeta$
τ_{zr}	2	1	$2\mu\varepsilon\xi\eta$	$\mu(\xi^2 + \zeta^2)$
τ	2	0	$(3\lambda + 2\mu)/3c^2$	0

In Table III τ is the mean normal stress,

$$\tau = \tfrac{1}{3}(\tau_{rr} + \tau_{\theta\theta} + \tau_{zz}) = (\lambda + \tfrac{2}{3}\mu)\nabla^2\phi. \tag{5.20}$$

The remaining two nonvanishing stresses are given by

$$\tau_{\theta\theta} = \frac{3\lambda}{3\lambda + 2\mu}\tau + 2\mu\frac{u_r}{r}, \qquad \tau_{rr} = 3\tau - (\tau_{\theta\theta} + \tau_{zz}). \tag{5.21}$$

B. Inversion of the Laplace Transforms

The key to completing the inverse Laplace transforms of $\bar{u}_i$ or $\bar{\tau}_{ij}$ in Eq. (5.19) is to replace the Bessel function by an integral representation (National Bureau of Standards, 1964, Eq. 9.1.21),

$$J_n(z) = \frac{i^{-n}}{\pi} \int_0^{\pi} e^{iz \cos \omega} \cos n\omega \, d\omega$$

$$= \begin{cases} \dfrac{2}{\pi} i^{-n} \operatorname{Re} \displaystyle\int_0^{\pi/2} e^{iz \cos \omega} \cos n\omega \, d\omega, & n = 0, 2, 4, \ldots, \\ \dfrac{2}{\pi} i^{-n+1} \operatorname{Im} \displaystyle\int_0^{\pi/2} e^{iz \cos \omega} \cos n\omega \, d\omega, & n = 1, 3, 5, \ldots. \end{cases} \tag{5.22}$$

We discuss in detail the cases of $\bar{u}_z$ and $\bar{u}_r$ in this section.

The $\bar{u}_z$ and $\bar{u}_r$ in Eq. (5.19) may be rewritten as ($q = 1$)

$$\bar{u}_z(\xi, z, s) = s^3 \bar{F}(s) \bar{I}_z(r, z, s), \qquad \bar{u}_r(\xi, z, s) = s^3 \bar{F}(s) \bar{I}_r(r, z, s), \tag{5.23}$$

where

$$\bar{I}_z = \int_0^{\infty} E_z(\xi) e^{-sh(\xi, z)} J_0(s\xi r) \xi \, d\xi, \tag{5.24}$$

$$\bar{I}_r = \int_0^{\infty} \xi E_r(\xi) e^{-sh(\xi, z)} J_1(s\xi r) \xi \, d\xi. \tag{5.25}$$

In the above, we have replaced $S(\xi)\pi(\xi)D(\xi)$ for $\bar{I}_z$ by $E_z(\xi)$, and that for $\bar{I}_r$ by $\xi E_r(\xi)$.

Substitution of the first of Eq. (5.22) into (5.24) gives rise to

$$\bar{I}_z(r, z, s) = \frac{2}{\pi} \operatorname{Re} \int_0^{\pi/2} d\omega \int_0^{\infty} E_z(\xi) e^{s[i\xi r \cos \omega - h(\xi, z)]} \xi \, d\xi. \tag{5.26}$$

Note that an interchange of order of integrations in ω and in ξ has been made since the integral in ξ is uniformly convergent for $0 < \omega < \pi/2$. Similarly,

$$\bar{I}_r(r, z, s) = \frac{2}{\pi} \operatorname{Im} \int_0^{\pi/2} d\omega \cos \omega \int_0^{\infty} \xi E_r(\xi) e^{s[i\xi r \cos \omega - h(\xi, z)]} \xi \, d\xi. \tag{5.27}$$

Inverse Laplace transform of an expression like $\bar{I}_z$ and $\bar{I}_r$ has been discussed in great detail in the original treatise by Cagniard (1939, Chapter 5). Note that the integral in ξ is analogous to that in Eq. (4.1). The entire exponent $i\xi r \cos \omega - h(\xi, z)$ is equivalent to $g(x, z; \xi)$ in Eqs. (4.1)–(4.3).

1. *Transformation of the Variable ξ to t*

Following the original method of Cagniard, we transform ξ to t by

$$-t = i\xi r \cos \omega - \sum_j z_j(\xi^2 + a_j^2)^{1/2}. \tag{5.28}$$

The t has an extremum at $t = t_M = t(\xi_M)$ where ξ_M is the root of

$$\frac{dt}{d\xi} = \sum_j \frac{z_j \xi}{(\xi^2 + a_j^2)^{1/2}} - ir \cos \omega = 0. \tag{5.29}$$

These are the same as Eqs. (4.3) and (4.4), respectively, when x is replaced by $r \cos \omega$. The original path of integration along the real ξ axis in Fig. 6 is transformed to one in the complex t plane (path $A'B'$) and then deformed to the real t axis in Fig. 7. The final result is

$$\bar{I}_z(r, z, s) = \frac{2}{\pi} \operatorname{Re} \int_0^{\pi/2} d\omega \int_{t_A}^{\infty} E_z[\xi(\omega, t)]\xi\left(\frac{\partial \xi}{\partial t}\right)_{\omega} e^{-st}\, dt, \tag{5.30}$$

where $t_A = \sum (z_i a_j)$.

Throughout the transformation from ξ to t, ω is held as a parameter. When $\omega = 0$, the path of integration is just that shown in Figs. 6 and 7 with $x = r$. When $\omega = \pi/2$, the mapping is the same as that in Eq. (4.3) with $x = 0$,

$$t]_{\omega=\pi/2} = \sum_j z_j(\xi^2 + a_j^2)^{1/2}.$$

With $x = 0$, the MC path is brought to coincide with the real ξ axis, and $A'B'$ with the real t axis. The transformation thus maps the AB path directly to $A'C'$ path along the real t axis and $t_A = t_M$.

Interchanging the order of integrations in t and ω in Eq. (5.30), one obtains

$$\bar{I}_z(r, z, s) = \int_{t_A}^{\infty} \left[\frac{2}{\pi} \operatorname{Re} \int_0^{\pi/2} E_z[\xi(\omega, t)]\xi\left(\frac{\partial \xi}{\partial t}\right)_{\omega} d\omega\right] e^{-st}\, dt. \tag{5.31}$$

This interchange is permissible even though $\partial\xi/\partial t$ is singular at t_M ($t_A < t_M < \infty$) because the singularity is of half-order. The inverse Laplace transform of $\bar{I}_z$ is then obtained by inspection:

$$I_z(r, z, t) = H(t - t_A)\frac{2}{\pi} \operatorname{Re} \int_0^{\pi/2} E_z[\xi(\omega, t; r, z)]\xi\left(\frac{\partial \xi}{\partial t}\right)_{\omega} d\omega. \tag{5.32}$$

Similarly, the inverse Laplace transform of $\bar{I}_r$ in Eq. (5.27) is

$$I_r(r, z, t) = H(t - t_A)\frac{2}{\pi} \operatorname{Im} \int_0^{\pi/2} \xi E_r[\xi(\omega, t; r, z)]\xi\left(\frac{\partial \xi}{\partial t}\right)_{\omega} \cos \omega\, d\omega. \tag{5.33}$$

The inverse Laplace transforms of $\bar{I}_z$ and $\bar{I}_r$ are thus completed.

2. *Transformation of the Variable ω to ξ*

The integrals in Eqs. (5.32) and (5.33) are inconvenient for numerical calculations because ξ is related to ω by an implicit function as defined Eq. (5.28). Although when j is summed from 1 to 2, an explicit expression for $\partial\xi/\partial t$ can be found, this is not possible when there are more than two radicals like $(\xi^2 + a_j^2)^{1/2}$. Thus it is desirable to transform the variable ω to ξ according to Eq. (5.28). This second change of variable is another main feature of the Cagniard method (Cagniard, 1939, Section 5-5).

From Eq. (5.28) we obtain, for $r \neq 0$,

$$\cos\omega = \frac{-t + \sum z_j(\xi^2 + a_j^2)^{1/2}}{i\xi r}. \tag{5.34}$$

When $\xi = 0$, the right-hand side of the above equation is indeterminate because $t = \sum z_j a_j$ at $\xi = 0$. Applying l'Hôpital's rule, we find $\cos\omega \to 0$ $(\omega \to \pi/2)$ as $\xi \to 0$. When $\omega = 0$, we find that $\xi = \xi_1(t)$ where ξ_1 is calculated from the following equation for a given t:

$$-t = i\xi_1 r - \sum_j z_j(\xi_1^2 + a_j^2)^{1/2}. \tag{5.35}$$

Hence, Eq. (5.32) is transformed to

$$I_z(r, z, t) = H(t - t_A)\frac{2}{\pi}\,\mathrm{Re}\int_{\xi_1}^{0} E_z(\xi)\xi\left(\frac{\partial\xi}{\partial t}\right)_\omega\left(\frac{\partial\omega}{\partial\xi}\right)_t d\xi.$$

It can be shown from Eq. (5.28) that

$$(\partial\xi/\partial t)_\omega(\partial\omega/\partial\xi)_t = -1/(i\xi r\sin\omega) = -1/iK,$$

where

$$K(\xi; r, z, t) = \left\{\xi^2 r^2 + \left[t - \sum_j z_j(\xi^2 + a_j^2)^{1/2}\right]^2\right\}^{1/2}. \tag{5.36}$$

The final answer is

$$I_z(r, z, t) = H(t - t_A)\frac{2}{\pi}\,\mathrm{Im}\int_0^{\xi_1(t)} E_z(\xi)\frac{\xi}{K(\xi; r, z, t)}\,d\xi. \tag{5.37}$$

Similarly,

$$I_r(r, z, t) = H(t - t_A)\frac{2}{\pi r}\,\mathrm{Im}\int_0^{\xi_1(t)} \xi E_r(\xi)\frac{t - \sum z_j(\xi^2 + a_j^2)^{1/2}}{K(\xi; r, z, t)}\,d\xi. \tag{5.38}$$

It should be noted that the integrations in Eqs. (5.37) and (5.38) are over the complex ξ plane along the path $AM\xi_1$ in Fig. 6. The $K(\xi)$ is multivalued and has a branch point at $\xi = \xi_1$. It is rendered single-valued by a branch cut from ξ_1 so that $K(\xi)$ is positive when Re ξ is positive. This branch cut can be any curve from ξ_1 toward the second quadrant of the ξ plane (branch cut not shown in Fig. 6).

We note that except for the choice of integration contours the derivations shown here adhere closely to those given by Cagniard (1939, Chapter 5 and Appendix 5-A). We prefer the paths of integration to lie in the first quadrant of the complex ξ plane since it is compatible with the complex analysis of most computer codes.

The answers for $u_z(t)$ and $u_r(t)$ are determined from Eq. (5.23) by convolution. For instance, when the source is a point explosion with time function $f(t)$ (Table I), we have

$$s^3\bar{F}(s) = s^2\bar{f}(s)/4\pi c^2. \tag{5.39}$$

The inverse transform is

$$[f''(t) + f'(0)\delta(t) + f(0)\delta'(t)]/4\pi c^2. \tag{5.40}$$

The displacements u_α $(\alpha = r, z)$ are

$$u_\alpha(r, z, t) = \frac{H(t - t_A)}{4\pi c^2}\left[\int_0^t f''(t-\tau)I_\alpha(r, z, \tau)\, d\tau + f'(0)I_\alpha(r, z, t) + f(0)\frac{\partial}{\partial t}I_\alpha(r, z, t)\right], \tag{5.41}$$

where $I_z(t)$ and $I_r(t)$ are given by Eqs. (5.37) and (5.38).

Equation (5.41) determines the transient displacements for waves traveling along a specific generalized ray path. The total response is then obtained by summing the solutions for waves along all possible paths. Details are given in Section VII.

C. Derivation of the Point Source Response from the Line Source Solutions

The similarity between the solutions for waves generated by a line source, and those by a point source as discussed in Section IV and this section is striking. In fact, Lamb (1904) noted that a point source could be generated by superposing line sources uniformly in all azimuths through the source point. However, the mathematical formula has not been available until recent years. (Thau and Pao, 1970; Dampney, 1971).

Thau and Pao (1970) showed that the axisymmetric point source solution in (r, z) coordinates can be derived by superposing, not directly from the two-dimensional solution in (x, z) coordinates, but the x derivative of the Hilbert transform of the corresponding solutions for a line source. By various mathematical manipulations, they deduced the two formulas

$$u_r(r, z, t) = -\frac{1}{\pi r}\int_r^\infty \frac{\partial u_x(x, z, t)}{\partial x}\frac{x}{(x^2 - r^2)^{1/2}}\,dx,$$

$$u_z(r, z, t) = -\frac{1}{\pi}\int_r^\infty \frac{\partial u_z(x, z, t)}{\partial x}\frac{1}{(x^2 - r^2)^{1/2}}\,dx. \tag{5.42}$$

In the above formulas u_r and u_z on the left-hand sides are displacements due to an axisymmetric point source; u_x and u_z inside the integrals are those due to a line source.

Applying the preceding formulas to Eqs. (4.9) for a line source, we obtain

$$u_r(r, z, t) = -\frac{2A}{\pi r}H(t - t_A)\int_{t_A}^{t} f^{(n+1)}(t - \tau) \times \operatorname{Im}\int_r^\infty \frac{x\xi E_x(\xi)}{(x^2 - r^2)^{1/2}}\frac{\partial \xi}{\partial x}\,dx\,d\tau,$$

$$u_z(r, z, t) = -\frac{2A}{\pi}H(t - t_A)\int_{t_A}^{t} f^{(n+1)}(t - \tau) \times \operatorname{Re}\int_r^\infty \frac{E_z(\xi)}{(x^2 - r^2)^{1/2}}\frac{\partial \xi}{\partial x}\,dx\,d\tau. \tag{5.43}$$

The key step in this derivation is to note that

$$\frac{\partial}{\partial x}\int_0^{\xi_1} \xi E(\xi)\,d\xi = [\xi E(\xi)]_{\xi=\xi_1}\frac{\partial \xi_1}{\partial x} = \xi_1 E(\xi_1)\frac{\partial \xi_1}{\partial x}. \tag{5.44}$$

The variable of integration x is then changed to ξ according to Eq. (4.3); the limits of integration are $\xi \to 0$ as $x \to \infty$, and $\xi \to \xi_1(r, t)$ as $x \to r$, where ξ_1 is determined by

$$ir\xi_1 = \sum_j z_j(\xi_1^2 + a_j^2)^{1/2} - t. \tag{5.45}$$

We thus find, from Eqs. (4.3),

$$\xi^2(x^2 - r^2) = -\xi^2 r^2 - \left[t - \sum_j z_j(\xi^2 + a_j^2)^{1/2}\right]^2 = -K^2(\xi; r, z, t). \tag{5.46}$$

The $K(\xi; r, z, t)$ is precisely that defined in Eq. (5.36). Substitution of the above results into Eq. (5.43) yields

$$u_r(r, z, t) = \frac{2A}{\pi r} H(t - t_A) \int_{t_A}^{t} f^{(n+1)}(t - \tau)$$

$$\times \operatorname{Im} \int_0^{\xi_1} \xi E_x(\xi) \frac{\partial K(\xi; r, z, \tau)}{\partial \tau} d\xi \, d\tau,$$

$$u_z(r, z, t) = \frac{2A}{\pi} H(t - t_A) \int_{t_A}^{t} f^{(n+1)}(t - \tau)$$

$$\times \operatorname{Im} \int_0^{\xi_1} \xi E_z(\xi) \frac{1}{K(\xi; r, z, \tau)} d\xi \, d\tau. \qquad (5.47)$$

These results are in total agreement with those in Eqs. (5.37)–(5.41) when $n = 1$ and $f'(0) = f(0) = 0$. This analysis also shows that $E_x(\xi) = E_r(\xi)$, and the $E_z(\xi)$ for both line and point source solutions are the same.

Comparing Eqs. (5.47) with Eqs. (4.9), we find that the response due to an axisymmetric point source and that due to a line source in the same medium can be calculated by the same computer code, the only difference being the extra factors $\partial K/\partial \tau$ and $1/K$. However, for asymmetric loadings, inverse transforms like those shown in Section V,B must be performed, and the responses cannot be easily deduced from those due to either a line source or an axisymmetric point source.

D. Solutions for a Point Source in an Unbounded Medium

It is instructive to apply the general solution derived in previous sections to the simple problem of a point source of explosion in an infinite medium [Eq. (2.32)]. The Laplace transformed solution for the potential $\bar{\phi}$ is [Eq. (2.35)]

$$\bar{\phi}(r, z, s) = \frac{1}{4\pi c^2} s\bar{f}(s)\bar{I}_\phi(r, z, s), \qquad (5.48)$$

$$\bar{I}_\phi(r, z, s) = \int_0^\infty \frac{1}{\eta} e^{-s\eta|z - z_0|} J_0(s\xi r)\xi \, d\xi. \qquad (5.49)$$

Comparing $\bar{I}_\phi$ with $\bar{I}_z$ in Eq. (5.24), we identify

$$E_z(\xi) = \eta^{-1} = (\xi^2 + 1/c^2)^{-1/2}, \qquad h(\xi, z) = \eta|z - z_0|. \qquad (5.50)$$

According to Eq. (5.37), the inverse transform of $\bar{I}_\phi$ is

$$I_\phi(r, z, t) = H(t - t_A)\frac{2}{\pi} \operatorname{Im} \int_0^{\xi_1(t)} \frac{\xi}{\eta K(\xi; r, z, t)} d\xi, \qquad (5.51)$$

where, according to Eq. (5.36),

$$K = [\xi^2 r^2 + (t - \eta|z - z_0|)^2]^{1/2} = [\eta^2 R^2 - 2t|z - z_0|\eta + t^2 - r^2/c^2]^{1/2} \tag{5.52}$$

and

$$R^2 = r^2 + (z - z_0)^2. \tag{5.53}$$

The upper limit of integration ξ_1 is determined from Eq. (5.35), which can be solved analytically to yield

$$\xi_1(t) = R^{-2}[|z - z_0|(t - R^2/c^2)^{1/2} + irt]^{1/2}. \tag{5.54}$$

The integration in Eq. (5.51) can be carried out as follows:

$$\begin{aligned}
\operatorname{Im} \int_0^{\xi_1} \frac{\xi}{\eta K(\xi; r, z, t)} d\xi \\
&= \tfrac{1}{2} \operatorname{Im} \int_0^{\xi_1} \frac{d(\xi^2 + c^{-2})}{\eta K(\xi; r, z, t)} \\
&= \operatorname{Im} \int_{\eta_0}^{\eta_1} \frac{d\eta}{(R^2\eta^2 - 2t|z - z_0|\eta + t^2 - r^2/c^2)^{1/2}} \\
&= \operatorname{Im} \left[\frac{1}{R} \sinh^{-1} \frac{R^2\eta - t|z - z_0|}{r(t^2 - R^2/c^2)^{1/2}}\right]_{\eta_0}^{\eta_1} = \frac{1}{R}\frac{\pi}{2} H\left(t - \frac{R}{c}\right).
\end{aligned} \tag{5.55}$$

Note that the lower limit $\eta_0 = \eta(0) = 1/c$ is real, which contributes nothing to the imaginary part of the solution. The upper limit $\eta_1 = (\xi_1^2 + c^{-2})^{1/2}$ is calculated from Eq. (5.54):

$$\eta_1 = R^{-2}[t|z - z_0| + ir(t^2 - R^2/c^2)^{1/2}]. \tag{5.56}$$

At this limit, the result of the integration is $\sinh^{-1}(i) = i\pi/2$. Substituting Eq. (5.55) into (5.51), we obtain

$$I_\phi(r, z, t) = \frac{1}{R} H(t - t_M) = \frac{1}{R} H\left(t - \frac{R}{c}\right). \tag{5.57}$$

For this source ray, $t_A < t_M = R/c$.

The inverse transform of $\bar{\phi}$ in Eq. (5.49) is then found by convolution. Since the inverse transform of $s\bar{f}(s)$ is $f'(t) + f(0)\delta(t)$, we obtain, if $f(0) = 0$,

$$\phi(r, z, t) = \frac{1}{4\pi c^2} \int_0^t f'(t - \tau) \frac{1}{R} H\left(\tau - \frac{R}{c}\right) d\tau = \frac{1}{4\pi c^2 R} \int_{R/c}^t f'(t - \tau)\, d\tau. \tag{5.58}$$

With $f(0) = 0$, the final answer is

$$\phi(r, z, t) = \frac{f(t - R/c)}{4\pi c^2 R}. \tag{5.59}$$

From $\phi(r, z, t)$, we can determine the radial displacement u_R:

$$u_R = \frac{\partial \phi}{\partial R} = -\frac{1}{4\pi c^2}\left[\frac{1}{cR} f'\left(t - \frac{R}{c}\right) + \frac{1}{R^2} f\left(t - \frac{R}{c}\right)\right]. \tag{5.60}$$

In cylindrical coordinates, $R = [r^2 + (z - z_0)^2]^{1/2}$,

$$u_r = \frac{r}{R} u_R, \qquad u_z = \frac{z - z_0}{R} u_R. \tag{5.61}$$

The result in Eq. (5.59) can, of course, be obtained directly from $\bar{\phi}(R, t)$ [see Eq. (6.8)]. This example shows that for slightly more complicated functions E_z and $h(\xi, z)$ than those in Eq. (5.50), a closed form integration like that in Eq. (5.55) would not be possible, and the only recourse is through numerical integration.

VI. General Solutions for a Point Source

Elastic waves generated by an obliquely applied concentrated force or any other asymmetric point source like a dipole, double force, etc., may be represented by all three potentials ϕ, ψ, χ in Eqs. (2.11)–(2.14). The boundary conditions are that the stresses τ_{zz}, τ_{zr}, $\tau_{z\theta}$ and displacements u_r, u_θ, and u_z are continuous across the interface $z = z_i$ $(i = 1, 2, \ldots)$ at all times.

Since these conditions must be satisfied for all values of r and θ, continuity of displacements and stresses across the entire boundary would require the continuity of the following six quantities (Chandra, 1968):

$$\begin{aligned} &\tau_{zz}: \quad \frac{\lambda}{c^2}\frac{\partial^2 \phi}{\partial t^2} + 2\mu\left(\frac{\partial^2 \phi}{\partial z^2} + \frac{\partial^2 \psi}{\partial z^2} - \frac{1}{C^2}\frac{\partial^3 \psi}{\partial z\, \partial t^2}\right), \\ &\tau_{zr}, \tau_{z\theta}: \quad \mu\left(2\frac{\partial \phi}{\partial z} + 2\frac{\partial^2 \psi}{\partial z^2} - \frac{1}{C^2}\frac{\partial^2 \psi}{\partial t^2}\right), \quad \mu\frac{\partial \chi}{\partial z}; \end{aligned} \tag{6.1}$$

$$\begin{aligned} &u_z: \quad \frac{\partial \phi}{\partial z} + \frac{\partial^2 \psi}{\partial z^2} - \frac{1}{C^2}\frac{\partial^2 \psi}{\partial t^2}, \\ &u_r, u_\theta: \quad \phi + \frac{\partial \psi}{\partial z}, \quad \chi. \end{aligned} \tag{6.2}$$

At a traction-free surface, the quantities in Eq. (6.1) must vanish.

According to Eq. (2.13), the displacement vector derived from the χ potential $[\partial\chi/r\,\partial\theta, -\partial\chi/\partial r, 0]$ is parallel to the planes $z =$ constant. The corresponding wave motion is designated as the SH wave. Furthermore, χ is uncoupled from the other two potentials in the boundary conditions (6.1) and (6.2). Hence, upon the incidence of an SH wave represented by χ at the interface of two layers, the reflected and transmitted waves are also horizontally polarized. The total wave motion may then be divided into two parts: the P and SV waves as represented by ϕ and ψ, and the SH wave as represented by χ. Each part is reflected and refracted along the layers independently of the other.

A. Generalized Rays for an Arbitrarily Oriented Force

Let a concentrated force with time function $f(t)$ be applied at an interior point $z = z_0$ in the direction of a unit vector $\mathbf{a}$. If $\mathbf{e}_i$ $(i = 1, 2, 3)$ indicate the unit vectors in x, y, z directions, and $\mathbf{e}_r$, $\mathbf{e}_\theta$, $\mathbf{e}_z$ the unit vectors in cylindrical coordinates, we have (Fig. 13)

$$\mathbf{a} = a_1\mathbf{e}_1 + a_2\mathbf{e}_2 + a_3\mathbf{e}_3 = a_r\mathbf{e}_r + a_\theta\mathbf{e}_\theta + a_z\mathbf{e}_z, \tag{6.3}$$

where

$$a_r = a_1 \cos\theta + a_2 \sin\theta, \qquad a_\theta = -a_1 \sin\theta + a_2 \cos\theta, \qquad a_z = a_3.$$

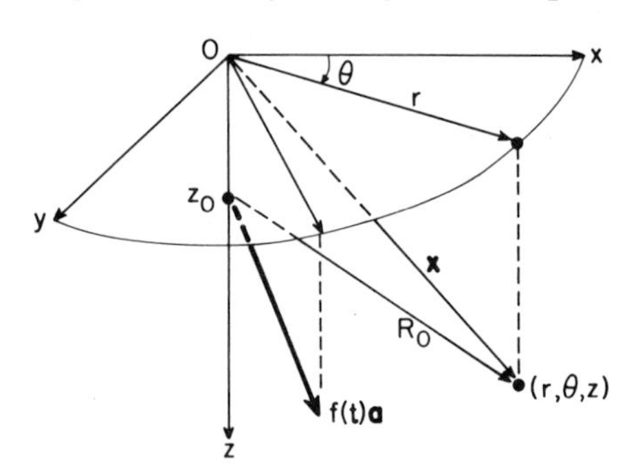

Fig. 13. Geometry of an oblique, concentrated force.

1. *The Source Functions*

When $f(t)$ is a harmonic function in time, that is, $f(t) = \exp(-i\omega t)$, the Green's dyadic for elastic waves is (Morse and Feshbach, 1953, p. 1783)

$$\mathbf{G}(\mathbf{x}\,|\,\mathbf{x}_0, \omega) = \frac{1}{4\pi\rho\omega^2} \times \left\{\frac{\omega^2}{C^2}\mathbf{I}g_S(\mathbf{x}\,|\,\mathbf{x}_0, \omega) - \nabla[g_P(\mathbf{x}\,|\,\mathbf{x}_0, \omega) - g_S(\mathbf{x}\,|\,\mathbf{x}_0, \omega)]\,\nabla\right\}, \tag{6.4}$$

where

$$g_P(\omega) = e^{i\omega R/c}/R, \qquad g_S(\omega) = e^{i\omega R/C}/R, \tag{6.5}$$

and

$$R = [(x - x_0)^2 + (y - y_0)^2 + (z - z_0)^2]^{1/2}.$$

Physically, the components of the dyadic $G_{mn}(\mathbf{x} \,|\, \mathbf{x}_0, \omega)$ represent the displacements at $\mathbf{x}$ in the direction of the x_n axis due to a harmonic, concentrated body force applied at the point $\mathbf{x}_0$ in the direction of the x_m axis. The displacement field generated by a single concentrated force in the direction $\mathbf{a}$ is then $\mathbf{u}_0(\omega)\exp(-i\omega t)$, where

$$\mathbf{u}_0(\omega) = \mathbf{G} \cdot \mathbf{a} = \frac{1}{4\pi\rho\omega^2}\left\{\mathbf{a}\frac{\omega^2}{C^2} g_\text{S}(\omega) - \nabla(\mathbf{a} \cdot \nabla)[g_\text{P}(\omega) - g_\text{S}(\omega)]\right\}. \tag{6.6}$$

The preceding solution can readily be transformed to the Laplace transformed displacement fields $\bar{\mathbf{u}}(\mathbf{x}, s)$ due to a concentrated force $f(t)\mathbf{a}$ applied at $(0, 0, z_0)$. We simply change ω to is and obtain

$$\bar{\mathbf{u}}_0(\mathbf{x}, s) = \frac{\bar{f}(s)}{4\pi\rho s^2}\left\{\nabla(\mathbf{a} \cdot \nabla) g_\text{P}(s) - \left[\nabla(\mathbf{a} \cdot \nabla) - \frac{s^2}{C^2}\mathbf{a}\right] g_\text{S}(s)\right\}, \tag{6.7}$$

where

$$g_\text{P}(s) = \frac{e^{-sR_0/c}}{R_0} = s\int_0^\infty \frac{\xi}{\eta} e^{-s\eta|z - z_0|} J_0(s\xi r)\, d\xi,$$

$$g_\text{S}(s) = \frac{e^{-sR_0/C}}{R_0} = s\int_0^\infty \frac{\xi}{\zeta} e^{-s\zeta|z - z_0|} J_0(s\xi r)\, d\xi,$$

$$R_0 = [x^2 + y^2 + (z - z_0)^2]^{1/2}. \tag{6.8}$$

The two integrals in Eq. (6.8) are the well-known Sommerfeld integral representation of a radial wave function. The derivation of the integral of $g_\text{P}(s)$ follows directly that of Eqs. (2.32) and (2.35).

Since in cylindrical coordinates

$$\mathbf{a} \cdot \nabla = a_r \frac{\partial}{\partial r} + a_\theta \frac{1}{r}\frac{\partial}{\partial \theta} + a_z \frac{\partial}{\partial z}, \tag{6.9}$$

and both functions $g_\text{P}(s)$ and $g_\text{S}(s)$ are independent of θ, we find the three components of $\bar{\mathbf{u}}_0(\mathbf{x}, s)$ as

$$\bar{u}_{0r} = \frac{\bar{f}(s)}{4\pi\rho s^2}\left[\left(a_r\frac{\partial^2}{\partial r^2} + a_z\frac{\partial^2}{\partial r\,\partial z}\right)(g_\text{P} - g_\text{S}) + a_r\frac{s^2}{C^2} g_\text{S}\right],$$

$$\bar{u}_{0\theta} = \frac{\bar{f}(s)}{4\pi\rho s^2} a_\theta\left[\frac{1}{r}\frac{\partial}{\partial r}(g_\text{P} - g_\text{S}) + \frac{s^2}{C^2} g_\text{S}\right],$$

$$\bar{u}_{0z} = \frac{\bar{f}(s)}{4\pi\rho s^2}\left[\left(a_r\frac{\partial^2}{\partial r\,\partial z} + a_z\frac{\partial^2}{\partial z^2}\right)(g_\text{P} - g_\text{S}) + a_z\frac{s^2}{C^2} g_\text{S}\right]. \tag{6.10}$$

The $\bar{\mathbf{u}}_0(\mathbf{x}, s)$ can be derived from the potentials $\bar{\phi}(\mathbf{x}, s)$, $\bar{\psi}(\mathbf{x}, s)$, and $\bar{\chi}(\mathbf{x}, s)$ by equating Eq. (6.10) to (2.13). The results are

$$\begin{aligned}
\bar{\phi}_0(s) &= \frac{\bar{f}(s)}{4\pi\rho s^2}(\mathbf{a}\cdot\nabla)g_{\mathrm{P}} \\
&= a_z s^2 \bar{F}(s)\int_0^\infty S_{\mathrm{P}}(\xi)e^{-s\eta|z-z_0|}J_0(s\xi r)\xi\, d\xi \\
&\quad + a_r s^2 \bar{F}(s)\int_0^\infty S'_{\mathrm{P}}(\xi)e^{-s\eta|z-z_0|}J_1(s\xi r)\xi\, d\xi, \\
\bar{\psi}_0(s) &= -a_z s\bar{F}(s)\int_0^\infty S_{\mathrm{V}}(\xi)e^{-s\zeta|z-z_0|}J_0(s\xi r)\, d\xi \\
&\quad - a_r s\bar{F}(s)\int_0^\infty S'_{\mathrm{V}}(\xi)e^{-s\zeta|z-z_0|}J_1(s\xi r)\, d\xi, \\
\bar{\chi}_0(s) &= -a_\theta s^2\bar{F}(s)\int_0^\infty S_{\mathrm{H}}(\xi)e^{-s\zeta|z-z_0|}J_1(s\xi r)\, d\xi,
\end{aligned} \tag{6.11}$$

where $\bar{F}(s) = \bar{f}(s)/4\pi\rho s^2$. The source functions S_{P}, S'_{P}, S_{V}, S'_{V}, and S_{H} are listed in Table IV. Note that when **a** coincides with the z axis, $a_z = 1$, and $a_r = a_\theta = 0$, and the preceding results reduce to the axisymmetric solutions in Eqs. (2.33) and (2.35).

TABLE IV

SOURCE FUNCTIONS FOR AN INCLINED CONCENTRATED FORCE [EQ. (6.11)]

Source functions		At the interior[a]	At the surface $z = 0$[b]
$S(\xi)$	P	$-\varepsilon$	$-1 + R^{\mathrm{PP}} + (\xi/\zeta)R^{\mathrm{SP}}$
	V	ξ/ζ	$(\xi/\zeta)(1 + R^{\mathrm{SS}}) + R^{\mathrm{PS}}$
$S'(\xi)$	P	$-\xi/\eta$	$-(\xi/\eta)(1 + R^{\mathrm{PP}}) - R^{\mathrm{SP}}$
	V	ε	$1 - R^{\mathrm{SS}} - (\xi/\eta)R^{\mathrm{PS}}$
$S_{\mathrm{H}}(\xi)$		$1/\zeta C^2$	$(1 + R^{\mathrm{H}})/(\zeta C^2)$

[a] $\varepsilon = \pm 1$ if the ray is in the $\pm z$ direction.
[b] $R^{\mathrm{PP}}, \ldots, R^{\mathrm{SS}}$ are given in Eq. (3.11); R^{H} in Eq. (6.14).

2. *The Reflected and Transmitted Rays*

The generalized rays for the reflected and transmitted waves may then be constructed like those in the two-dimensional case (Section III). In analogy to Eqs. (3.3) and (3.4), we obtain three potentials along each ray path (Müller, 1969):

$$\bar{\phi}(s) = a_z s^2 \bar{F}(s) \int_0^\infty S\pi e^{-sh(\xi, z)} J_0(s\xi r)\xi \, d\xi$$

$$+ a_r s^2 \bar{F}(s) \int_0^\infty S'\pi e^{-sh(\xi, z)} J_1(s\xi r)\xi \, d\xi,$$

$$\bar{\psi}(s) = -a_z s\bar{F}(s) \int_0^\infty S\pi e^{-sh(\xi, z)} J_0(s\xi r) \, d\xi$$

$$-a_r s\bar{F}(s) \int_0^\infty S'\pi e^{-sh(\xi, z)} J_1(s\xi r) \, d\xi,$$

$$\bar{\chi}(s) = -a_\theta s^2 \bar{F}(s) \int_0^\infty S_H \pi_H e^{-sh(\xi, z)} J_1(s\xi r) \, d\xi. \tag{6.12}$$

The phase function $h(\xi, z)$ is defined in Eq. (5.18), and it may be rewritten as

$$h(\xi, z) = \sum_{k=1}^{n} h_k(\xi^2 + a_k^2)^{1/2}, \tag{6.13}$$

where n is the number of segments of the ray, h_k the projection of the kth segment of the ray on the z axis, a_k the slowness (reciprocal of the velocity) along the kth segment of the ray, and $\bar{F}(s) = \bar{f}(s)/4\pi\rho s^2$.

For the coupled P-SV waves, the S and S' may be either a P source (S_P and S'_P) or an SV source (S_V and S'_V), depending on the mode of the first segment of a generalized ray. The π is the continuous product of $n - 1$ reflection and transmission coefficients as given in Section III,B. Whether the ray is a P wave (ϕ) or an SV wave (ψ) depends on the mode of the last segment as specified by $h(\xi, z)$.

For the SH wave, S_H is the source function in Eq. (6.11). The π_H is the product of $n - 1$ reflection and transmission coefficients of SH waves, which are

$$R^H = \frac{\mu_2\zeta_2 - \mu_1\zeta_1}{\mu_1\zeta_1 + \mu_2\zeta_2}, \qquad \zeta_1 = (\xi^2 + C_1^{-2})^{1/2},$$

$$T^H = \frac{2\mu_2\zeta_2}{\mu_1\zeta_1 + \mu_2\zeta_2}, \qquad \zeta_2 = (\xi^2 + C_2^{-2})^{1/2}. \tag{6.14}$$

The subscript 2 for μ and C denotes shear modulus and shear wave velocity respectively in the source (incident wave) medium, and the subscript 1 is for the adjacent medium. For a traction free surface, $R^H = 1$.

Displacements may be calculated from Eq. (6.12) according to

Eq. (2.13). Despite the difference in the $\bar{\phi}$ and $\bar{\psi}$ potentials, the displacements for P or SV wave may be cast in the same integral form with a different receiver function (Müller, 1969):

P or SV wave:

$$\bar{u}_r(s) = a_z s^3 \bar{F} \int_0^\infty S\pi D e^{-sh} \xi J_1 \, d\xi,$$

$$- a_r s^3 \bar{F} \int_0^\infty S'\pi D e^{-sh} \xi J_0 \, d\xi - \frac{a_r}{a_\theta} \bar{u}_\theta(s),$$

$$\bar{u}_\theta(s) = -a_\theta \frac{s^2}{r} \bar{F} \int_0^\infty S'\pi D e^{-sh} J_1 \, d\xi,$$

$$\bar{u}_z(s) = a_z s^3 \bar{F} \int_0^\infty S\pi D e^{-sh} \xi J_0 \, d\xi$$

$$+ a_r s^3 \bar{F} \int_0^\infty S'\pi D e^{-sh} \xi J_1 \, d\xi; \tag{6.15}$$

SH wave:

$$\bar{u}_r(s) = a_r \frac{s^2}{r} \bar{F} \int_0^\infty S_{\mathrm{H}} \pi_{\mathrm{H}} D_{\mathrm{H}} e^{-sh} J_1 \, d\xi,$$

$$\bar{u}_\theta(s) = -a_\theta s^3 \bar{F} \int_0^\infty S_{\mathrm{H}} \pi_{\mathrm{H}} D_{\mathrm{H}} e^{-sh} \xi J_0 \, d\xi + \frac{a_\theta}{a_r} \bar{u}_r(s),$$

$$\bar{u}_z(s) = 0. \tag{6.16}$$

The receiver functions D of the three displacements for either P or SV wave, and D_{H} for SH wave are collected in Table V.

It is seen from Eqs. (6.15) and (6.16) that the ray integrals for all

TABLE V

RECEIVER FUNCTIONS D AND D_{H} FOR DISPLACEMENTS [EQS. (6.15) AND (6.16)]

	Mode	At the interior	At the free surface $z = 0$
u_r	P	$-\xi$	$-\xi(1 + R^{\mathrm{PP}}) - \zeta R^{\mathrm{PS}}$
or u_θ	SV	$-\varepsilon\zeta$	$\zeta(1 - R^{\mathrm{SS}}) - \xi R^{\mathrm{SP}}$
	SH	1	$1 + R^{\mathrm{H}} = 2$
u_z	P	$-\varepsilon\eta$	$\eta(1 - R^{\mathrm{PP}}) - \xi R^{\mathrm{PS}}$
	SV	$-\xi$	$-\xi(1 + R^{\mathrm{SS}}) - \eta R^{\mathrm{SP}}$

displacements are either in the form of $\bar{I}_z$ in Eq. (5.24), or $\bar{I}_r$ in Eq. (5.25). The inverse Laplace transforms of $\bar{u}_i(s)$ can thus be found as in Section V,B.

B. Additional Source Functions

In addition to a center of dilatation [Eq. (2.32)], a vertical force [Eq. (2.33)], and an oblique concentrated force [Eq. (6.11)], there are many other body forces that are often used to model material defects, cracks, or faults. The cases of a double force without moment, a single body couple, and an oblique force on the surface will be discussed in this section. The cases of an impulsive stress applied to the wall of a small cylindrical cavity (Brekhovskikh, 1960, p. 305) and of a spherical cavity (Ghosh, 1973; Gajewski, 1977) inside a layered medium are also of interest, but will not be discussed here.

1. *An Oblique Force at the Free Surface*

In Section II,B,3, the cases of a vertical force in an infinite medium and a vertical force at the free surface of a half-space were treated as two different boundary value problems. Actually, the solution of the latter can be derived from that of the former.

The source functions for a vertical force are given by Eqs. (2.33), (2.35), and Table I. The P wave with the source function S_P is reflected at the surface of a half-space to form a PP and a PS ray (see Figs. 4 and 11). Similarly, the incident shear wave with the source function S_V is reflected as SP and SS rays. When z_0 approaches zero, the incident P ray coalesces with the reflected PP and SP rays to form a direct P ray from the surface source to a receiver. The three ray integrals can then be combined into a single integral as given by $\bar{\phi}$ in Eq. (2.35) with $z_0 = 0$ and S_P is replaced by $S_P + S_P R^{PP} + S_V R^{SP}$. From Table I we find that the new surface source function for the P wave should be

$$S_P(\text{surface}) = -1 + R^{PP} + \frac{\xi}{\zeta} R^{SP}. \tag{6.17}$$

Similarly, the incident S ray is combined with the reflected SS and PS rays to form a new surface source function for the S wave represented by $\bar{\psi}$:

$$S_V(\text{surface}) = \frac{\xi}{\zeta} + \frac{\xi}{\zeta} R^{SS} + R^{PS}. \tag{6.18}$$

Substituting the reflection coefficients in Eq. (3.11) into the previous equations, we obtain

$$S_P(\text{surface}) = 2(\xi^2 + \zeta^2)/(C^2\Delta_r), \qquad S_V(\text{surface}) = -4\eta\xi/(C^2\Delta_r). \tag{6.19}$$

Since $C^2 = \mu/\rho$, these results are in total agreement with those shown in Table I for a vertical force at the surface.

The source functions for an inclined force applied at the surface of a half-space can be derived analogously from those in Eq. (6.11). They are listed in Table IV.

2. *Double Forces and Body Couples*

The construction of the solutions for a double force without moment or a single couple (a double force with moment) along a coordinate axis from those for a concentrated force is well known (Love, 1944, Section 213; White, 1965). The solutions for oblique double forces may be similarly derived.

Let $\mathbf{u}(\mathbf{x}, t)$ be the displacement generated by a concentrated force f_0 at $(0, 0, z_0)$ in the direction of a unit vector $\mathbf{a}$ (Fig. 14). The $\mathbf{a}$ forms an orthog-

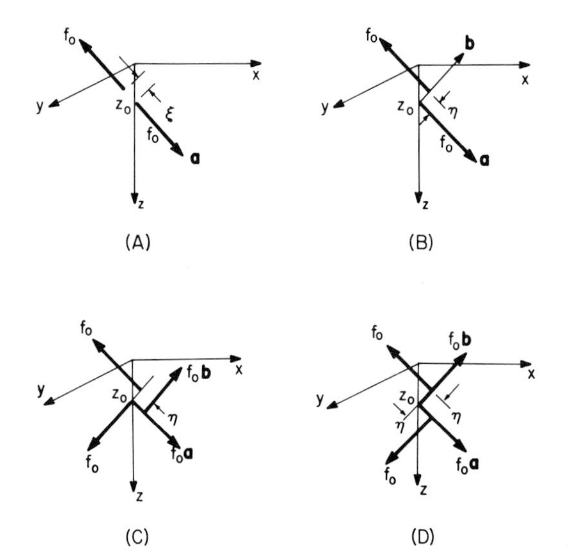

FIG. 14. Point sources in an unbounded solid: (a) double force; (b) single couple; (c) center of rotation; (d) double couple without moment.

onal triad with two other unit vectors $\mathbf{b}$ and $\mathbf{c}$. The solutions for the double-forces are as follows:

(a) A double force (double force in $\mathbf{a}$ without moment):

$$\mathbf{u}^{a}(\mathbf{x}, t) = -D(\mathbf{a} \cdot \nabla)\mathbf{u}(\mathbf{x}, t). \tag{6.20}$$

(b) A single couple (one double force with moment about $\mathbf{c}$):

$$\mathbf{u}^{b}(\mathbf{x}, t) = M(\mathbf{b} \cdot \nabla)\mathbf{u}(\mathbf{x}, t). \tag{6.21}$$

(c) A center of rotation (two double forces with moment about **c**):

$$\mathbf{u}^{c}(\mathbf{x}, t) = M\mathbf{c} \cdot [\nabla \times \mathbf{G}(\mathbf{x}, t)]. \tag{6.22}$$

In the preceding results D and M are the strengths of the double force and single couple, respectively,

$$D = \lim_{\xi \to 0} (f_0 \xi), \qquad M = \lim_{\eta \to 0} (f_0 \eta), \tag{6.23}$$

where ξ is the distance separating the two forces along the **a** direction, and η is that along the **b** direction. $\mathbf{G}(t)$ is Green's dyadic for elastic waves (Achenbach, 1973, Sect. 3.8; Pao and Varatharajulu, 1976),

$$\mathbf{u} = \mathbf{a} \cdot \mathbf{G} = \mathbf{G} \cdot \mathbf{a}; \qquad \mathbf{G}(t) = \int_{-\infty}^{\infty} \mathbf{G}(\omega) e^{-i\omega t}\, d\omega.$$

$\mathbf{G}(\omega)$ is given in Eq. (6.4).

From Eq. (6.21) we can construct the solution for a double couple without moment (Fig. 14). This source is similar to the center of rotation except that the senses of f_0 along the directions of **b** and $-$**b** are reversed.

(d) A double couple without moment:

$$\mathbf{u}^{d}(\mathbf{x}, t) = M(\mathbf{b} \cdot \nabla)\mathbf{u}(\mathbf{x}, t; \mathbf{a}) + M(\mathbf{a} \cdot \nabla)\mathbf{u}(\mathbf{x}, t; \mathbf{b}). \tag{6.24}$$

By properly choosing the orientations of **a** and **b**, this source is often used to represent the fault motion of earthquakes (Harkrider, 1976). Other type of sources due to inelastic strains may be constructed from an integration of Green's dyadics (Willis, 1965).

Based on the above equations, cylindrical components for **u** can easily be derived. For example, the components for a single couple with unit strength ($M = 1$) are:

$$u_r^{\mathrm{b}} = b_r \frac{\partial u_r}{\partial r} + b_\theta \left(\frac{1}{r} \frac{\partial u_r}{\partial \theta} - \frac{u_\theta}{r} \right) + b_z \frac{\partial u_r}{\partial z},$$

$$u_\theta^{\mathrm{b}} = b_r \frac{\partial u_\theta}{\partial r} + b_\theta \left(\frac{1}{r} \frac{\partial u_\theta}{\partial \theta} + \frac{u_r}{r} \right) + b_z \frac{\partial u_\theta}{\partial z},$$

$$u_z^{\mathrm{b}} = b_r \frac{\partial u_z}{\partial r} + b_\theta \frac{1}{r} \frac{\partial u_z}{\partial \theta} + b_z \frac{\partial u_z}{\partial z}. \tag{6.25}$$

The b_r, b_θ, b_z are cylindrical components of the unit vector **b** as defined in Eq. (6.3).

Since $\bar{u}_r(s)$, $\bar{u}_\theta(s)$, and $\bar{u}_z(s)$ for a concentrated force in the **a** direction are given in Eqs. (6.15) and (6.16), and their inverse Laplace transforms can be found, we can then determine $\mathbf{u}^{\mathrm{b}}(t)$ directly from Eq. (6.25). Details are given by Müller (1969).

C. Approximate Analysis of the Ray Integrals

The generalized ray method requires that the contribution of all arriving rays be included for the complete solution. For many problems, this means the calculation of a large number of ray integrals plus a large amount of work of numerical integration. For a certain application, an approximate solution coupled with an efficient computing scheme may be sufficient.

We discuss in this section two approximate methods of analysis, one is the *plane wave approximation* (Pekeris *et al.*, 1965; Helmberger, 1972; Wiggins and Helmberger, 1974); the second is the *first motion approximation* (Knopoff and Gilbert, 1959; Helmberger, 1968; Müller, 1969). The former method approximates a spherical wave generated by a point source by plane waves, which is accomplished by replacing the Bessel functions in the ray integrals by their high frequency asymptotic expressions. The latter evaluates the predominant part of the wave motion near the arrival time of a ray.

Further simplification which bypasses the use of Cagniard's method altogether is possible. Note that for plane strain solutions [Eqs. (3.16) and (3.17)] or for the plane wave approximation of axisymmetric solutions [Eq. (5.23)], the Laplace-transformed displacement is of the form

$$\bar{u}_z(x, z, s) = A\bar{f}(s) \operatorname{Re} \int_0^\infty E_z(\xi) e^{s[g(x,\, z;\, \xi) - \varepsilon]}\, d\xi. \tag{6.26}$$

In the above A is an amplitude constant, and ε a phase constant. This integral can be evaluated by either the *method of stationary phase* or the *method of equal phase*. The former method evaluates analytically the dominant part of the integral when the phase is stationary, that is, when $dg(x, z; \xi)/d\xi = 0$. The latter method replaces the condition of stationary phase by the condition of equal phase, that is, the finding of values of ξ such that $t = -g(x, z; \xi)$. After the ξ integral is evaluated approximately, the inverse Laplace transform can often be carried out in closed form (Wiggins, 1976; Chapman, 1976a; Wiggins and Madrid, 1974).

1. *Plane Wave Approximation*

The plane wave approximation assumes that the source function is either of short time duration or of high frequency content. In this method the Bessel function is replaced with the first term of its asymptotic expansion.

To be specific, consider the problem of a point source in Section V. Instead of using Eq. (5.22), the first term of the asymptotic expansion of Bessel function for large arguments is used (National Bureau of Standards, 1964, Eq. 9.2.1),

$$J_0(s\xi r) = (2/\pi s\xi r)^{1/2} \cos(s\xi r - \tfrac{1}{4}\pi).$$

It can be rewritten as

$$J_0(s\xi r) = \text{Re}(2/\pi is\xi r)^{1/2}e^{is\xi r}. \tag{6.27}$$

where we have taken $(-i) = \exp(-i\pi/4)$.

This approximation then reduces $\bar{I}_z(s)$ in Eq. (5.24) to a form like $\bar{u}(s)$ in Eq. (4.1), except for the factor $(is\xi r)^{-1/2}$. Applying the transformation

$$-\tau = i\xi r - \sum_j z_j(\xi^2 + a_j^2)^{1/2}, \tag{6.28}$$

the inversion of $I_z(r, z, s)$ without the factor $s^{-1/2}$ is found by inspection. Since the inverse transform of $s^{-1/2}$ is $(\pi t)^{-1/2}$, the complete answer for the inverse Laplace transform of $\bar{I}_z(r, z, s)$ is found by convolution:

$$I_z(r, z, t) = H(t - t_A)\left(\frac{2}{\pi r}\right)^{1/2} \text{Im} \int_{t_A}^{t} E_z(\xi)\left(\frac{i\xi}{\pi}\right)^{1/2} \frac{d\xi}{d\tau} \frac{d\tau}{(t-\tau)^{1/2}}. \tag{6.29}$$

The ξ in the integrand is a function of τ, which is found by solving Eq. (6.28).

If the variable τ is transformed back to ξ according to Eq. (6.28), we obtain, like the change of Eq. (4.7) to (4.9),

$$I_z(r, z, t) = H(t - t_A)\left(\frac{2}{\pi r}\right)^{1/2} \text{Im} \int_0^{\xi_1(t)} E_z(\xi)\left(\frac{i\xi}{\pi}\right)^{1/2} \frac{1}{[t - \tau(\xi)]^{1/2}}\, d\xi, \tag{6.30}$$

where $\xi_1(t)$ is obtained by solving ξ for $\tau = t$ in Eq. (6.28).

Comparing Eq. (6.30) with the exact answer in Eq. (5.37), we note that the plane wave approximation is equivalent to the following approximation of the factor K:

$$K(\xi; r, z, t) \cong (-2i\xi r)^{1/2}(t - \tau)^{1/2}. \tag{6.31}$$

This can directly be derived by noticing first that the K in Eq. (5.36) is

$$K(\xi; r, z, t) = \left[\left(t - \sum_j z_j\eta_j - i\xi r\right)\left(t - \sum_j z_j\eta_j + i\xi r\right)\right]^{1/2}, \tag{6.32}$$

where $\eta_j = (\xi^2 + a_j^2)^{1/2}$. In view of Eq. (6.28) we find

$$K(\xi; r, z, t) = \{[t - \tau(\xi) - 2i\xi r][t - \tau(\xi)]\}^{1/2}. \tag{6.33}$$

In dealing with problems of short duration of observation and with high frequency components of waves ($s\xi r \gg 1$), the following condition is valid:

$$|t - \tau| \ll |2i\xi r|. \tag{6.34}$$

Hence Eq. (6.33) reduces to (6.31) in this approximation.

This method has several advantages: (1) The solution is an integral over time, allowing the integrands for each ray to be computed before doing the final convolution. (2) As the integrands are being computed, the relative contribution of each ray may be compared and those with small values left out. (3) The convolution factor $(t - \tau)^{-1/2}$ could be combined analytically with the source time function, and the number of numerical integrations is thus reduced by one.

The method has successfully been applied to approximate the exact results of Pekeris *et al.* (1965) by Helmberger (1972).

2. *First Motion due to a Refracted Ray*

The first motion approximations depend on separating the rapidly varying parts from the slowly varying parts in the integration of Eq. (6.29). For the first motion of a head wave, the critical point is $\xi_k = ia_k$ where $a_k = 1/c_k$ and c_k is the P wave speed in the refracting layer k (Section IV,B,2). Then the rapidly varying function in Eq. (6.29) is $\eta_k = (\xi^2 + a_k^2)^{1/2}$. To first order, all other functions in the integrand can be then considered as constants evaluated at $\xi_k = ia_k$.

Let $R(\xi)$ be the reflection coefficient off the refracting layer. We decompose E_z in Eq. (6.29) as

$$E_z(\xi) = E_z'(\xi)R(\xi), \tag{6.35}$$

where $E_z'(\xi)$ does not include any factor that contains η_k. Based on Eq. (3.9), we write

$$R(\xi) = \frac{A(\xi_k) - \eta_k B(\xi_k)}{A'(\xi_k) + \eta_k B'(\xi_k)}, \tag{6.36}$$

where A, B, A', B' are real constants (at $\xi = \xi_k$) which are determined from the actual reflection coefficient (Helmberger, 1968).

The exact transient response in Eq. (5.41) can be expressed as

$$u_z(r, z, t) = \frac{H(t - t_A)}{4\pi c^2} \int_0^t f'(t - \tau) \frac{\partial}{\partial \tau} I_z(r, z, \tau)\, d\tau + f(0) \frac{\partial}{\partial t} I_z(r, z, t). \tag{6.37}$$

From Eq. (6.29) we find

$$\frac{\partial I_z(r, z, t)}{\partial t} = H(t - t_A) \frac{1}{\pi} \left(\frac{2}{r}\right)^{1/2} \times \operatorname{Im} \int_{t_A}^t \frac{\partial}{\partial \tau} \left[E_z(\xi)(i\xi)^{1/2} \frac{d\xi}{d\tau} \right] \frac{d\tau}{(t - \tau)^{1/2}}. \tag{6.38}$$

However, the only rapidly varying part in the integrand is $R(\xi)$. Thus

$$\frac{\partial}{\partial \tau} \left[E_z(i\xi)^{1/2} \frac{d\xi}{d\tau} \right] \cong E_z'(i\xi)^{1/2} \frac{d\xi}{d\tau} \frac{\partial R}{\partial \tau}.$$

To first order in η_k,

$$R(\xi) = \frac{1}{A'}\left[A - \eta_k \frac{BA' + B'A}{A'}\right]; \tag{6.39}$$

we have, with $\eta_k = (\xi^2 + a_k^2)^{1/2}$,

$$\frac{\partial R}{\partial t} = -\frac{AB' + A'B}{A'^2}\frac{\xi}{\eta_k}\frac{d\xi}{dt}. \tag{6.40}$$

To calculate η_k, we expand ξ about ξ_k,

$$\xi = ia_k + \frac{d\xi}{dt}(t - t_k) + \cdots, \tag{6.41}$$

and obtain

$$\eta_k \cong \left(2ia_k \frac{d\xi}{dt}\right)^{1/2}(t - t_k)^{1/2}. \tag{6.42}$$

From Eq. (6.28), we find

$$\left(\frac{dt}{d\xi}\right)_{\xi=\xi_k} = -ir + \sum_j \frac{z_j\xi_k}{(\xi_k^2 + a_j^2)^{1/2}} = -iL_k, \tag{6.43}$$

where

$$L_k = r + \sum \frac{iz_j\xi_k}{(\xi_k^2 + a_j^2)^{1/2}}.$$

The L_k may be interpreted as the length of wave travel in the refractor [see Eq. (4.18)].

Substituting Eqs. (6.43) and (6.42) in Eq. (6.40) and then in (6.38) we obtain

$$\frac{\partial I_z(r, z, t)}{\partial t} = H(t - t_k)\frac{1}{\pi}\left(\frac{2}{r}\right)^{1/2} \operatorname{Im} \int_{t_k}^{t} \frac{A'B + AB'}{A'^2} \times \frac{\xi(-i\xi)^{1/2}E'(\xi)}{(2a_k L_k^3)^{1/2}} \frac{d\tau}{(\tau - t_k)^{1/2}(t - \tau)^{1/2}}. \tag{6.44}$$

In the above t_A has been replaced by t_k since it is the time of first arrival of the head wave.

The integration of the rapidly varying part is

$$\int_{t_k}^{t} \frac{d\tau}{(\tau - t_k)^{1/2}(t - \tau)^{1/2}} = \pi.$$

The remaining part is constant when ξ is evaluated at $\xi = \xi_k = ia_k$. The result is

$$\frac{\partial I_z(r, z, t)}{\partial t} = H(t - t_k)\left(\frac{1}{rL_k^3}\right)^{1/2}\left(\frac{A'B + AB'}{A'^2}\right)E'(ia_k)a_k. \qquad (6.45)$$

From Eq. (6.37), we finally obtain, if $f(0) = 0$,

$$u_z(r, z, t) = H(t - t_k)f(t - t_k)\left(\frac{1}{rL_k^3}\right)^{1/2}\left(\frac{A'B + AB'}{A'^2}\right)a_k E'(ia_k). \quad (6.46)$$

Thus, the head wave displacement first motion has the same time behavior as the source function.

There might be another head wave arrival due to the fast S wave speed in the refractor. Let C_l be this wave speed. We calculate L_l from Eq. (6.43) by setting $\xi = \xi_l = i/C_l$. If $L_l > 0$, there will be a refracted S wave arrival, and the rapidly varying function is $\eta_l = (\xi^2 + C_l^{-2})^{1/2}$.

The analysis follows in a similar manner except that A, B, A', B', and E_z' are now generally complex constants evaluated at $\xi = \xi_l$. We must also evaluate the following integral

$$I_l = \int_{t_k}^{t} \frac{d\tau}{(t - \tau)^{1/2}(\tau - t_l)^{1/2}}, \qquad (6.47)$$

where t_k is the time value used for Eq. (6.44). From a table of integrals we find

$$I_l = -i \log \frac{|t - t_l|}{t_k^*} + \pi H(t - t_l), \qquad (6.48)$$

where the integration constant is

$$t_k^* = 2[t_k^2 - t_k(t + t_l) - tt_l]^{1/2} + 2t_k - t - t_l. \qquad (6.49)$$

Hence, for the refracted S wave,

$$\frac{\partial I_z(r, z, t)}{\partial t} = \frac{a_l}{(rL_l^3)^{1/2}}\left\{H(t - t_l)\,\mathrm{Re}\left[\left(\frac{A'B + AB'}{A'^2}\right)_l E'(ia_l)\right]\right.$$
$$\left. + \frac{1}{\pi}\log\frac{|t - t_l|}{t_k^*}\,\mathrm{Im}\left[\left(\frac{A'B + AB'}{A'^2}\right)_l E'(ia_l)\right]\right\}. \qquad (6.50)$$

Equation (6.50) differs from (6.45) with the additional logarithmic singularity.

3. *First Motion due to a Direct Ray*

The arrival of a direct wave is determined at $\xi = \xi_M$ such that (Section IV,B,1)

$$(dt/d\xi)_{\xi = \xi_M} = 0. \qquad (4.4)$$

Thus, the only rapidly varying function is $d\xi/dt$, and the rest of the integrand in Eq. (6.29) can be treated as a constant evaluated at $\xi = \xi_M$.

To evaluate $d\xi/dt$, expand t in a power series about t_M:

$$t - t_M = \left(\frac{dt}{d\xi}\right)_{\xi_M} (\xi - \xi_M) + \frac{1}{2}\left(\frac{d^2t}{d\xi^2}\right)_{\xi_M} (\xi - \xi_M)^2 + \cdots. \tag{6.51}$$

With $(dt/d\xi)_{\xi_M} = 0$, it can be solved for ξ:

$$\xi = \xi_M + [2(t - t_M)(d^2t/d\xi^2)_{\xi_M}^{-1}]^{1/2} \tag{6.52}$$

and

$$d\xi/dt = [2(t - t_M)(d^2t/d\xi^2)_{\xi_M}]^{-1/2}. \tag{6.53}$$

From Eq. (6.43), we obtain

$$\frac{d^2t}{d\xi^2} = \sum_j \frac{z_j a_j^2}{(\xi^2 + a_j^2)^{3/2}}. \tag{6.54}$$

Therefore, the first motion approximation of Eq. (6.29) is

$$I_z(r, z, t) \cong \left(\frac{1}{r}\right)^{1/2} \frac{1}{\pi} H(t - t_A) \operatorname{Im} \int_{t_A}^{t} E_z(\xi_M) \times (i\xi_M)^{1/2} \left[\frac{d^2t}{d\xi^2}\right]_{\xi_M}^{-1/2} \frac{d\tau}{(\tau - t_M)^{1/2}(t - \tau)^{1/2}}. \tag{6.55}$$

For the direct ray, the lower limit of integration t_A is replaced by t_M. However, when the wave along a direct ray follows that along the refracted ray, the lower limit is replaced by t_d where t_d ($<t_M$) is the upper limit of integration in the first motion approximation of the refracted waves as discussed in Section VI,C,2. The time-dependent part of the integral in Eq. (6.55) is the same as Eq. (6.47). Upon the completion of the integration, we find

$$I_z(r, z, t) = \left[\frac{|\xi_M|}{r}\right]^{1/2} \left[\sum_j \frac{z_j a_j^2}{(\xi_M^2 + a_j^2)^{3/2}}\right]^{-1/2} \times \left\{H(t - t_M) \operatorname{Re}[E_z(\xi_M)] + \frac{1}{\pi} \log \frac{|t - t_M|}{t_d^*} \operatorname{Im}[E_z(\xi_M)]\right\}, \tag{6.56}$$

when t_d^* is given by Eq. (6.49) if the t_k in it is replaced by t_d, and t_l by t_M.

Again, there is a logarithmic singularity in the result if $\operatorname{Im}[E_z(\xi_M)] \neq 0$, which happens only when there is a head wave preceding the direct wave.

VII. Transient Responses and Numerical Examples

The solutions of displacement field for pulses traveling along a specific generalized ray path are given in Eq. (4.9) for plane strain and in Eq. (5.41) for axisymmetric responses. The transient response at $\mathbf{x}$ is then determined by summing solutions for pulses along all possible ray paths. It may be expressed compactly as

$$u_\alpha(\mathbf{x}, t) = A_\alpha \sum_{j=1}^{\infty} \left\{ H(t - t_{Aj}) \int_{t_A}^{t} f^{(n+1)}(t - \tau) I_{\alpha j}(\mathbf{x}, \tau)\, d\tau \right.$$
$$\left. + f^{(n)}(0) I_{\alpha j}(\mathbf{x}, t) + f^{(n-1)}(0) \frac{\partial}{\partial t} I_{\alpha j}(\mathbf{x}, t) \right\}. \tag{7.1}$$

The index j pertains to the jth generalized ray.

For plane strain cases, $\alpha = x$ or z, and we have from Eq. (4.9),

$$I_{xj} = 2 \operatorname{Im} \int_0^{\xi_1(\tau)} \xi E_{xj}(\xi)\, d\xi, \qquad I_{zj} = 2 \operatorname{Re} \int_0^{\xi_1(\tau)} E_{zj}(\xi)\, d\xi. \tag{7.2}$$

For axisymmetric cases, $\alpha = r$ or z, and we obtain from Eq. (5.37) or (5.47),

$$I_{rj} = \frac{2}{\pi r} \operatorname{Im} \int_0^{\xi_1(\tau)} \xi E_{xj}(\xi) \frac{\partial K_j(\xi; \mathbf{x}, \tau)}{\partial \tau}\, d\xi,$$

$$I_{zj} = \frac{2}{\pi} \operatorname{Im} \int_0^{\xi_1(\tau)} \xi E_{zj}(\xi) \frac{1}{K_j(\xi; \mathbf{x}, \tau)}\, d\xi. \tag{7.3}$$

The integration with respect to ξ is along the complex contour AMC shown in Fig. 9.

The coefficients A_α and index n depend on the nature of source functions (Table I). For a source of explosion, we have $n = 1$ and $A_\alpha = (4\pi c^2)^{-1}$; for a concentrated force inside a layer, $n = 0$ and $A_\alpha = (4\pi\rho)^{-1}$; for a concentrated force at the free surface, $n = 0$ and $A_\alpha = (2\pi\mu)^{-1}$. When $n = 0$, the $(n - 1)$th derivative of the time function $f^{(n-1)}(t)$ is zero.

Since, in general, the duration of observation $(0, t)$ at any given station $\mathbf{x}$ is finite, the infinite upper limit of summation may be replaced by a finite number N. We thus arrange all rays in ascending order according to the number of reflections because the arrival times of the rays increase with the number of reflections. All rays with arrival times t_j less than the maximal times of observation must be included in the summation. The t_A that is the lower limit of integration in Eq. (7.1) is the minimum of all t_{Aj} ($j = 1, 2, \ldots, N$). For a medium with a single layer, the matrix formulation of ray groups in Section V,A,1 provides a systematic method for ordering all rays. Calculations of t_j are discussed in Section VII,A.

Since each ray integral $I_{\alpha j}$ has its own contour of integration (Figs. 6 and 9), it must be evaluated separately. The natural approach then is to calculate first each $I_{\alpha j}(\tau)$ at a preselected time τ between 0 and t, and then to collect all rays at each value of τ:

$$I_\alpha(\mathbf{x}, \tau) = \sum_{j=1}^{N} I_{\alpha j}(\mathbf{x}, \tau). \tag{7.4}$$

Having calculated $I_\alpha(\tau)$ for many discrete but closely spaced values of τ, we then determine $u_\alpha(t)$ by a single convolution:

$$u_\alpha(\mathbf{x}, t) = H(t - t_A) \int_{t_A}^{t} f^{(n+1)}(t - \tau) I_\alpha(\mathbf{x}, \tau)\, d\tau + f^{(n)}(0) I_\alpha(\mathbf{x}, t) + f^{(n-1)}(0) \frac{\partial}{\partial t} I_\alpha(\mathbf{x}, t). \tag{7.5}$$

Various methods have been proposed to evaluate the integrals like $I_{\alpha j}$ (Cagniard, 1939; Abramovici and Alterman, 1965; Müller, 1968, Part II). We discuss in Section VII,B a numerical scheme for integrating $I_{\alpha j}$ in the complex ξ plane. Numerical results will be presented in Section VII,C for two examples: a point source of explosion [$n = 1$ in Eqs. (7.1) and (7.5)] inside a plate with a receiver on its surface, and the same source in a layer overlying a half-space of different material. These results may be compared with those obtained by Knopoff (1958a) for a plate, and by Perkeris *et al.* (1965) and Helmberger (1972) for a single layer. Numerical results are also given by Garvin (1956) for a line source of explosion inside a half-space, by Mooney (1976) for a point force on the surface of a half-space, and by Schmuely (1975) for a line of forces on the surface of a plate.

A. Arrival Times of Generalized Rays

In Eqs. (7.2) and (7.3) $\xi_1(t)$ for the jth ray is determined from Eq. (4.3) or (5.45):

$$(t)_j = -ir\xi_{1j}(t) + \sum_k z_{kj}[\xi_{1j}^2(t) + a_k^2]^{1/2}. \tag{7.6}$$

Before calculating $\xi_{1j}(t)$ for a given value of t, we must calculate the location of the stationary point M from Eq. (4.4):

$$\left(\frac{dt}{d\xi}\right)_j = -ir + \sum_k \frac{z_{kj}\xi_j}{(\xi_j^2 + a_k^2)^{1/2}} = 0. \tag{7.7}$$

When there are one or two radicals ($k = 1, 2$), this equation can be solved exactly [e.g., Eq. (4.25)]. Otherwise it must be solved numerically. We found that even for the case of two radicals, it is more convenient to find the root by applying Müller's method (Conte and de Boor, 1972), a general purpose root solving algorithm. As in Section IV, the root of Eq. (7.7) is denoted as $\xi = \xi_{Mj} = ib_{Mj}$.

Müller's method achieves quadratic convergence, and the convergence is rapid if a set of initial values closely bracketing the correct answer is made. ξ_{Mj} is always on the positive imaginary ξ axis, and it lies below the minimum value of a_k $(k = 1, 2, \ldots)$ if $z_{kj} \neq 0$ in Eq. (7.6). A preliminary search for ξ_{Mj} starts at the minimum of ia_k and proceeds down the imaginary axis at large intervals until a change of sign for the values of $dt/d\xi$ occurs. The large interval bracketing the root is reduced by successively halving the interval and keeping the two values of ξ at which $dt/d\xi$ are of opposite signs. After a couple iterations of this binary chopping, the final two values of ξ are used as the initial values for Müller's algorithm. The binary chopping is necessary in order to make the bracketing interval sufficiently small. Otherwise, the singularity at $\xi = \min(ia_k)$ in Eq. (7.7) makes the Müller's method totally ineffective. With a set of closely bracketing initial values for Müller's method, an accurate value for the root was found in six iterations.

After the finding of ξ_M Eq. (7.6) is used to calculate t_{Mj} by setting $\xi_{1j}(t) = \xi_{Mj}$. t_{Mj} is the arrival time of the jth ray along the direct ray path (Section IV,B,1). For direct rays, $t_j = t_{Mj}$ in Eq. (7.1).

The ξ_{Mj} is then compared with the branch points ia_f associated with head waves (Sections IV,B,2 and IV,2,C). If $|\xi_{Mj}| > |ia_f|$, we must then calculate t_f in Eq. (4.16) because waves traveling along the refracted ray path arrive earlier than those along the direct ray path. The arrival time t_{fj} for the refracted ray is calculated by substituting $\xi_{1j} = ia_f$ into Eq. (7.6). For refracted rays, $t_j = t_{fj}$ in Eq. (7.1).

The values of $\xi_{1j}(t)$ in Eq. (7.6) for $t > t_j$ are also calculated with Müller's method. The initial values are not critical in this calculation so ξ_{Mj} or ia_f is used as the first upper limit $\xi_{1j}(t_1)$. The values for $\xi_{1j}(t)$ is pure imaginary between ia_f and ξ_{Mj} for head waves, and is complex in the first quadrant after the arrival of direct rays (Fig. 9).

As an example, we show in Table VI the minimum arrival times $(\tau = C_1 t/h_1)$ for the first 20 generalized rays of a three-layered medium in Fig. 2. The data of the layers are as follows:

$$\begin{array}{lll} h_1 = 1.0, & h_2 = 2.0, & z_0 = 2.0, \\ c_1 = \sqrt{3}, & C_1 = 1.0, & \rho_1 = 1.21, \\ c_2 = 1.1c_1, & C_2 = 1.1C_1, & \rho_2 = 2.00, \\ c_3 = 2.2c_1, & C_3 = 2.2C_1, & \rho_3 = 3.00. \end{array}$$

TABLE VI

MINIMUM ARRIVAL TIMES FOR THE FIRST TWENTY RAYS AT THREE STATIONS

	A ($r = 2$)		B ($r = 5$)		C ($r = 10$)	
1	PP_1	1.557	PP_1	2.948	$pP_3^*P_2P_1$	4.502
2	PS_1	2.064	$pP_3^*P_2P_1$	3.190	$pP_3^*P_2S_1$	4.952
3	pP_2P_1	2.405	pP_2P_1	3.441	$pP_3^*S_2P_1$	5.334
4	PP_1^3	2.523	PS_1	3.534	$PP_3^*P_2^2P_1$	5.411
5	pP_2S_1	2.853	PP_1^3	3.608	PP_1	5.522
6	$PP_1^2S_1$	2.974	$pP_3^*P_2S_1$	3.641	$pP_3^*P_2P_1^3$	5.531
7	$pP_3^*S_2P_1$	3.234	pP_2S_1	3.957	$pS_3^*P_2P_1$	5.689
8	pS_2P_1	3.240	$pP_3^*S_2P_1$	4.022	pP_2P_1	5.771
9	$PP_2^2P_1$	3.375	$PP_3^*P_2^2P_1$	4.099	$pP_3^*S_2S_1$	5.784
10	$PP_1S_1^2$	3.432	$PP_1^2S_1$	4.138	$PP_3^*P_2^2S_1$	5.862
11	$pP_2P_1^3$	3.485	$PP_2^2P_1$	4.166	$pP_3^*P_2P_1^2S_1$	5.981
12	PP_1^5	3.596	$pP_3^*P_2P_1^3$	4.219	PP_1^3	6.036
13	$pP_3^*S_2S_1$	3.685	pP_2P^3	4.301	PS_1	6.128
14	pS_2S_1	3.706	PP_1^5	4.439	$PS_3^*P_2^2P_1$	6.21359
15	$PP_2^2S_1$	3.809	$pS_3^*S_2P_1$	4.466	$PP_2^2P_1$	6.21362
16	PS_1^3	3.901	$pP_3^*S_2S_1$	4.472	$pS_3^*P_2S_1$	6.223
17	$pP_2P_1^2S_1$	3.920	pS_2P_1	4.477	$PP_3^*P_2S_2P_1$	6.243
18	$PP_1^4S_1$	4.032	$PP_3^*P_2^2S_1$	4.550	pP_2S_1	6.345
19	$PP_2S_2P_1$	4.171	$PP_2^2S_1$	4.643	$pS_3^*P_2P_1^3$	6.401
20	$pS_2P_1^3$	4.284	$pP_3^*P_2P_1^2S_1$	4.669	$pP_2P_1^3$	6.407

There are three observation stations A, B, and C at the surface, their ranges being

$$r_A = 2.0, \qquad r_B = 5.0, \qquad r_C = 10.0.$$

In the table each ray is identified by the mode of each segment (P or S) and the layer (subscript 1, 2, or 3). The first segment from an explosive source in layer 2 is either a downward p mode or upward P mode where the subscript 2 is omitted. With reference to Fig. 8A there should be two rays for a two-segment path, PP_1 and PS_1; and four rays for a three-segment direct path, pP_2P_1, pP_2S_1, pS_2P_1, and pS_2S_1. In Fig. 8B and 8C there might be eight rays for a three-segment refracted path, $pP_3^*P_2P_1, \ldots, pS_3^*S_2S_1$.

However, we note that at station A, PP_1^3 ray (number 4, upward P mode and then three segments of P mode in layer 1) arrives earlier than many three-segment rays, and only two of the eight three-segment refracted rays exist. More refracted rays occur as r increases. At station B, six refracted rays exist. At station C, all eight rays exist but only five are shown in the table. Note also because of higher P-wave speed, the PP_1^3 ray (number 12) is faster than the PS_1 ray (number 13); and the PP_2^2P ray (15) is faster than the pP_2S_1 ray (18) at station C.

B. Numerical Evaluation of the Complex Integrals

1. *Integration in the Complex ξ Plane*

The path of integration for integrals in Eqs. (7.2) and (7.3) is shown in Fig. 9 along the contour $AM\xi_1$. The most complicated case (case C-b) is repeated in Fig. 15 and the integration starts from the point S ($\xi_S = ia_6$ and $a_6 < a_5 < b_M$). Since the integrand is analytic to the right of this contour, we may replace the path $SM\xi_1$ by $QP\xi_1$ as shown in Fig. 15. The point Q can be any point between A and S, and P is the *pivotal point* somewhere in the complex ξ plane. As a start, we chose $\xi_Q = 0.8\xi_S$ and $\xi_P = \xi_S + 0.2|\xi_S|$. The values for ξ_P will be changed at selected times in the calculation so that the lengths of the two segments QP and $P\xi_1$ are approximately equal.

There are several reasons for selecting the new path $QP\xi_1$. First, it lies entirely in the first quadrant, avoiding the branch points (ia_6, ia_5, etc.), the stationary point (ib_M), and other poles of the integrand along the imaginary axis (Rayleigh pole, Stoneley poles, etc.). These singular points cause great difficulty in numerical integration. Secondly, standard numerical methods of integration, such as Gaussian quadrature may be used along these straight paths. Thirdly, since the segment QP only changes occasionally, the time independent part of the integrand, $S(\xi)\pi(\xi)D(\xi)$, along QP is calculated only once for a wide range of t.

Gaussian quadrature is used to numerically evaluate these integrals. The complex interval (Q, P) is mapped onto the real interval $(-1, +1)$ by a change of integration variable, and the ten-point Gaussian formula is used.

The integration of the interval (P, ξ_1) for Eq. (7.2) can be carried out analogously. However, for Eq. (7.3), it is not as straightforward as above

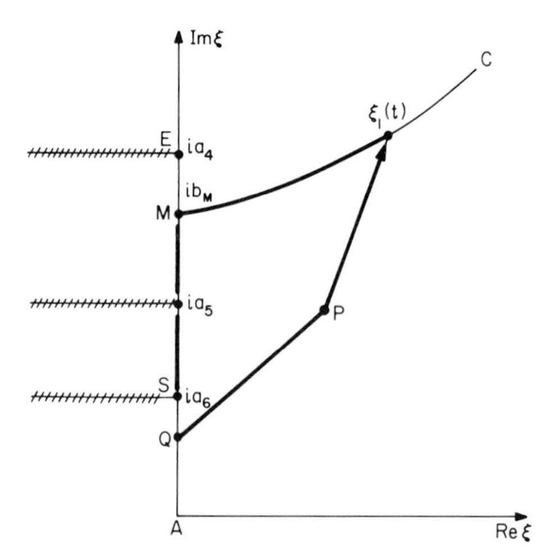

Fig. 15. Change of the integration contour $SM\xi_1$ to $QP\xi_1$.

because the function $K_j(\xi_j; \mathbf{x}, t)$ is zero at $\xi = \xi_1(t)$. This is a half-order integrable singularity which requires special numerical treatment.

First, the integration variable is changed from ξ to u:

$$u = (\xi^2 - \xi_1^2)^{1/2} \tag{7.8}$$

and

$$du = \xi \, d\xi/(\xi^2 - \xi_1^2)^{1/2}. \tag{7.9}$$

Completing this change of variable introduces a factor $[\xi(u)^2 - \xi_1^2]^{1/2}$ in the numerator, which matches the half-order singularity of K in the denominator. A power series expansion is then made about the point $u = 0$ to analytically remove the common factor and facilitate numerical evaluation of the integrand for u near 0. The complex integration interval for u is $(0, (\xi_p^2 - \xi_1^2)^{1/2})$ which can now be mapped to the real interval $(-1, +1)$ and the ten-point Gaussian integration formula can be used. Note that Gaussian quadrature uses an open interval not requiring evaluation of the integrand at the endpoints.

The integration of Eq. (7.3) can thus be done for each value of τ. The results for the jth generalized ray are to be added to the results of the other contributing rays to yield $I_\alpha(\tau)$ in Eq. (7.4).

2. *Convolution Integration*

The results I_α calculated at time interval $d\tau$ must now be convoluted with a time function of the source to complete the solution. Equation (7.5) provides the form of this convolution.

Since data for I_α are provided at a fixed interval $d\tau$, the integral in Eq. (7.5) is evaluated using Simpson's rule. If the limits of integration do not match values of time at which I_α was computed, a cubic interpolation formula is used to approximate I_α for integration in these gaps. In addition the cubic interpolation formula is also used just prior to and after sharp wave front arrivals in order to minimize the numerical errors associated with such wave fronts.

The source function used for the examples in Section VII,C is a parabolic ramp function. It is composed of several quadratic time functions $f_2(t) = (t^2/2)H(t)$ according to the second-order differences. The superposed result is (Pekeris *et al.*, 1965)

$$\begin{aligned} f(t) &= f_2(t) - 2f_2(t - \Delta) + f_2(t - 2\Delta) \\ &= 0, && t < 0, \\ &= \tfrac{1}{2}t^2, && 0 < t < \Delta, \\ &= \tfrac{1}{2}t^2 - (t - \Delta)^2, && \Delta < t < 2\Delta, \\ &= \tfrac{1}{2}t^2 - (t - \Delta)^2 + \tfrac{1}{2}(t - 2\Delta)^2 = \Delta^2, && 2\Delta < t. \end{aligned} \tag{7.10}$$

2Δ is the rise time of the parabolic ramp function.

For this source function, $f(0) = 0$ and $f'(0) = 0$; only the integral in Eq. (7.5) contributes to the answer.

Since $f''(t)$ is piecewise continuous, a special weighting and interpolation subroutine is used to calculate this function. First, the integral

$$\hat{I}_\alpha(\mathbf{x}, t) = H(t - t_A) \int_{t_A}^{t} I_\alpha(\mathbf{x}, \tau)\, d\tau \tag{7.11}$$

is computed at time interval $2\,dt$ using Simpson's rule. Then u_α is calculated with the weighting formula

$$u_\alpha(\mathbf{x}, t) = \hat{I}_\alpha(\mathbf{x}, t) - 2\hat{I}_\alpha(\mathbf{x}, t - \Delta) + \hat{I}_\alpha(\mathbf{x}, t - 2\Delta). \tag{7.12}$$

A cubic interpolation is used if the rise time 2Δ is not an even integral multiple of the data interval $2\,dt$.

C. Transient Responses for a Plate and a Layered Solid

Numerical calculations were done for the plate and the one-layer solid shown in Fig. 16. The material constants of the two examples are chosen to be the same:

$$\lambda_1 = \mu_1, \qquad \rho_1 = 1.21, \qquad C_1 \equiv 1.0, \qquad c_1 = \sqrt{3}\,C_1,$$
$$\mu_2 = 2\mu_1, \qquad \rho_2 = 2.0, \qquad C_2 \equiv 1.1, \qquad c_2 = \sqrt{3}\,C_2.$$

The minimum arrival times ($\tau = C_1 t/h$) along various paths for a surface receiver at $r = 5h$ in a plate are shown in Table VII. The results are presented by groups (Fig. 16), and each group consists of rays having equal number of segments. Rays are labeled by the number of P or S modes. Thus

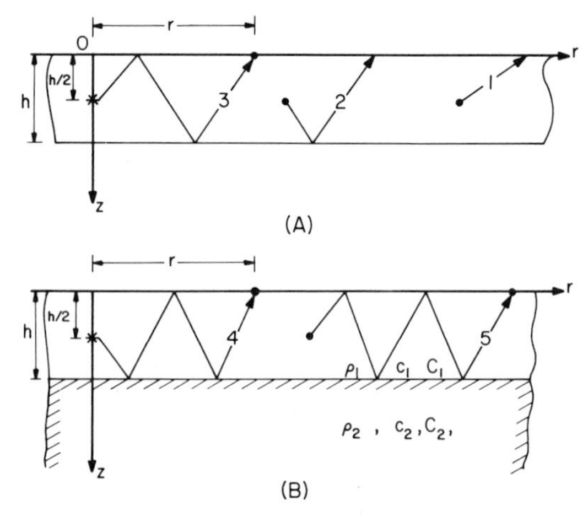

Fig. 16. Five ray groups in a single layer: (a) a plate; (b) a layer overlaying a half-space.

TABLE VII

MINIMUM ARRIVAL TIMES FOR RAYS IN A PLATE AND A LAYER ($r = 5h$)

Group	Direct ray (plate and layer)		Arrival time $\tau = C_1 t/h$	Refracted ray (layer)	Arrival time $\tau = C_1 t/h$
1	P		2.901		
2	P^2		3.014	pP*P	2.985
	pS		3.720	pP*S	3.596
3	P^3	(1)	3.228		
	P^2S	(2)	3.849	PpP*S	3.836
	PS^2	(1)	4.540	PsP*S	4.447
4	P^4	(1)	3.524		
	P^3S	(3)	4.088		
	P^2S^2	(3)	4.690	pP^*PS^2	4.688
	pS^3	(1)	5.361	pP^*S^3	5.298
5	P^5	(1)	3.884		
	P^4S	(4)	4.409		
	P^3S^2	(6)	4.957		
	P^2S^3	(4)	5.539		
	PS^4	(1)	6.185	PP^*S^4	6.149
6	P^6	(1)	4.292		
	P^5S	(5)	4.791		
	P^4S^2	(10)	5.304		
	P^3S^3	(10)	5.837		
	P^2S^4	(5)	6.397		
	pS^5	(1)	7.014	pP^*S^5	7.001
7	p^7	(1)	4.735		
	P^6S	(6)	5.217		
	P^5S^2	(15)	5.707		
	P^4S^3	(20)	6.209		
8	P^8		5.204		
	P^7S		5.674		
	P^6S^2		6.149		
9	P^9		5.694		
	P^8S		6.154		
10	P^{10}		6.198		

P^4 represents four P segments, the first one being downward; P^3S represents three P segments and one S segment and includes the following three combinations: pPpS, pPsP, and pSpP. Similarly, P^2S^2 represents also three rays; pPsS, pSpS, and pSsP. All rays in group 7 or higher with an arrival time larger than 6.5 are not listed in the table.

For the overlaying layer, the direct rays are the same as those in a plate (Table VII). However, the lower medium has higher wave speeds and the rays may be refracted along the interface. This is checked out in each group and the additional refracted rays are also listed in Table VII.

The source function in both examples is a center of dilatation defined by Eq. (2.32), and the time function $f(t)$ is specified by Eq. (7.10). In view of Eq. (5.59), the potential for the wave field generated by this point source in an unbounded medium is

$$\phi(r, z, t) = \frac{1}{4\pi c_1^2 R}\frac{q}{\Delta^2}[f_2(t) - 2f_2(t-\Delta) + f_2(t-2\Delta)], \tag{7.13}$$

where

$$f_2(t) = \frac{1}{2}\left(t - \frac{R}{c_1}\right)^2 H\left(t - \frac{R}{c_1}\right). \tag{7.14}$$

In the solution for ϕ we have inserted the factor q/Δ^2, where q is an unspecified coefficient, depending on the strength of the point source; and the factor Δ^{-2} renders the time function dimensionless.

Based on Eq. (7.13), the radial displacement u_R in Eq. (5.60) reaches peak value at $t = R/c_1 + \Delta$, and remains constant after $t = R/c_1 + 2\Delta$,

$$\begin{aligned} u_R &= 0, && t < R/c_1, \\ &= \frac{-q}{4\pi c_1^2}\left(\frac{1}{2R^2} + \frac{1}{c_1 R\Delta}\right), && t = R/c_1 + \Delta, \\ &= \frac{-q}{4\pi c_1^2}\frac{1}{R^2}, && t \geq R/c_1 + 2\Delta. \end{aligned} \tag{7.15}$$

Following Pekeris *et al.* (1965), we set

$$q = (-4\pi c_1^2 \Delta)Q. \tag{7.16}$$

The scalar factor Q, which carries with it the proper dimensions, will be set equal to unity in all numerical calculations. Thus, the peak value for u_R at large R is approximately $1/c_1 R$, which is independent of Δ.

The graphs for $u_r = ru_R/R$ and $u_z = zu_R/R$ are shown in Fig. 17 for

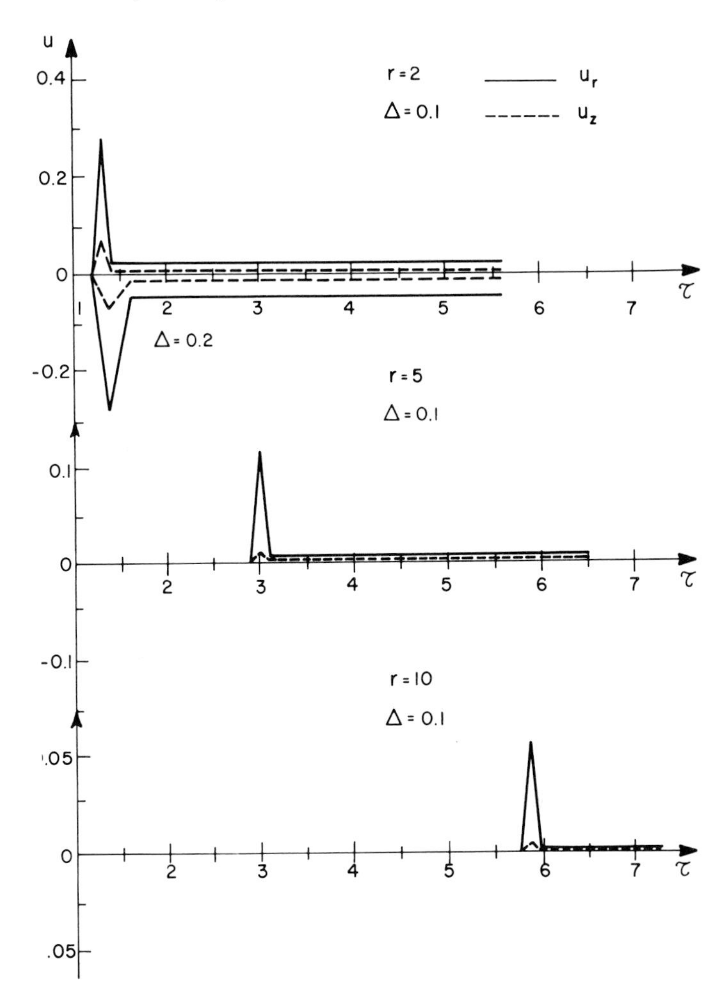

FIG. 17. Displacements at $z = 0$ generated by a point source (center of dilatation at $z_0 = 0.5$, $r = 0$).

three values of r ($r = 2h$, $5h$, $10h$) and $z = h/2$. The half-rise time is taken as $\Delta = 0.1h/C_1$. For purpose of comparison, displacements at $r = 2h$ for $\Delta = 0.2h/C_1$ are also shown in the figure, in negative magnitude.

If an observation point is in the interior of the medium, the results shown in Fig. 17 will be the solutions due to the source ray. For an observation point at the surface, this ray has been combined with the two reflected rays when we adopted the surface receiver function like that in Table II.

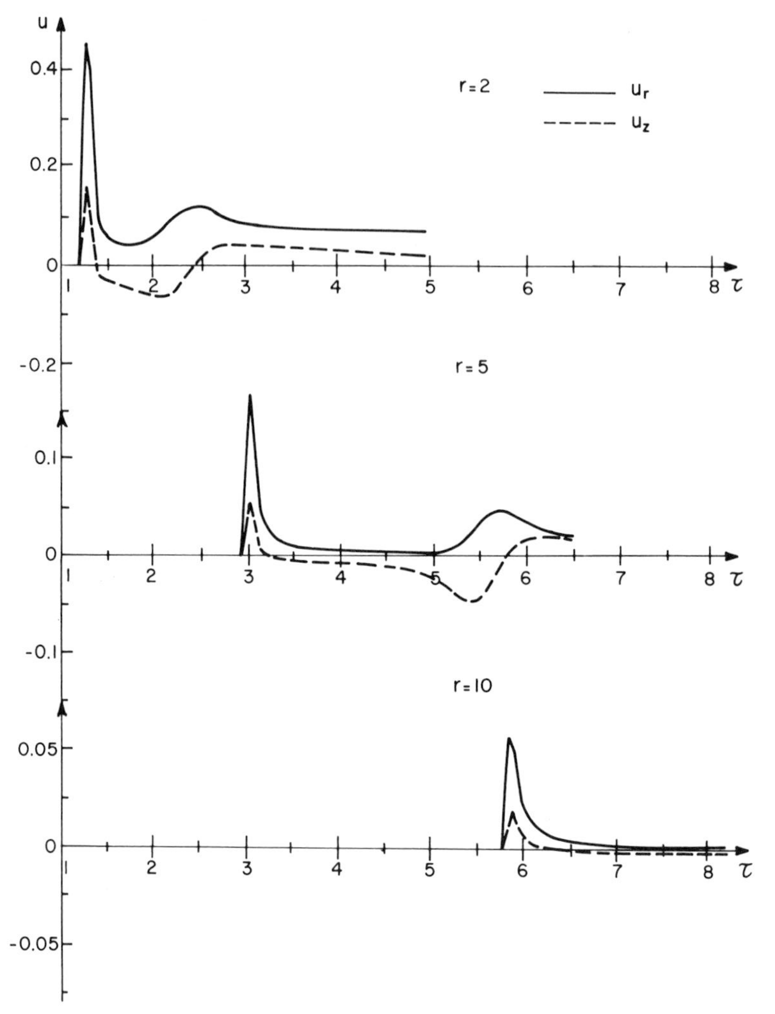

FIG. 18. Displacements at the surface of a half-space due to a point source (center of dilatation at $z_0 = 0.5$, $\Delta = 0.1$).

1. *Displacements at the Surface of a Plate*

Displacements generated by various groups of rays are shown in Figs. 18–21. We use the same source function in Eq. (7.16) with $Q = 1$. Like those shown in Fig. 17, the displacements u are undetermined by a scale factor. The normalized time $\tau = t/t_0$ where $t_0 = h/C_1$ is the time required for an S wave to travel a thickness of the plate. The half-duration of the source pulse Δ is also normalized by t_0.

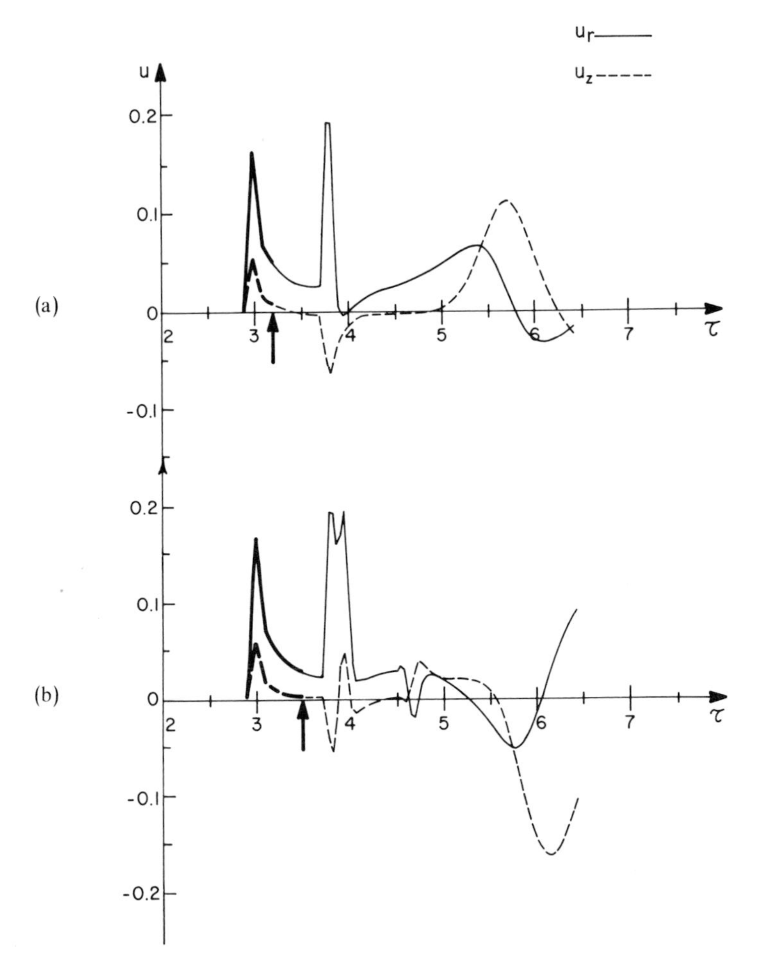

FIG. 19. Displacements at the surface of a plate due to a point source ($r = 5, \Delta = 0.1$): (a) two ray groups; (b) three ray groups; (c) four ray groups; (d) five ray groups; (e) six ray groups; (f) all rays prior to $\tau = 6.5$. (c)–(f) on pp. 258, 259.

The first group (Fig. 18) represents the waves from the point source interacting with the upper free surface. This solution completely ignores the existence of the lower free surface, and thus represents the solution for a half-space infinite in extent in the $+z$ direction. These half-space results include the source ray as well as P and S wave reflections off the upper free surface to the receiver located an infinitesimal distance below that surface. Note the amplification of the motion, shown in Fig. 17 particularly in the vertical direction, and the additional late time response, the Rayleigh surface wave. The solution of generalized rays includes the Rayleigh wave contribu-

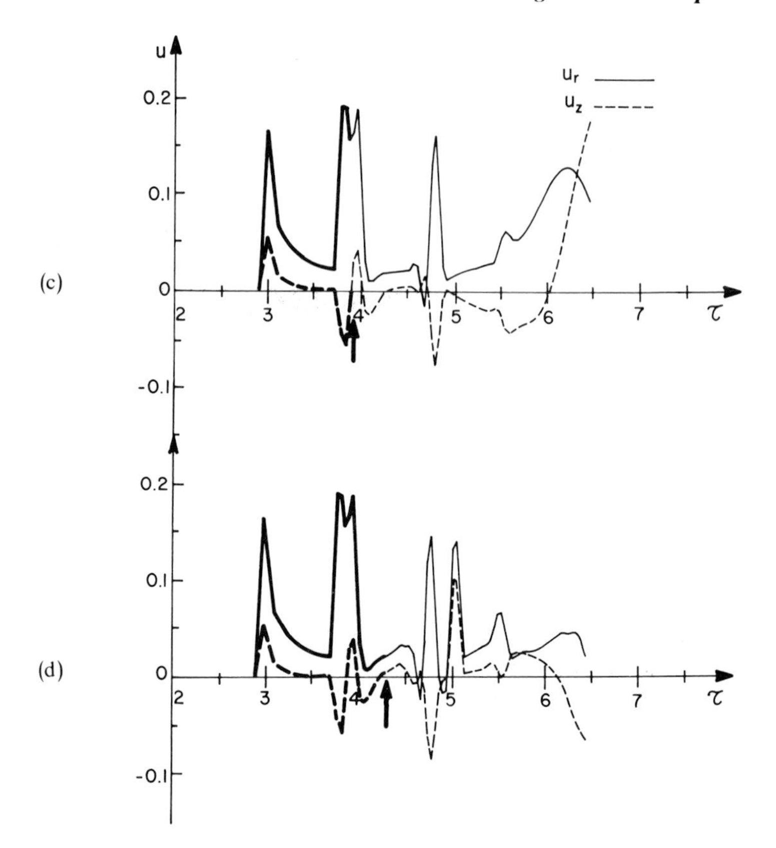

FIG. 19—*continued*

tion of each ray. When $z \cong 0$ the Cagniard contour $M\xi_1$ in Fig. 6 passes close to the Rayleigh pole on the imaginary axis, which has a predominant effect on the surface disturbance.

The next graph (Fig. 19a) shows the results of adding the second group of rays. The second group represents a semispherical P wave traveling down from the source and interacting with the bottom surface, producing P and S waves in an upward reflection (no head waves in this plate problem). This ray includes the additional downward P and S wave reflections off the upper surface (the surface receiver effect).

The solution shown in Fig. 19a has no corresponding physical analog such as the first group. It is valid only before the third group arrives, as denoted by the arrow at $\tau = 3.228$. However, it does show the individual ray

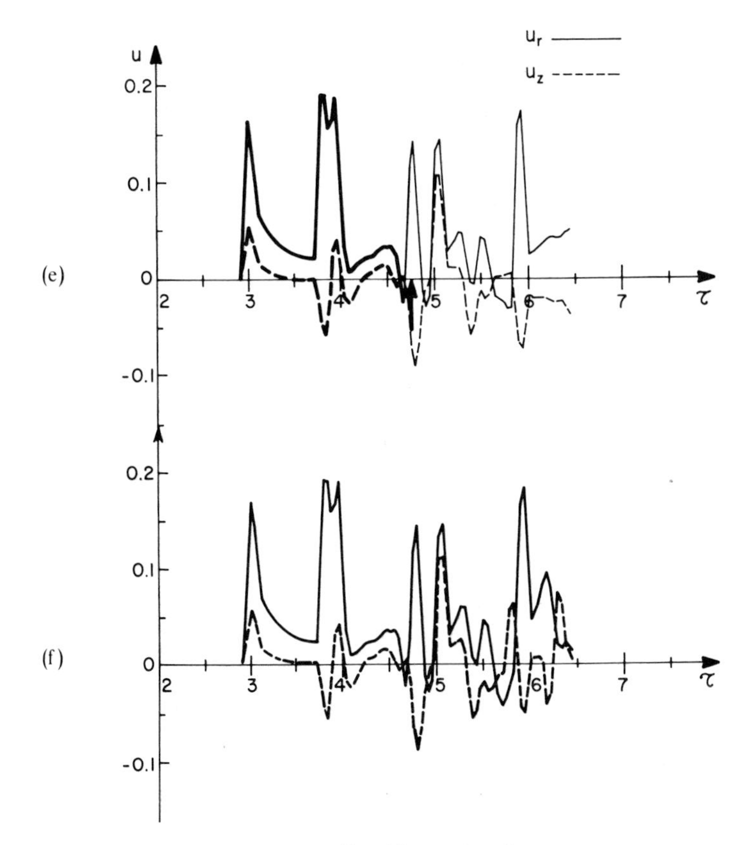

FIG. 19—*continued*

contribution. The pP ray, arriving just after the source ray changes slightly the amplitude, but the pS ray at $\tau = 3.720$ produces a major disturbance. Also note that the Rayleigh pole contributions of these rays have changed the late time portion of the response.

Figure 19b shows the addition of the third group. The significant wave arrivals are the P^2S at 3.849 and PS^2 at 4.540. This solution is valid up to $\tau = 3.524$ (P^4 ray).

In the next ray group (Fig. 19c), the significant ray is the P^2S^2 of the fourth group arriving at 4.690. This ray is a good example of a multiple arrival ray or degenerate ray (Dainty and Dampney, 1972). All rays in this group start with a p segment, and there are three ways to permute the remaining one P and two S segments: pPsS, pSpS, and pSsP. All of them

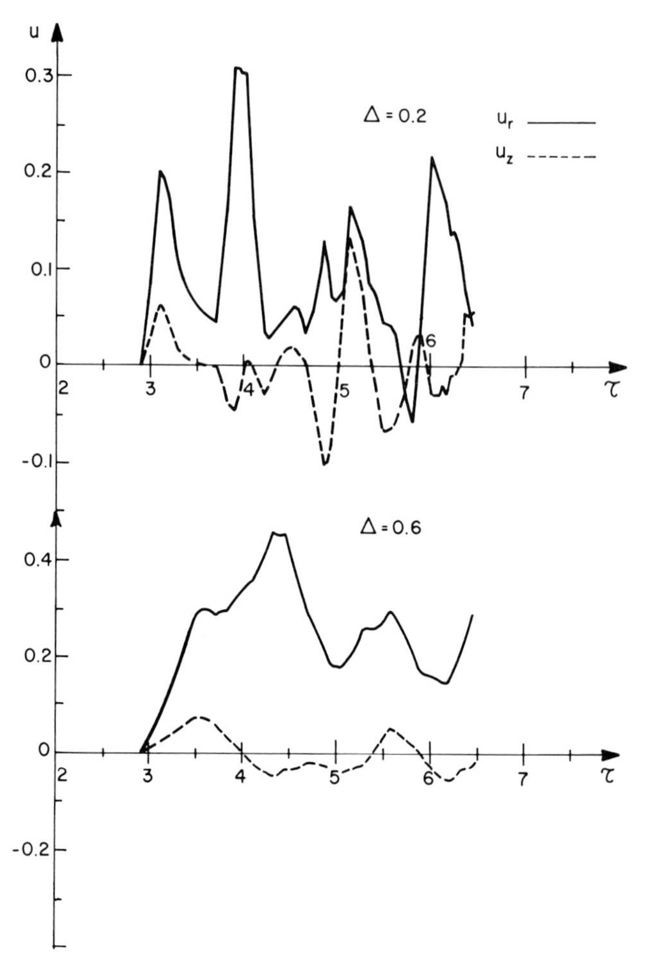

FIG. 20. Effect of the duration of the source on the surface displacements of a plate ($r = 5$).

arriving at the same time must be included in the solution. The computer code calculates the results of these rays simultaneously, using the same Cagniard contour but selecting different reflection coefficients and receiver functions. The solution shown in this graph is valid to $\tau = 3.884$ (P^5 ray).

The effects of the fifth group are shown in Fig. 19d, valid until $\tau = 4.292$ (P^6 ray); and the sixth group in Fig. 19e, valid until $\tau = 4.735$ (P^7 ray). Note that despite the increasing number of reflections, each group is contributing significant wave arrivals at late times.

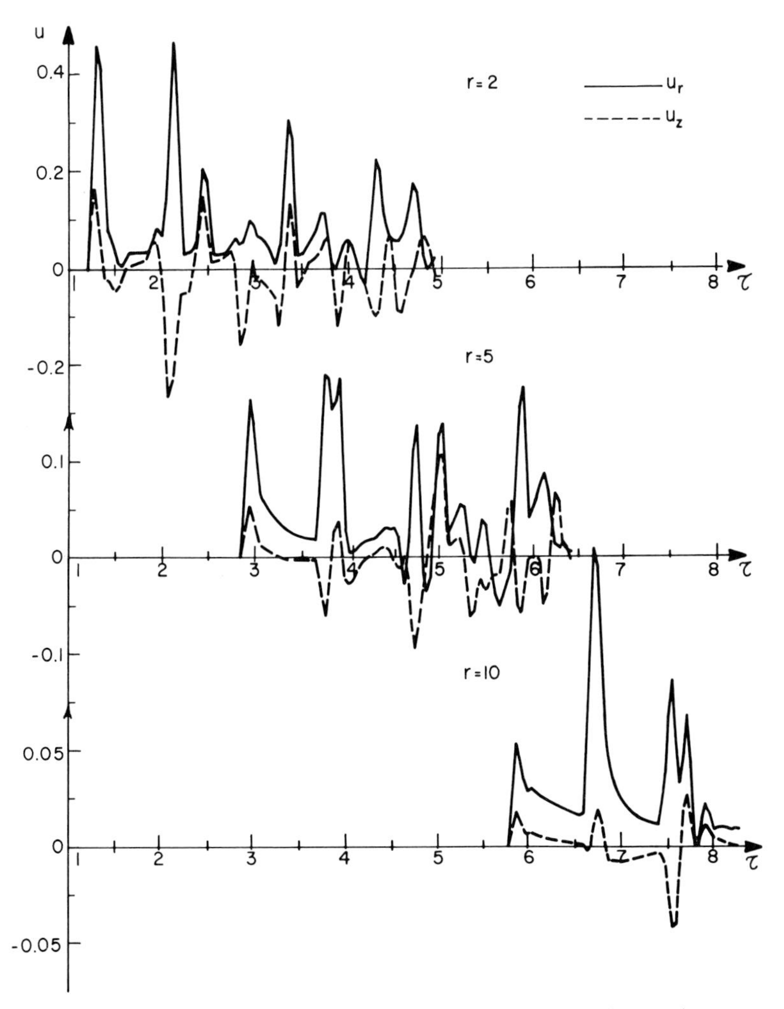

FIG. 21. Surface displacements of a plate at various ranges ($\Delta = 0.1$).

The complete plate solution up to $\tau = 6.5$ is shown in Fig. 19f. Its late time response differs from the previous graphs. The complete plate solution represents 30 distinct wave arrivals with 98 separate ray computations since it includes many simultaneous ray arrivals. It is interesting to note that the Rayleigh surface wave effects, so pronounced in the first five groups, are not apparent now because of the strong rays with late arrival times.

To examine the effects of other input parameters on the plate response, we show in Fig. 20 the results when the pulse duration is increased to $\Delta = 0.2$ and 0.6, the other parameters being the same as in Fig. 19. It is seen that the longer duration [Eqs. (7.13) and (7.16)] magnifies slightly the displacements, and spreads out the disturbance over a longer time.

Figure 21 shows the effect of the receiver range. The arrival times for $r = 2h$, $10h$, as well as $r = 5h$ are listed in Table VIII for the first four groups. At all ranges the rays with mode conversion (PS, P^2S, PS^2, P^3S, and P^2S^2) make significant contributions to the total response.

2. *Displacements at the Surface of the Overlaying Layer*

The solution to the one layer over a half-space problem is shown in Fig. 22. The results for vertical displacement can be compared to prior calculations of Pekeris *et al.* (1965) and of Alterman and Karal (1968).

In contrast to the plate, this solution shows a stronger initial P-wave reflection off the lower surface, increasing the magnitude of the first peak. Subsequent reflections off the lower surface dissipate energy into the half-space, thus reducing the peaks associated with later arriving rays.

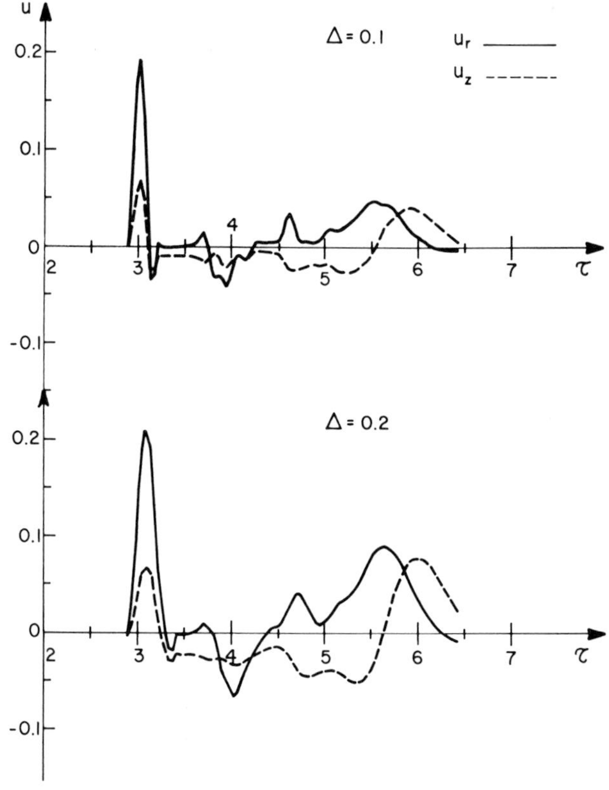

FIG. 22. Surface displacements due to a point source in a layer overlaying a half-space ($r = 5$).

TABLE VIII

ARRIVAL TIMES τ AT A SURFACE POINT IN A PLATE

	τ		
Rays	$2h$	$5h$	$10h$
P	1.190	2.901	5.781
P^2	1.443	3.014	5.832
PS	2.024	3.720	6.598
P^3	1.848	3.228	5.951
P^2S	2.342	3.849	6.659
PS^2	2.876	4.540	7.415
P^4	2.327	3.523	6.117
P^3S	2.789	4.088	6.780
P^2S^2	3.263	4.690	7.481
PS^3	3.761	5.361	8.232

The comparatively large disturbances after $\tau = 5.0$ is similar to the Rayleigh surface wave effect shown in Fig. 18 but quite different in nature. As discussed by Pekeris *et al.* (1965), these disturbances correspond to the Airy phase of the Rayleigh normal modes [Eq. (2.38) of Section II,C]. Since the Airy phases decay in the order of $r^{-1/3}$, while other modes decay in the order of $r^{-1/2}$, they become dominant at large distances. This is evident in the graphs shown in the aforementioned reference for $r = 10h$, and $20h$.

Additional results are presented in Fig. 22 showing the effect of doubling the pulse duration from $\Delta = 0.1$ to $\Delta = 0.2$. Recall that the displacement u has been normalized by the factor $\Delta Q/\Delta^2$ [Eqs. (7.13) and (7.16)]. For the early time behavior, the peak magnitude of the response remains nearly constant due to the normalization factor, while the pulse duration of the response doubles. But for the late time behavior (the Airy phase of Rayleigh modes), the duration remains constant while the peak magnitude doubles. Thus the magnitude of the early time behavior is inversely proportional to the rise time of the source as well as directly proportional to the magnitude of the source, whereas the magnitude of the late time behavior is proportional only to the magnitude of the source.

ACKNOWLEDGMENT

The research of the senior author (Y.H.P.) is supported by grants from the National Science Foundation to the College of Engineering, and the Materials Science Center of Cornell University.

References

Abramovici, F., and Alterman, Z. (1965). *Methods Comput. Phys.* **4**, 349.

Achenbach, J. D. (1973). "Wave Propagation in Elastic Solids." North-Holland Publ., Amsterdam.

Alterman, Z., and Karal, F. C. (1968). *Bull. Seismol. Soc. Am.* **58**, 367.

Berry, M. J., and West, G. F. (1966). *Geophys. Monogr., Am. Geophys. Union,* **10**, 464.

Born, M., and Wolf, E. (1975). "Principles of Optics." Pergamon, Oxford.

Bortfeld, R. (1962). *Geophys. Prospect.* **10**, 519.

Brekhovskikh, L. M. (1960). "Waves in Layered Media." Academic Press, New York.

Bromwich, T. J. I'A. (1916). *Proc. London Math. Soc.* **15**, 401.

Cagniard, L. (1939). "Reflexion et réfraction des Ondes séismiques progressives." Gauthier-Villars, Paris ["Reflection and Refraction of Progressive Seismic Waves" (English transl. by E. A. Flynn and C. H. Dix). McGraw-Hill, New York, 1962].

Chandra, U. (1968). *Bull. Seismol. Soc. Am.* **58**, 993.

Chao, C.-C. (1960). *J. Appl. Mech.* **27**, 559.

Chapman, C. H. (1974). *Geophys. J. R. Astron. Soc.* **36**, 673.

Chapman, C. H. (1976a). *Geophys. Res. Lett.* **3**, 153.

Chapman, C. H. (1976b). *Geophys. J. R. Astron. Soc.* **46**, 201.

Chen, P. (1977). Ph.D. Thesis, Cornell University, Ithaca, New York.

Conte, S. D., and de Boor, C. (1972). "Elementary Numerical Analysis," 2nd ed. McGraw-Hill, New York.

Dainty, A. M., and Dampney, C. N. G. (1972). *Geophys. J. R. Astron. Soc.* **28**, 147.

Dampney, C. N. G. (1971). *Bull. Seismol. Soc. Am.* **61**, 1583.

Davids, N. (1959). *J. Appl. Mech.* **26**, 651.

Davids, N., and Lawhead, W. (1965). *J. Mech. Phys. Solid* **13**, 199.

Debye, P. (1909). *Ann. Phys. (Leipzig)* [4] **30**, 57.

Dunkin, J. W., and Corbin, D. G. (1970). *Bull. Seismol. Soc. Am.* **60**, 167.

Ewing, M., Worzel, J. L., and Pekeris, C. L. (1948). Mem., *Geol. Soc. Am.* **27**, 1.

Ewing, W. M., Jardetzky, W. S., and Press, F. (1957). "Elastic Waves in Layered Media." McGraw-Hill, New York.

Fuchs, K. (1966). *Bull. Seismol. Soc. Am.* **56**, 75.

Gajewski, R. R. (1977). Ph.D. Thesis, Cornell University, Ithaca, New York.

Garvin, W. W. (1956). *Proc. R. Soc. London, Ser. A* **234**, 528.

Ghosh, S. K. (1973). *Pure Appl. Geophys.* **105**, 781.

Gilbert, F., and Backus, G. E. (1966). *Geophysics* **31**, 326.

Gilbert, F., and Helmberger, D. B. (1972). *Geophys. J. R. Astron. Soc.* **27**, 57.

Harkrider, D. G. (1964). *Bull. Seismol. Soc. Am.* **54**, 627.

Harkrider, D. G. (1970). *Bull. Seismol. Soc. Am.* **60**, 1937.

Harkrider, D. G. (1976). *Geophys. J. R. Astron. Soc.* **47**, 97.

Haskell, N. A. (1953). *Bull. Seismol. Soc. Am.* **43**, 17.

Helmberger, D. V. (1968). *Bull. Seismol. Soc. Am.* **58**, 179.

Helmberger, D. V. (1972). *Bull. Seismol. Soc. Am.* **62**, 325.

Knopoff, L. (1958a). *J. Appl. Phys.* **29**, 661.

Knopoff, L. (1958b). *J. Geophys. Res.* **63**, 619.

Knopoff, L., and Gilbert, F. (1959). *J. Acoust. Soc. Am.* **31**, 1161.

Knopoff, L., Gilbert, F., and Pilant, W. L. (1960). *J. Geophys. Res.* **65**, 265.

Lamb, H. (1904). *Philos. Trans. R. Soc. London, Ser. A* **203**, 1.

Lapwood, E. R. (1949). *Philos. Trans. R. Soc. London, Ser. A* **242**, 63.

Love, A. E. H. (1944). "Mathematical Theory of Elasticity," 4th ed. Dover, New York.

Mencher, A. G. (1953). *J. Appl. Phys.* **24**, 1240.
Mooney, H. M. (1976). *Geophysics* **41**, 243.
Morse, P. M., and Feshbach, H. (1953). "Methods of Theoretical Physics." McGraw-Hill, New York.
Müller, G. (1968). *Z. Geophys.* **34**, 147.
Müller, G. (1969). *Z. Geophys.* **35**, 347.
National Bureau of Standards. (1964). *In* "Handbook of Mathematical Functions" (M. Abramowitz and I. A. Stegun, eds.), p. 358. US Govt. Printing Office, Washington D.C.
Newlands, M. (1952). *Philos. Trans. R. Soc. London, Ser. A* **245**, 213.
Norwood, F. R. (1975). *Int. J. Solids Struct.* **11**, 33.
Pao, Y. H., and Ceranoglu, A. (1977). *J. Appl. Mech.* (submitted for publication).
Pao, Y. H., and Kaul, R. K. (1974). *In* "R. D. Mindlin and Applied Mechanics" (G. Hermann, ed.), p. 149. Pergamon, Oxford.
Pao, Y. H., and Varatharajulu, V. (1976). *J. Acoust. Soc. Am.* **59**, 1361.
Pao, Y. H., Gajewski, R. R., and Thau, S. A. (1971). "Analysis of Ground Wave Propagation in Layered Media," Rep. No. DASA 2697. Defense Nuclear Agency, Washington, D.C.
Pekeris, C. L. (1940). *Proc. Natl. Acad. Sci. U.S.A.* **26**, 433.
Pekeris, C. L. (1955). *Proc. Natl. Acad. Sci. U.S.A.* **41**, 469.
Pekeris, C. L., and Longman, I. M. (1958). *J. Acoust. Soc. Am.* **30**, 323.
Pekeris, C. L., Longman, I. M., and Lifson, H. (1959). *Bull. Seismol. Soc. Am.* **49**, 247.
Pekeris, C. L., Alterman, Z., and Abramovici, F. (1963). *Bull. Seismol. Soc. Am.* **53**, 39.
Pekeris, C. L., Alterman, Z., Abramovici, F., and Jarosch, H. (1965). *Rev. Geophys.* **3**, 25.
Richards, P. G. (1971). *Geophysics* **36**, 798.
Schmuely, M. (1975). *Int. J. Solids Struct.* **11**, 679.
Sneddon, I. N. (1951). "Fourier Transforms." McGraw-Hill, New York.
Spencer, T. W. (1960). *Geophysics* **25**, 625.
Spencer, T. W. (1965a). *Geophysics* **30**, 363.
Spencer, T. W. (1965b). *Geophysics* **30**, 369.
Stokes, G. G. (1849). *Trans. Cambridge Philos. Soc.* **9**, 1.
Thau, S. A., and Pao, Y. H. (1970). *Int. J. Eng. Sci.* **8**, 207.
Thomson, W. T. (1950). *J. Appl. Phys.* **21**, 89.
Van der Pol, B., and Bremmer, H. (1937). *Philos. Mag.* [7] **24**, 141 and 824.
White, J. E. (1965). "Seismic Waves: Radiation, Transmission, and Attenuation." McGraw-Hill, New York.
Wiggins, R. A. (1976). *Geophys. J. R. Astron. Soc.* **46**, 1.
Wiggins, R. A., and Helmberger, D. V. (1974). *Geophys. J. R. Astron. Soc.* **37**, 73.
Wiggins, R. A., and Madrid, J. A. (1974). *Geophys. J. R. Astron. Soc.* **37**, 423.
Willis, J. R. (1965). *J. Mech. Phys. Solids* **13**, 377.

Author Index

C

D

E

F

G

Subject Index

B

C

D

E

F

G

H

I

L

M

Contents of Previous Volumes

Volume I, Part A—Methods and Devices

Volume I, Part B—Methods and Devices

Volume II, Part A—Properties of Gases, Liquids and Solutions

Volume II, Part B—Properties of Polymers and Nonlinear Acoustics

Volume III, Part A—Effect of Imperfections

Volume III, Part B—Lattice Dynamics

Volume IV, Part A—Applications to Quantum and Solid State Physics

Volume IV, Part B—Applications to Quantum and Solid State Physics

Volume V

Volume VI

Volume VII

Volume VIII

Volume IX

Volume X

Volume XI

Volume XII

A
B 7
C 8
D 9
E 0
F 1
G 2
H 3
I 4
J 5